T0312082

SUSTAINABLE PRODUCTION AND LOGISTICS

SUSTAINABLE PRODUCTION AND LOGISTICS
Modeling and Analysis

Edited by Eren Özceylan and Surendra M. Gupta

CRC Press
Taylor & Francis Group
Boca Raton London New York

CRC Press is an imprint of the
Taylor & Francis Group, an **informa** business

First edition published 2021
by CRC Press
6000 Broken Sound Parkway NW, Suite 300, Boca Raton, FL 33487-2742

and by CRC Press
2 Park Square, Milton Park, Abingdon, Oxon, OX14 4RN

© 2021 Taylor & Francis Group, LLC

CRC Press is an imprint of Taylor & Francis Group, LLC

Library of Congress Cataloging-in-Publication Data

ISBN: 978-0-367-43130-3 (hbk)
ISBN: 978-0-367-71499-4 (pbk)
ISBN: 978-1-003-00501-8 (ebk)

Typset in Times
by MPS Limited, Dehradun

Dedication

Dedicated to our families:

Ayca Özceylan, and Duru Özceylan
-Eren Özceylan

Sharda Gupta, Monica Gupta, and Neil M. Gupta
-Surendra M. Gupta

Table of Contents

Preface

Sustainable production and logistics deal with the management of environmental, economic, and social impacts, and the encouragement of good governance practices, throughout the lifecycles of products and services. Sustainability is the ability to continue a defined behavior indefinitely. The objective of sustainability in production and logistics is to create, protect, and grow long-term environmental, economic, and social value for all stakeholders involved in bringing products and services to the market as well as product recovery. Thus, it is the management of material and information flow in and between facilities—(re)manufacturing and assembly/disassembly plants, forward/reverse logistics, supplier selection, transportation, and scheduling. Sustainable production and logistics related issues have found a large following in industry and academia who aim to find solutions to the problems that arise in this rapidly emerging research area. Problems are widespread, ranging from remanufacturing to disassembly; from logistics to vendor selection. This book addresses several important issues faced by strategic, tactical, and operational planners of sustainable production and logistics, using efficient models in a variety of decision-making situations, providing easy-to-use mathematical and/or simulation modeling-based solution methodologies.

This book is organized into 18 chapters. The first chapter, by Pandit and Gupta, addresses the issue of preventing warranty fraud arising in the remanufacturing sector initiated by the customer. The study examines strategies available to the warranty provider to mitigate fraud originating from the customer in the consumer electronics remanufacturing sector. Discrete event simulation is employed to compare how well a sensor-embedded product may improve upon the existing system by comparing relevant fraud statistics. The sensor-embedded scenario was able to preemptively stop more frauds, but further research into sensor cost would be required to truly ascertain the value of sensor implementation.

The second chapter by Zhou and Gupta creates a value-based pricing model for new and remanufactured products belonging to different generations under various conditions. It discusses the optimal wholesale and retail prices in a vertical Nash equilibrium model. The piecewise demand functions are generated based on customers' perceived value on technology and quality in two segments: quality-conscious and technology-savvies. The authors found that the optimal pricing decision is modified when the proportion of the two customer categories is changed. It was also revealed how value differentiation between the two groups influence the prices and the profits of the retailer, manufacturer, remanufacturer, and the joint supply chain.

In the third chapter, Eligüzel, Özceylan, and Mete applied an integer programming approach of the disassembly line, balancing model to the end-of-life wind turbines. They performed various scenarios by altering the cycle times to observe its effect on the number of open workstations. The results showed that end-of-life wind turbines can be disassembled systematically and optimally using this method.

The fourth chapter, by Kucukkoc, Keskin, and Li, applied Data Envelopment Analysis (DEA) to disassembly line balancing problems. The objective was to maximize line efficiency. The workings of the approach were demonstrated through numerical examples. Computational tests were conducted to measure the performance of the proposed approach. The results demonstrated the viability of the approach.

In the subsequent chapter (Chapter 5), Kenger, Koç, and Özceylan introduced an integrated collection-disassembly-distribution problem within a reverse supply chain, consisting of a disassembly center, collection centers, and remanufacturing centers. The integrated problem sought to optimize the reverse supply chain network, cooperating with the disassembly line, balancing together with the vehicle routing problem. The authors proposed two mixed-integer mathematical formulations for several problem cases. The first model dealt with the case in which one type component was demanded, and the second model dealt with the multi-components case. The chapter conducted extensive scenario analyses on formulations to provide managerial and policy insights.

Chapter 6, by Tozanlı, Kongar, and Gupta, introduces the concept of digital twins in disassembly-to-order (DTO) systems to develop a predictive trade-in pricing policymaking strategy. The authors constructed a discrete event simulation model to gain insights regarding the behavior of the returned devices as well as the expected overall cost of DTO operations. Following a design-of-experiments study, a quantitative assessment on acquiring the optimal trade-in incentive was presented from an original equipment manufacturer's perspective by proposing an engaging price from the perspectives of all parties involved in the transaction.

Chapter 7, by McGovern and Gupta, reviews the requirements for justifying the use of metaheuristics. Using disassembly-line balancing problem as a case study, the authors discussed metaheuristic design, coefficient selection, and stopping criteria to be used while solving such problems. The discussion included the creation of a performance evaluation data set to compare solution-generating techniques.

In Chapter 8, Demirel, Demirel, and Gökçen addressed the problem of managing the recovery of packaging waste efficiently. To that end, they presented a modeling framework for packaging waste recycling network design based on a case study considering the packaging and textile waste treatment in Kayseri, which is one of the largest cities of Turkey.

Chapter 9, by Dulman, Gupta, and Yamada, examined the use of sensors in a closed-loop supply chain system, including maintenance of the products and their end-of-life processing. To determine the impact that the sensors have on the system, regular dishwasher and dryer and sensor-embedded dishwasher and dryer systems were developed. These systems were modeled using discrete event simulation. The models were subsequently tested through experiments. The results indicated that sensor-embedded systems performed better than regular systems in terms of disassembly, inspection, and maintenance costs.

In Chapter 10, Ilgin used DEMATEL (Decision Making Trial and Evaluation Laboratory) for a detailed analysis of the factors that hinder the wide-spread and effective use of industry 4.0 technologies in reverse logistics.

Chapter 11 by Aldoukhi and Gupta used a robust mathematical model with multi-objectives for a closed-loop supply chain network design. In their model, because of uncertainties in demand, substitution of new or remanufactured products is considered. This can avoid expensive backorders and long lead times. The authors presented an illustrative application from the National Industrial Clusters Development Program in Saudi Arabia about starting a tire manufacturing factory.

In Chapter 12, Alqahtani, Rajkhan, and Gupta presented a two-stage repair-full refund warranty policy analysis for an Advanced Remanufacturing-To-Order system for sensor-embedded products. The goal of the proposed approach was to determine how to predict a two-stage repair-full refund warranty period for the disassembled components and remanufactured products using sensor information about the age of every end-of-life retrieved product on-hand to meet remanufactured product and component demands while minimizing the cost associated with warranty and maximizing manufacturer's profit. Different simulation scenarios were considered, and a case example was presented for illustration of the model applicability.

Chapter 13, by Kesen and Yağmur, examined the mixed pallet order picking process in a distribution center of a logistics firm. They first developed a new layout plan for the distribution center by considering the demands and weights of the products. Then, they proposed a new mathematical model to minimize the total distance for picking the orders in a mixed pallet. They accomplished this by including an additional constraint to the well-known traveling salesman problem. By doing this, they were able to achieve about 28% improvement in the order picking process.

In Chapter 14, Çetinkaya, Özceylan, and Baytur discussed the problem of supplier selection to make a green purchase in a five-star accommodation business that has an ISO 14001 certificate. The authors used Analytical Network Process with BOCR (benefit, opportunity, cost, and risk) criteria to choose the best supplier among the alternatives. This multi-criterion decision-making problem had four main criteria and 14 sub-criteria. The task was to choose the most suitable supplier out of the three suppliers, considering their environmental sensitivity for purchasing cleaning materials.

Chapter 15, by Duman, Kongar, and Gupta, proposed a green supplier evaluation and selection method that evaluated suppliers according to their green competencies and environmental performances. The authors suggested a hybrid multi-criteria decision-making approach capable of handling imprecise quantitative and qualitative data. To demonstrate the functionality of the approach, a case study on a US-based company that manufactures and distributes plastic closures and dispensing systems internationally was considered. The results of the study along with a discussion were provided.

In Chapter 16, Subulan and Baykasoğlu presented an improved extension of the weighted hierarchical fuzzy axiomatic design approach to select the best transport options by considering several sustainability factors. From the methodological perspective, the classical crisp axiomatic design method was extended by generating negative valued information contents (overachievement) as well as positive valued information contents (underachievement). Additionally, the classical weighted hierarchical fuzzy axiomatic design method was also modified by incorporating the criteria hierarchy and negative valued information contents. Finally, an evaluation procedure was also proposed by considering both total positive and negative valued information contents. To show the practicality and applicability of the proposed extensions, a real-life application of an international logistics company was presented.

In Chapter 17, Alim, Koç, and Kesen addressed the problem of delays in railway transportation. In particular, the authors proposed a mathematical formula to minimize the delays due to idle waits to meet with other trains in a single line. They considered an instance with real data and solved it using mathematical formula. The authors reported and analyzed the results by comparing them with the current schedule. The results showed that the delays could be reduced by up to 38%.

In the final chapter (Chapter 18), Akyurt considered a flight scheduling problem in flight academies. The author used a mathematical model to solve this problem by taking economic, environmental, and social constraints into account. Considering the special circumstances of the aviation industry, a sustainable schedule was sought that took the social constraints (ergonomic conditions, mental conditions, etc.) caused by the human factor—in addition to the classical flight scheduling problem—into account. The proposed approach was applied to a sample case problem in a flight academy and the results were presented.

This work would not have been possible without the devotion and commitment of the contributing authors. They have been very patient in preparing their manuscripts. We would also like to express our appreciation to Taylor and Francis and its staff for providing seamless support in making this timely and important book possible.

Eren Özceylan, Ph.D.
Gaziantep, Turkey

Surendra M. Gupta, Ph.D., P.E.
Boston, Massachusetts, U.S.A.

About the editors

Eren Özceylan, Ph.D., is an Associate Professor of Industrial Engineering at Gaziantep University in Turkey. Dr. Özceylan holds BE and MS degrees with first honors from the Industrial Engineering Department of Selçuk University and a PhD degree in Computer Engineering from the Selçuk University, Konya, Turkey. He also had academic experience as a postdoctoral fellow at Northeastern University in Boston. His research interests are in the areas of sustainable production and distribution planning operations, including reverse logistics, closed-loop supply chains, and disassembly line balancing. He is also studying geographic information system-based, multi-criteria decision-making problems. Dr. Özceylan holds the "Science Academy Young Scientific Researchers Award" (BAGEP 2016) and the "Young Scientist Award" awarded by Gaziantep University (2017). Dr. Özceylan is the author of numerous journal and conference papers, the recipient of several research grants within his scholarly areas, and has presented his work at various national and international conferences. Dr. Özceylan is also one of the members of the Turkish Operational Research Society and INFORMS.

Surendra M. Gupta, Ph.D., is a Professor of Mechanical and Industrial Engineering and the Director of the Laboratory for Responsible Manufacturing at Northeastern University in Boston, Massachusetts, USA. He received his BE in Electronics Engineering from Birla Institute of Technology and Science, MBA from Bryant University, and MSIE and PhD in Industrial Engineering from Purdue University. He is a registered professional engineer in the State of Massachusetts, USA. Dr. Gupta's research interests span the areas of production/manufacturing systems and operations research. He is mostly interested in environmentally conscious manufacturing, reverse and closed-loop supply chains, disassembly modeling, and remanufacturing. He has authored or co-authored twelve books and over 600 technical papers published in edited books, journals, and international conference proceedings. His publications have received over 14,800 citations (with an h-index of 63) from researchers all over the world in journals, proceedings, books, and dissertations. He has traveled to all seven continents—Africa, Antarctica, Asia, Australia, Europe, North America, and South America—and presented his work at international conferences on six continents. In addition, he has delivered keynote speeches in international conferences in several countries, including Spain, The Netherlands, Denmark, France, Japan, Korea, Thailand, India, Taiwan, China, Saudi Arabia, and Turkey. Dr. Gupta has taught over 150 courses in areas such as operations research, inventory theory, queuing theory, engineering economy, supply chain management, and production planning and control. Among the many recognitions received, he is the recipient of outstanding research award and outstanding industrial engineering professor award (in recognition of teaching excellence) from Northeastern University, a distinguished professor award from IEOM Society International, and an outstanding doctoral dissertation advisor award from the American Society for Engineering Management.

Contributors

İbrahim Zeki Akyurt, Turkish Airlines Flight Academy, Aydın, Turkey.

Murtadha Aldoukhi, Department of Mechanical and Industrial Engineering, Northeastern University, Boston, MA, USA.

Muzaffer Alım, Technology Faculty, Batman University, Batman, Turkey.

Ammar Y. Alqahtani, Faculty of Engineering, Department of Industrial Engineering, King Abdulaziz University, Jeddah, Saudi Arabia.

Adil Baykasoğlu, Department of Industrial Engineering, Dokuz Eylül University, Izmir, Turkey.

Büşra Baytur, Industrial Engineering Department, Gaziantep University, Gaziantep, 27010, Turkey.

Cihan Çetinkaya, Management Information Systems Department, Adana Alparslan Türkeş Science and Technology University, Adana, Turkey.

Eray Demirel, 12th Regional Directorate of State Hydraulic Works, Kayseri, Turkey.

Neslihan Demirel, Faculty of Applied Sciences, Department of International Trade and Logistics, Kayseri University, Kayseri, Turkey.

Mehmet Talha Dulman, Department of Mechanical and Industrial Engineering, Northeastern University, Boston, MA, USA.

Gazi Murat Duman, Trefz School of Business, University of Bridgeport, Bridgeport, CT, USA.

İbrahim Miraç Eligüzel, Department of Industrial Engineering, Gaziantep University, Gaziantep, Turkey.

Hadi Gökçen, Faculty of Engineering, Department of Industrial Engineering, Gazi University, Ankara, Turkey.

Surendra M. Gupta, Department of Mechanical and Industrial Engineering, Northeastern University, Boston, MA, USA.

Mehmet Ali Ilgin, Department of Industrial Engineering, Manisa Celal Bayar University, Yunusemre, Manisa, Turkey.

Zülal Diri Kenger, Department of Industrial Engineering, Gaziantep University, Gaziantep, Turkey.

Saadettin Erhan Kesen, Industrial Engineering Department, Konya Technical University, Konya, Turkey.

Gulsen Aydin Keskin, Department of Industrial Engineering, Balikesir University, Balikesir, Turkey.

Çağrı Koç, Department of Business Administration, Social Sciences University of Ankara, Ankara, Turkey.

Elif Kongar, Departments of Mechanical Engineering and Technology Management, University of Bridgeport, Bridgeport, CT, USA.

Ibrahim Kucukkoc, Department of Industrial Engineering, Balikesir University, Balikesir, Turkey.

Zixiang Li, Engineering Research Centre for Metallurgical Automation and Measurement Technology, Wuhan University of Science and Technology, Wuhan, China.

Seamus M. McGovern, U.S. DOT National Transportation Systems Center, Cambridge, MA, USA.

Süleyman Mete, Department of Industrial Engineering, Gaziantep University, Gaziantep, Turkey.

Eren Özceylan, Department of Industrial Engineering, Gaziantep University, Gaziantep, Turkey.

Aditya Pandit, Department of Mechanical and Industrial Engineering, Northeastern University, Boston, MA, USA.

Albraa A. Rajkhan, Faculty of Engineering, Department of Industrial Engineering, King Abdulaziz University, Jeddah, Saudi Arabia.

Kemal Subulan, Faculty of Engineering, Department of Industrial Engineering, Dokuz Eylul University, Izmir, Turkey.

Özden Tozanlı, Center for Transportation and Logistics, Massachusetts Institute of Technology, Cambridge, MA, USA.

Ece Yağmur, Industrial Engineering Department, Konya Technical University, Konya, Turkey.

Tetsuo Yamada, Department of Informatics, The University of Electro-Communications, Chofu, Tokyo, Japan.

Liangchuan Zhou, Department of Mechanical and Industrial Engineering, Northeastern University, Boston, MA, USA.

1 Mitigating Customer-Initiated Warranty Fraud for Remanufactured Products in Reverse Supply Chain
A Study on the Effects of Fraud in the Remanufacturing Sector and an Attempt to Curtail them

Aditya Pandit and Surendra M. Gupta
Department of Mechanical and Industrial Engineering, Northeastern University, 334 Snell Engineering Center, 360 Huntington Avenue, Boston, MA 02115, USA

1 INTRODUCTION

A remanufactured product may perform as well as a new product; however, the consumer may not agree. This shows a level of uncertainty in the mind of the consumer regarding the quality of the remanufactured product and might lead to a decision to decide against buying it. Remanufacturers, therefore, feel the need to provide additional assurance to the consumer through product warranties. In systems that involve multiple parties, each with their own goals, motivations, and competing interests working together, it is inevitable to avoid fraud. Warranty fraud frequently occurs when a component part—not within the manufacturer's warranty period—is placed in a product within the manufacturer's warranty period. The product is returned to the manufacturer for replacement of the allegedly defective component to pursue a warranty. The absence of an effective means to control the fraudulent substitution of warranted parts imposes substantial costs on the manufacturers and consumers, and the products there are used (Hayes, 2005).

This type of fraud can also be seen in larger products such as automobiles, to smaller products like consumer electronics. This chapter attempts to improve existing methods of fraud prevention/detection by considering the implementation of sensors. The success or failure of sensor implementation will depend on any improvements in the proposed systems' fraud statistics, which include the number of frauds that go unchecked, the total cost of inspection, revenue lost from fraud, and the number of falsely charged claims. Using discrete event simulation, the study was able to conclude that sensor implementation has positive benefits when it comes to fraud detection and prevention, and this work acts as a demonstration for more complex fraud scenarios (that may involve two or more parties).

2 LITERATURE REVIEW

Over the past few decades, there has been an interest in environmentally conscious manufacturing which arose partly due to the implementation of government legislation and the dwindling of natural resources. The field of Environmentally Conscious Manufacturing and Product Recovery (ECMPRO) has produced many research papers that are of interest in setting up this study. Gungor and Gupta (1999) presented a state-of-the-art survey paper covering those published through 1998 in the field of ECMPRO. A later paper by Ilgin and Gupta (2010) presented another updated survey in the same area. These reviews showed building interest and while certain issues may not currently be prevalent (due to lack of product volume), preemptive steps should be taken to deal with problems that will acerbate with time—such as fraud.

One problem that needs to be confronted in remanufacturing operation is the high degree of variability in operation processing times. According to Ilgin and Gupta (2010), the level of uncertainty in the quantity, quality, and timing of returned products further complicates the analysis of remanufacturing systems. Therefore, it is also important for manufacturers to balance their sales strategies in response to the introduction of remanufacturing. Some large manufacturers have adopted remanufacturing marketing programs to promote recycling and selling of the remanufactured or refurbished products (Wu, 2012).

Another study by Guide and Li (2010) carried out auctions to understand the willingness of a consumer to pay for both new and remanufactured consumer and commercial products. Additionally, Vadde, Kamarthi, and Gupta (2006) focused on the pricing strategies with obsolescence. They regarded two models to counter the prospect of product obsolescence gradually and suddenly, considering the risk in demand decrease and inventory increase.

In the past, warranty has been used for competitive marketing. Many manufacturers and researchers have explored a host of ways to make a product more appealing by experimenting with warranty policies—adding additional services, extending the periods, offering favorable terms, etc. Podolyakina (2017) identified the relative level of cost incurred by the manufacturer to fully satisfy the consumers' warranty expectation—there is evidence that more expensive complex products involve extended decision making, thereby enhancing the probability that the warranty plays a significant role (Wilkes and Wilcox, 1981). The consumer relies more often on the manufacturer than they do on the retailer when it comes to the nature and quality of the goods—this fact has strongly influenced courts in warranty cases (Southwick, 1963). Most of the extant literature was focused on warranties in the new product industry; many of the same issues have been tackled in the remanufacturing sector as well. Alqahtani and Gupta (2017) considered a two-dimensional warranty policy to maximize consumer confidence and minimize cost for a remanufactured product (washing machine).

Warranty providers who outsource their product maintenance claims to external service agents who charge between 1-5% for their services (Murthy and Jack, 2016). Warranty fraud is a significant problem, affecting motor vehicles and other consumer products with multiple components that are the subject of warranty. These frauds are typically only between the manufacturer and the service agent. A study by Jack and Murthy (2017) used game theory to tackle service agent fraud in cases where the latter exaggerates the value of the claim to the manufacturer. The concept of warranty fraud in the remanufacturing industry was introduced and reviewed by Pandit and Gupta (2018a). Another paper by Pandit and Gupta (2019a) reviewed the issue of service agent remanufacturing fraud and its effect on the warranty provider by modeling the problem using discrete event simulation. In this paper, they discussed possible methods to alleviate fraud. Similarly, the issue of remanufacturing product fraud was modeled using game theory (Pandit and Gupta, 2019b), where they established the relationships between several factors such as fraud size, inspection frequency, and penalty amount. A study by Pandit and Gupta (2018b) first outlined the issue of remanufactured warranty fraud originating from the customer using a discrete event simulation model that served as the starting point for this study.

There are methods and systems to obtain and analyze data using embedded sensors in electronic products for warranty management. A data collection unit in an electronic product collects and reports data about environmental factors that are relevant to a warranty agreement and transmits the data over the communications link to a data interpretation unit. The use of sensors in consumer electronics products has been suggested in the past to consider matters both in EOL (Ilgin and Gupta, 2011, for washing machines) and during a product's (laptop) lifetime (Dulman and Gupta, 2018), suggesting that sensor-embedded products (SEP) could be an effective way to catch fraud. Sensors have many possibly useful applications to combat warranty fraud (Figure 1.1).

Based on the literature review, we can see that fraud is still a prevalent issue in the consumer product industry. However, most extant literature focuses on new products; little research is focused on frauds pertaining to remanufactured ones. While several techniques (such as game theory) have been used in the past, many models consider fraud and do not incorporate factors such as prior history and party motivation. Previous remanufacturing fraud studies have focused on fraud from warranty service agents, warranty administrators, and product parts providers but these studies have neglected an especially important entity in the warranty service chain—the customer. Additionally, the trend of sensor implementation has proven to be a boon in solving other problems, and literature shows that it has the potential to deal with fraud. These points influenced the direction and methodology followed in this study.

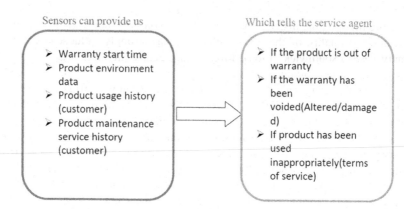

Sensors can provide us

> Warranty start time
> Product environment data
> Product usage history (customer)
> Product maintenance service history (customer)

Which tells the service agent

> If the product is out of warranty
> If the warranty has been voided(Altered/damaged)
> If product has been used inappropriately(terms of service)

FIGURE 1.1 Sensor Application in Warranty Service.

3 FRAUD OVERVIEW

Apart from costs such as revenue loss, warranty fraud has many indirect consequences. Repair data received from service agents are often used for research and development. Manufacturers use information acquired from field data to address product quality issues and take corrective action. This idea works well in principle but if this data includes fictitious or claims of dubious nature, separating real quality problems from fraudulent billing can be difficult. As a result, unreliable product quality feedback may delay corrective actions (Kurvinen, Töyrylä, and Murthy, 2016). This is an escalating problem, as it might also result in product recalls which are unnecessary and overly expensive, and reduce the effect of predictive maintenance. This also leads to incorrect decision-making, as the manufacturer might terminate profitable products that are incorrectly assumed to be generating loss. Warranty fraud can also affect the warranty provider's reputation and damage its brand. This is the case, for example, when customers are wrongly denied warranty service or receive poor service due to fraud (Kurvinen, Töyrylä, and Murthy, 2016). While many studies assume the customer as the victim of fraud, in this instance we try to examine how they attempt to defraud other parties.

The occurrence of customer-driven fraud is not confined to just the warranty period, as it often takes place after the warranty has expired. Fraudulent warranty claims may also be associated with products that have never been purchased by the customer. Customer fraud typically relates to cases where a customer tries to have a repair or replacement done although there is no warranty coverage—the warranty has expired, the customer has damaged the product, or the service action is not really covered by warranty. Usually, the fraud done by a customer is limited to one or a few cases per purchased product, although large-scale fraud also exists (Kurvinen, Töyrylä, and Murthy, 2016). Several different instances of warranty fraud are compiled in Tables 1.1 and 1.2 (summarized by Kurvinen, Töyrylä, and Murthy, 2016).

We attempt to narrow the scope of the customer warranty fraud by examining a specific scenario of fraud, namely the case of component substitution. For instance,

TABLE 1.1
Table of Frauds Originating from Customer

Fraud Victim	Motivation	Method
Warranty provider	Refund or replacement	Unjustified return or replacement of an item that is not faulty or is fake.
	Service cost avoidance	Getting out of warranty products repaired under warranty.
	Extra products or earning	Claiming and reselling parts or replacement items.
	Service level improvement	Claiming better service than entitled for.

TABLE 1.2
Table of Frauds Where Customer Is Victim

Fraud Source	Motivation	Method
Service agent	Extra revenue	Overselling extended warranties.
	Extra revenue	Selling nonexistent extended warranties.
	Extra revenue	Charging the customer for in-warranty service.
	Extra revenue	Upselling a replacement product when the customer is entitled to a warranty service.
	Extra revenue	Purposeful lack of first-time resolution.
Warranty provider	Extra income	Overselling extended warranties.
	Extra income	Selling nonexistent warranties.
	Extra income	Charging the customer for in-warranty service.
Sales channel	Extra revenue	Overselling extended warranties.
	Extra revenue	Selling nonexistent warranties.
	Inventory refreshment	Getting a new item from OEM and using a secondhand item for replacement.

having bought multiple numbers of similar products at separate times (e.g., LED light bulbs or tablets), the customer may try to utilize the warranty of the newest product to have the old one replaced. In a comparable way, defective parts in an out-of-warranty product can be changed to an in-warranty product and claimed under warranty. If the equipment or the proof of purchase does not have a serial number or the parts used are not serial number tracked, it is hard to control whether the proof of purchase relates to the faulty product. The same applies to large installations that consist of multiple equipment installed at various times and contain in-warranty and out-of-warranty products. In this situation, gaps in the warranty provider's entitlement process or data can be used to obtain service also for out-of-warranty products. In this chapter, we attempt to model an instance of substitution type customer-driven warranty fraud for a consumer electronic product.

3.1 DESCRIPTION OF THE SYSTEM

Based on relative importance, the different parties can be subdivided into three categories described in Table 1.3 (assume that the remanufacturer is the primary warranty provider). Primary parties refer to those chiefly involved in the warranty service chain, secondary parties assist primary parties in warranty operations, and tertiary parties have no direct role in the warranty operations, but they are invested in the performance of the warranty service.

Component substitution fraud mainly involves the primary parties—the consumer is the fraud's perpetrator and the service agents (and, by extension, the remanufacturer) are the victims.

In a typical warranty service system, when a product is rendered nonfunctional, it is inspected to determine the cause of failure. The information about such failure is transmitted to the service personnel who conducts the required service operations (e.g., replacing the failed components). After this process, deficient products are transferred to the service facility. After the maintenance process, the products are brought back to working condition. Once the maintenance service operations are complete, the products are returned to the customers. The generalized activity flow chart for all types of warranty maintenance is shown in Figure 1.2.

The case presented assumes that the claim is true and that the customer is acting in good faith. However, when introducing the concept of fraud into the scenario, it is assumed that for certain cases, at least one component in the product originates from another product no longer under warranty. The service of such a product would violate the warranty agreement between the customer and the remanufacturer. In addition to the extra service cost incurred, there would also be additional inspection and logistics costs to account for. The next section describes two simulation scenarios that were considered to model the problem further.

TABLE 1.3
Parties Involved in a Warranty Servicing Scenario

Category	Parties
Primary	Remanufacturer, service agents, customers
Secondary	Parts manufacturer, sales channel, warranty administrator
Tertiary	Leasers, inspectors, logistics companies, underwriters and insurers, government, shareholders

---

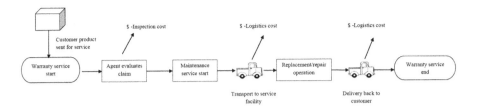

FIGURE 1.2 Warranty Service Operation for Products.

4 METHODOLOGY

This section discusses the process flow and model parameters that are pertinent to this study. The warranty service is assumed to be performed by a single service maintenance agency. Kurvinen, Töyrylä, and Murthy (2016) outlined the various service and return channels that are commonly employed. For regular claims, service begins when the item/product under warranty is submitted to an authorized service center/retailer (customer service). Customer service responsibilities include the return merchandise authorization process (RMA), the entitlement to verify customer's eligibility for warranty repair within the warranty period. Upon maintenance completion, the product is returned to the customer (return channel).

4.1 BASIC FLOW CHART

Figure 1.3 gives the generalized process flow of warranty customer claims. Customer (or customer service) passes the claim to the service agent after claim entitlement is conducted to ensure that the product is in warranty. The maintenance activities under consideration fall under repair or replacement. When the repair cost exceeds the replacement cost, the product is replaced instead. The decision to investigate a service claim from a customer after passing through the entitlement process is assumed to take place at random (random inspection). Additionally, we consider the inclusion of certain warning flags that might result in an inspection. For the purposes of the study, it is assumed that the claims (both fraudulent and true) pass through the RMA and entitlement processes without issue, as the study focuses on what comes after (the claims investigation). Table 1.4 below outlines the numerous factors that affect the customer claim, investigation process, and sensor behavior (for sensor-embedded systems).

4.2 SIMULATION MODEL SCENARIOS

The study uses discrete event simulation to model customer-driven fraud. The model was simulated using Rockwell arena software; the input to the simulation model are customer claims which arrive at a rate governed by distribution functions. The logistics costs were assumed at a fixed rate per trip (to and from the repair center), while the inspection cost was assumed on an hourly basis. We consider the following two cases for our research.

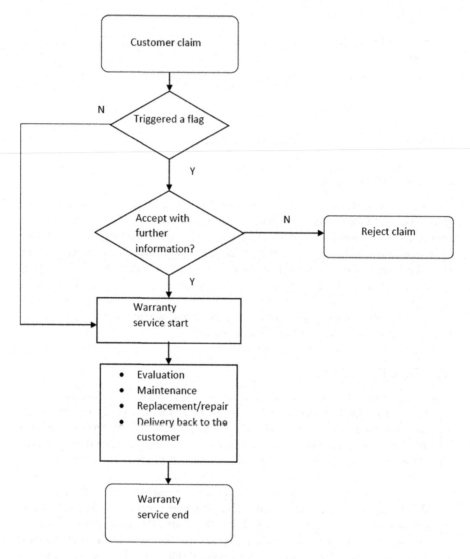

FIGURE 1.3   Basic Flow Chart (Warranty).

Scenario 1 – Regular Systems

In a regular system (RL), when a warranty claim is made by the customer, it is the service agent's responsibility to first verify that the claim is true and conduct any services to fulfill the warranty contract. Built-in systems exist to catch substitution frauds. For instance, in electronic products, each part possesses serial numbers that can be cross-checked to determine if the component is from the same product or series. This process is fairly time intensive and cannot catch all types of substitution fraud (components that may be used across multiple product series such as batteries might be substituted between products of different generations without the service

TABLE 1.4
Simulation Model Factors

Category	Subcategory	Assumptions and justifications
Factors that push the customer into committing fraud	*Prior fraud history*	It is also assumed that a customer's prior experience with the investigation process will bias them either toward or away from fraud. For example, if a customer gets away with committing fraud once, then the probability that they will commit fraud next time increases. Conversely, if a customer is caught, then the probability of committing fraud again is similarly reduced.
	Confidence in the investigation process	If the customer can commit fraud in multiple successive instances without being caught by either the audit or secondary inspection, the probability that the customer would commit fraud would increase with each successive escape.
Factors that affect claim processing	*Prior criminal record*	The prior record of the customer (only of caught fraud not committed fraud) will influence the investigator's decision to accept the claims as true.
	Warning flags	There are certain indicators that might set warning flags when a claim is examined. (an unusual number of claims in a fixed time period).
	Inspection error	There are several reasons that the inspection claims can be imperfect. Human error can occur due to a number of reasons, though most often by inexperience or negligence.
Factors that affect sensor behavior	*Claim data on record*	As the sensor can provide better records (dates, times, location data, etc.), it will more accurately determine if the submitted claims are true.
	Sensor damage	There is always the possibility that failure of a sensor may occur, either deliberately or by accident, and it is often not easy to determine if it was done maliciously to cover up fraud, or just an accident.

agent being any wiser). In cases of fraud, we consider multiple forms of cost to the remanufacturer. Productivity loss occurs when service agents try to rectify fraudulent claims while putting true claims on hold. In the system, productivity loss time is calculated by determining the time between receiving the faulty claim and the service completion time. Maintenance costs include the costs that are incurred because of inspecting the product failure. Since a corrective maintenance strategy is chosen and the failed components are replaced, the material cost associated with the replacement of the components is added to the overall maintenance cost. Finally, logistics costs are also taken into consideration within maintenance costs.

Scenario 2 – With Sensors

In sensor-embedded scenarios (SE), we assume that sensors can make up for some of the shortfalls that exist in regular systems. To highlight this effect, we consider the existence of human error in the model and consider claims to be reviewed by multiple inspectors that may fall into one of four categories, each with slightly different process times and fraud detection capabilities. The model assumes that the customer's past fraud history will influence the decision to commit fraud in the future. For instance, the model assumes that if a customer has been previously successful in committing fraud, the truthfulness of the next claim will be different if they were previously caught. In addition, sensors are assumed to retain data on parts that have been replaced in previous inspections which theoretically should prevent a repeat of fraud using the same product. In theory, sensors would shorten the time required to verify all the components from the same product and prevent additional substitution scenarios (the battery example). In addition to costs that are incurred in the regular scenario, we also assume the cost of sensor implementation. A product recall is a request from a manufacturer to return a product after the discovery of safety issues or product defects that might endanger the consumer or put the maker/seller at risk of legal action. We consider the possibility of fraud resulting in an eventual product recall. This will allow us to determine if the cost of sensor implementation is worth any potential fraud catching benefits. The next section outlines the basic assumptions that were made to simplify our model.

4.3   SIMULATION MODEL PARAMETERS

Model assumptions. The simulation model simplifies many of the complexities of the warranty service chain to narrow the studies' focus on the fraudulent claim investigation process. The assumptions that have been made are as follows:

- The warranty service starts when the defective product is brought to the local service center, assuming customers can clear RMA online.
- The product (supposedly) experiences either a partial failure (Intermittent) or complete failure (nonfunctional).
- Maintenance activities are primarily corrective in nature.
- Do not assume a product with an extended warranty.
- Assume that some amount of disassembly will be required for all claims.
- More than one instance of fraud ends the warranty provider's obligation to the customer (benefit of the doubt).
- If found fraudulent, the customer is charged for transportation costs.
- Assume that the SA and WP are behaving honestly.

With the model assumptions laid out, the next section details how the distinct stages of the warranty are modeled.

4.4 MODEL FORMULATION

4.4.1 The Investigation Processes

The claims investigation process can occur due to one of two reasons, the first being random inspection and the second being the existence of a suspicious claim resulting in an inspection. Figure 1.4 models the probability of a customer claim being selected for a random inspection; the factors that determine a claim to be suspicious were previously outlined in Table 1.4 (simulation model factors). The model makes an allowance for inspection errors (Type I and Type II) and for errors due to experience (job experience). Table 1.5 categorizes the different investigators into one of four categories: competent, expert, incompetent, and novice.

4.4.2 Warranty Claims

a Normal Claims

A single customer may raise zero, one, or more warranty claims over the warranty period. White goods (i.e. refrigerators, ovens, and other large household appliances), consumer electronics, and watches, for example, typically have little to no repairs during their warranty period. More complex products like cars or large industrial installations will have a higher number of claims over the warranty period. Customer warranty claims occur in an uncertain manner over time. The cumulative count of claims follows a stochastic (unpredictable) counting process over time (Kurvinen, Töyrylä, and Murthy 2016).

b Fraud Claims

The underlying reasons for fraud are vast and complex; we attempt to model the issue of fraud from an engineering perspective; therefore, we isolate the two

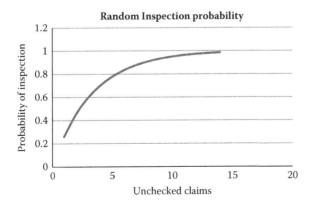

FIGURE 1.4 Probability of Random Inspection.

TABLE 1.5
The Investigation Process

Type	Probability of Detecting Fraud	Probability of Falsely Charging Customer
Inspector 1	80	10
Inspector 2	90	5
Inspector 3	65	15
Inspector 4	70	15

primary factors (from an engineering standpoint)—they both stem from the customer's confidence in the investigation process (Table 1.6).

4.4.3   Fraud and Logistics Costs

Many costs that are associated with warranty service operation exists; this study focuses on those that pertain/highlight the fraud-related aspect of this problem. Table 1.7 lists these costs.

4.4.4   Remanufacturer

Arguably the worst possible consequence of warranty fraud is a product recall since an artificial inflation in the number of returned products can be misconstrued by the product remanufacturer as a genuine product flaw and lead to mass withdrawal of the product. The exact nature of this process is dependent on the type of product, product flaw (if the latter causes bodily harm to consumers), and so on. For the model, we assume a cutoff percentage that triggers the initial stages of a product recall as the starting point. We simulate the model for different cutoff percentages (Table 1.8) to observe how it affects fraudulent claim processing.

TABLE 1.6
Factors That Influence Service Agent Fraud

Factors	Rationale
Previously escaped	If the customer has previously committed a fraud and was caught, it is assumed that the customer will be more timid (less likely to commit fraud) the next time they make a claim (Figure 1.5) and vice versa.
Total number of claims	It is assumed that the probability of fraud increases proportionally with the number of claims (especially if the customer has never been caught in the past) (Figure 1.6).

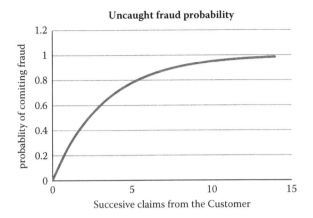

FIGURE 1.5 Probability of Customer Committing Fraud after Being Caught.

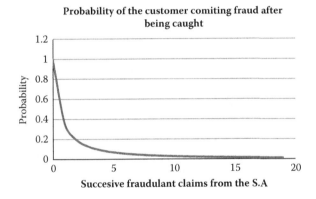

FIGURE 1.6 Probability of Customer Committing Fraud as a Factor of the Overall Number of Claims Submitted.

4.4.5 Sensor Behavior Assumptions

Based on work by Dulman and Gupta (2018) and Ondemir and Gupta (2014), we assume the following relationship (Figure 1.7) between the cost of sensor implementation and the confidence (probability) that the sensors accurately determined a fraudulent claim. The aim of sensor implementation is to free auditors to look at claims that are more likely to be fraudulent and reduce the need to look at a random selection of claims.

4.5 SIMULATION MODEL OUTPUT

4.5.1 Desired Output Metrics

In order to assess the efficacy of the RL and SE system, a number of output statistics were deemed necessary. Table 1.9 lists the parameters of relevance to this chapter.

TABLE 1.7
Costs Pertaining to the Model

Cost	Explanation
Logistics costs	These include the costs to transport the item from the customer to the maintenance center, and back to the customer.
Investigation costs	The cost of investigating the fraudulent claim.
Uncaught Fraud costs	This refers to the loss to the manufacturer in being defrauded which, in this instance, is equal to the cost of the component being replaced for free.
Total cost	The net cost to the remanufacturer.

5   NUMERICAL EXAMPLE

5.1   CASE 1 – LAPTOP

We consider the number of warranty claims by customers to a service agent for a small-scale electronic appliance. The laptop has an average lifespan of four years, the warranty length is 14 months. There is one component that is substituted (memory RAM card). The data presents the number of claims per month that the service agent receives over three years (Figure 1.8, Table 1.10).

5.2   CASE 2 – MOBILE PHONES

We consider the number of warranty claims by customers to a service agent for a small-scale electronic appliance. The laptop has an average lifespan of four years, the warranty length is 14 months. There are two components that are being substituted (battery and memory card). The data presents the number of claims per month that the service agent receives over three years (Figure 1.9, Table 1.11).

TABLE 1.8
Product Recall Scenarios

Product Recall Cases	Percentage of Failed Products of the Total
1	5
2	10
3	11
4	12
5	13
6	14
7	15

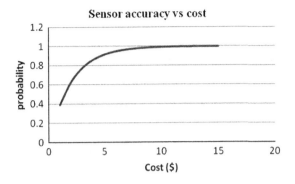

FIGURE 1.7   Relationship between Sensor Accuracy and Sensor Costs.

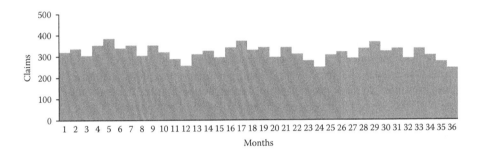

FIGURE 1.8   Number of Monthly Laptop Warranty Claims over Three Years.

TABLE 1.9
Desired Output Statistics

Output Statistics	Regular Systems	Sensor Embedded Systems
Escaped fraud cases	✓	✓
Fraud cases that escape the sensor	–	✓
Falsely charged	✓	✓
Successfully caught fraud	✓	✓
Number of claims processed	✓	✓
Number of claims flagged by sensor	–	✓
Claims found to be valid	✓	✓
Time spend processing claim	✓	✓
Number of falsely flagged true claims	–	✓
Logistics costs	✓	✓

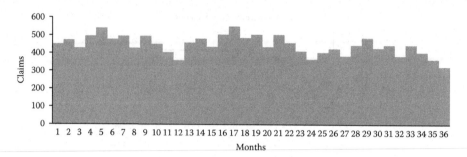

FIGURE 1.9   Number of Monthly Mobile Phone Warranty Claims over Three Years.

6   RESULTS AND DISCUSSIONS

For Case 1, each system (RS and SEP), experiments were carried out and the data pertaining to the total profit, inspection cost, and penalty and fraud costs were tracked. Table 1.12 contrasts the difference between the base scenario and the SEP scenario by describing statistics pertinent to fraud detection and inspections.

The data indicated that the use of sensors significantly reduced the cost of the inspection process but there were certain parameters that did not follow this pattern—for example, in normal systems, the inspection cost for the simulation period was $81738 while this increased to $86677 for the SEP systems. Table 1.13 presents the average values of the performance measures mentioned above, as well as the total cost for both systems.

Results (the results of the pairwise t-tests, including mean differences and p-values, are presented in Table 1.14.) show that the sensor embedded system shows a statistically significant improvement over the base scenario with respect to fraud

TABLE 1.10
Laptop Warranty Claims Over Three Years

Month\Total Claims per Month	Year 1	Year 2	Year 3
Jan	320	310	305
Feb	336	326	320
Mar	304	295	290
Apr	352	341	336
May	384	372	366
Jun	339	329	323
July	352	341	336
Aug	304	295	290
Sep	352	341	336
Oct	320	310	305
Nov	288	279	275
Dec	256	248	244

TABLE 1.11
Mobile Phone Warranty Claims over Three Years

Month\Total Claims per Month	Year 1	Year 2	Year 3
Jan	450	458	405
Feb	473	481	425
Mar	428	435	385
Apr	495	504	446
May	540	550	486
Jun	477	485	429
July	495	504	446
Aug	428	435	385
Sep	495	504	446
Oct	450	458	405
Nov	405	412	365
Dec	360	366	324

TABLE 1.12
Fraud Detection Statistics for Laptop

Service Agent Claims Statistics	Base Case (RS)	With Sensors (SEP)	Percentage Improvement (%)
Uncaught false claims	1438.92	1140	25.797
Uncaught false claims (only manual)	1438.92	889.2	N/A
Uncaught false claims (with sensor)	–	250.8	N/A
Falsely charged	548.16	524.4	4.267
Caught fraud (from investigation)	913.6	980.4	7.454
Total general audits	11420	11400	–
Total claims flagged by sensors	–	1664.4	N/A
Total inspection	5458.76	5768.4	5.877(decrease)
True claims (from investigation)	3745.76	4058.4	8.294
Average time in system	35.071 hrs	29.10 hrs	3.164 (decrease)
No. of SA claims that fools sensors	–	296.4	N/A
No. of true SA claims that are flagged by sensors	–	1231.2	N/A
Approved claims	9958.24	9895.2	0.499
Max queue length	2	1	50

TABLE 1.13

Fraud Detection Statistics

Measure	Base System ($)	Sensor-Embedded System ($)
Inspection costs	81738	86677
Uncaught fraud costs	50274	39970
Total costs	189012	183747

TABLE 1.14

Pairwise *t*-Test Results for Mean Difference

Measure	Mean Difference ($)	p-Value
Inspection costs	−4939.8	<0.0001
Uncaught fraud costs	10304	<0.0001
Total costs	5264.2	<0.0001

detection. For Case 2, experiments were carried out for each system (RS and SEP), and the data pertaining to the total profit, inspection cost, and penalty and fraud costs were tracked. Table 1.15 contrasts the difference between the base scenario and the SEP scenario by describing statistics pertinent to fraud detection and inspections.

The data indicated that the use of sensors significantly reduced the cost of the inspection process but there were certain parameters that did not follow this pattern— for example, in normal systems, the inspection cost for the simulation period was $118305 while this reduced to $125235 in the SEP systems. Table 1.16 presents the average values of the performance measures mentioned above, as well as the total cost for both systems.

Preliminary results (the results of the pairwise *t*-tests, including mean differences and p-values, are presented in Table 1.17.) show that the sensor embedded system shows a statistically significant improvement over the base scenario with respect to fraud detection. A number of correlations between different factors were also observed. If subsequent frauds are considered to be unbiased by previous claims, we see that the propensity for committing higher types (in terms of cost) of fraud exists. If, however, we consider that there is a link between previous claims and subsequent fraud, we see the number of larger-sized frauds decrease; likewise, the overall sensor value also goes down. There is a negative correlation between sensor value and the total number of frauds committed, while there is a positive correlation between inspection cost and the total cost. In summary, it is possible to use sensors to not just combat, but also better track fraud. Sensors also provide additional

TABLE 1.15
Fraud Detection Statistics for Mobile Phone

Service Agent Claims Statistics	Base Case (RS)	With Sensors (SEP)	Percentage Improvement (%)
Uncaught false claims	2020.41	1603.5	26
Uncaught false claims (only manual)	2020.41	1250.73	N/A
Uncaught false claims (with sensor)	-	352.77	N/A
Falsely charged	769.68	737.61	4.3
Caught fraud (from investigation)	1282.8	1379.01	7.5
Total general audits	16035	16035	–
Total claims flagged by sensors		2341.11	N/A
Total inspection	7664.73	8113.71	5.9 (decrease)
True claims (from investigation)	5259.48	5708.46	8.34
Average time in system	300.2971	309.7994	3.164 (decrease)
No. of SA claims that fools Sensors	–	416.91	N/A
No. of true SA claims that are flagged by sensors	–	1731.78	N/A
Approved claims	13982.52	13918.38	0.5
Max queue length	4	1	

TABLE 1.16
Fraud Detection Statistics

Measure	Base System ($)	Sensor-Embedded System ($)
Inspection costs	118305	125235
Uncaught fraud costs	72765	57750
Total costs	273570	265485

TABLE 1.17
Pairwise *t*-Test Results for Mean Difference

Measure	Mean Difference($)	p-Value
Inspection costs	−6930	<0.0001
Uncaught fraud costs	15015	<0.0001
Total costs	8085	<0.0001

benefits because they can be used to gain an economic advantage in a closed-loop supply chain system.

7 CONCLUSIONS

Results show that the sensor embedded system performed better than the regular system in terms of its ability to catch and deter frauds. For Case 2, there was a 22% decrease in the number of fraudulent claims that escaped, a 17% decrease in the average time it takes to process claims, and an 18.6% decrease in total logistics costs. Based on early results, it was judged that the cost savings from fraud and maintenance justified the price of sensor implementation. This chapter described the literature surrounding the issues of fraud and warranties. This review served to show the importance of tackling fraud in the remanufacturing service industry. The different ways in which customer-driven fraud affects the warranty service industry were elaborated. Lastly, the problem was modeled using simulation to better understand and combat warranty fraud. Only one potential application of the sensor dealing with fraud was modeled, but there are many more that have yet to be tackled. The issue of warranty fraud has been an overlooked problem in the area of remanufacturing, and further exploration would prove beneficial in promoting remanufacturing as a whole.

REFERENCES

Alqahtani, A. Y., & Gupta, S. M. 2017. "Warranty as a marketing strategy for remanufactured products." *Journal of Cleaner Production*, 161, 1294–1307.

Dulman, M. T., & Gupta, S. M. 2018. "Evaluation of maintenance and EOL operation performance of sensor-embedded laptops." *Logistics*, 2(3), 1–22.

Guide, V. D. R., & Li, J. 2010. "The potential for cannibalization of new products sales by remanufactured products." *Decision Sciences*, 41(3), 547–572.

Gungor, A., & Gupta, S. M. 1999. "Issues in environmentally conscious manufacturing and product recovery: a survey." *Computers & Industrial Engineering*, 36(4), 811–853.

Hayes, D. A. 2005. Vehicle control system with radio frequency identification tag. Pyper Products Corporation, Patent.

Ilgin, M. A., & Gupta, S. M. 2010. "Environmentally conscious manufacturing and product recovery (ECMPRO): A review of the state of the art." *Journal of Environmental Management*, 91(3), 563–591.

Ilgin, M. A., & Gupta, S. M. 2011. "Recovery of sensor embedded washing machines using a multi-kanban controlled disassembly line." *Robotics and Computer Integrated Manufacturing*, 27(2), 318–334.

Jack, N., & Murthy, D. N. P. 2017. "Game theoretic modelling of service agent warranty fraud." *Journal of the Operational Research Society* 68(11), 1399–1409.

Kurvinen, M., Töyrylä, I., & Murthy, D. N. P. 2016. *Warranty Fraud Management: Reducing Fraud and Other Excess Costs in Warranty and Service Operations.* Hoboken, New Jersey: Wiley.

Murthy, D. N. P., & Jack, N. 2016. "Game theoretic modelling of service agent warranty fraud." School of Mechanical and Mining Engineering, The University of Queensland, St Lucia, QLD 4072, Australia; and 2Springfield, Fife, Scotland KY15 5SA, UK.

Ondemir, O., & Gupta, S. M. 2014. "Quality management in product recovery using the Internet of Things: An optimization approach." *Computers in Industry*, 65(3), 491–504.

Pandit, A., & Gupta, S. M., 2018a. "Warranty fraud in remanufacturing." Proceedings of the Global Interdisciplinary Conference: Green Cities, June 27-30, Nancy, France.

Pandit, A., Gupta S. M. 2018b. "Mitigating customer driven fraud for remanufactured products in reverse supply chain." Proceedings of 16th International Logistics and Supply Chain Congress, Denizli, Turkey.

Pandit, A. & Gupta, S. M., 2019a. "Warranty fraud in a remanufacturing environment," in *Responsible Manufacturing – Issues Pertaining to Sustainability*, Edited by A. Y. Alqahtani, E. Kongar, K. K. Pochampally, & S. M. Gupta, CRC Press, 11, 241–261, ISBN: 978-0815375074.

Pandit, A., & Gupta, S. M. 2019b. "Impact of warranty fraud on remanufactured products." Proceedings of International Conference on Remanufacturing, Amsterdam, Netherlands.

Podolyakina, N. 2017. "Estimation of the relationship between the products reliability, period of their warranty service and the value of the enterprise cost." *Procedia Engineering*, 178, 558–568.

Southwick, A. F. 1963. "Mass marketing and warranty liability." *Journal of Marketing*, 27(2), 6–12

Vadde, S., Kamarthi, S. V., & Gupta, S. M. 2006. "Pricing of end-of-life items with obsolescence." Proceedings of the 2006 IEEE International Symposium on Electronics and the Environment, Volume 11(pp. 156–160), San Francisco, California.

Wilkes, R. E., & Wilcox, J. B. 1981. "Limited versus full warranties: the retail perspective." *Journal of Retailing*, 57, 65–76.

Wu, C. H. 2012. "Price and service competition between new and remanufactured products in a two-echelon supply chain." *International Journal of Production Economics*, 140(1), 496–507.

2 Price Decision for New and Remanufactured High-Technology Products Under Vertical Nash Model

Liangchuan Zhou and Surendra M. Gupta
Department of Mechanical and Industrial Engineering,
Northeastern University, 334 Snell Engineering Center, 360
Huntington Avenue, Boston, MA 02115, USA

1 INTRODUCTION

The concept of environmentally conscious manufacturing and product recovery (ECMPRO) is interpreted as the manufacturers' responsibility regarding take-back legislation, customers' awareness of green products, and economics (Andrew-Munot et al., 2015). Based on the literature surveys in the year 1999 (Gungor and Gupta, 1999) and 2010 (Ilgin and Gupta, 2010), much attention was focused on environmentally conscious design and production, material recycling, remanufacturing, reverse, and closed-loop supply chains. However, only a few of them discussed the marketing-related issues for remanufactured products, especially for short lifecycle items. Hi-tech products such as smartphones and laptops are short lifecycle items. Manufacturers launch new generation models intensively to meet customer's requirements on fashion design and technology innovation. Customers tend to purchase new models even when the old one has much remaining value. Remanufacturing practices make use of old products and avoid the massive use of new materials—it is a comprehensive and rigorous industrial process where a previously sold, worn, or non-functional product or component is returned to a "like-new" or "better-than-new" condition and warranted in performance level and quality (Council). These products also target low-end and green customers and are sold at optimal prices, which are decided by the customer's perceived value.

Heterogeneous customers vary on perceived values on different products. For example, quality-conscious customers prefer brand-new products, so they are assumed to have higher perceived value on quality and lower perceived value on technology than that of technology-savvies. However, value-based price decision is also influenced by market competition. The market situation of hi-tech products is complicated. On the side of the manufacturer, a new model is released when the old product is still in the inventory; on the side of customers, they purchased the latest model and throw away the old one with much residual value. The end-of-use products become the supply of remanufactured items. Therefore, new and re-manufactured products belonging to different generations coexist in the market. To manage the overall profit for all kinds of products well, manufacturers need to identify the market position of several types of items, selling them at different prices to different customers, and avoid cannibalization with new models.

In the view of a supply chain system, manufacturers and retailers need to find optimal pricing policies for different types of products. A successful marketing strategy for new and remanufactured products can address the customer's perceived value and keep the equilibrium in profit from new and remanufactured products of different generations for all supply chain members.

This chapter will address the decision making on the wholesale and retail price decisions for the new product belonging to the earlier generation and the remanufactured products belonging to the latest generations in a Vertical Nash Equilibrium environment. In the following sections, previous papers regarding marketing policies for heterogeneous customers, supply chain structure po-licies for the system members, and competition among supply chain members will be presented. Then, the customer value-based price model will be given. Numerical examples will help understand the impact of the population pro-portion of each customer segment on the optimal profit, and how perceived value differentiation between two segments influence price decision and profit.

2   LITERATURE REVIEW

Many researchers focused on pricing decisions for new and remanufactured pro-ducts. In the hi-tech industry or apparel industry, manufacturers release new gen-eration models or seasonal products regularly. Prices of commodities differ for various generations and seasons and are also dependent on product condition. Remanufactured and used products usually have lower prices than the new products due to the apparent inferior quality. Researchers explore the optimal price for high-technology or seasonal products in three ways: first, the price can be decided by customers' reservation value in diverse segments for several types of products. Second, the price can be impacted by customers' purchase preferences in different sales channels and bargaining behavior. Third, the price can be adjusted by vertical and horizontal competition and cooperation among different manufacturers and retailers.

Finding an optimal price helps balance the tradeoff between generating more revenue and losing business opportunities. In many cases, an optimal price is based on the customer's perception. Customer's acceptance of the products and will-ingness to pay (WTP) is critical in pricing strategy. To understand people's purchase behavior, Kuo and Huang examined the dynamic retail price decisions for the products from two different generations, considering inventory level and remaining selling periods (Kuo and Huang, 2012). They generalized the Nash bargaining solution model in two scenarios—posted-pricing-first and negotiation-first—to find the optimal price. Customers are classified into two segments—price-takers and bargainers. Authors suggested an increased posted price when bargainers are the majorities, while negotiable price is the optimal strategy for an earlier generation of electronic products. Zhu and Yu also conducted a case study on new, remanufactured, and refurbished electronic products to analyze customers' buying behavior, but they paid more attention to the service level (Zhu and Yu, 2018). Under the dynamic game model, they found that three products' service differentiation is reduced as consumer preference for service increases. Liu, Guo, Guo, and Lei examine the customers' WTP and acceptance level for the remanufactured products by deciding the independent price for the new product in the first cycle period and the joint price for both new and remanufactured products in the second cycle (Liu et al., 2016). They extended the study into a dual-channel system with one manufacturer, one retailer, and one e-retailer, discussing the influence of bargaining power between the manufactured and the retailer. The results show that all supply chain members receive benefits from selling remanufactured products. Customer's acceptance level for remanufactured products is positively correlated with the manufacturer's profit growth, significantly more than that of the retailer and e-retailer. Gan discusses the same phenomenon (Gan et al., 2017). They established a two-time period model to find optimal prices for new and remanufactured products in a separate sales-channel—remanufactured products are sold through a direct channel, while the new products are sold via a retailer. They pointed out that lower remanufacture acceptance contributes to higher profit for the retailer but lower profit for the manufacturer. However, if the acceptance level is high, the manufacturer can level the power as the Stackelberg leader to gain more benefits than the retailer. They also found that a dual-channel brings more profit to the total supply chain than a single-channel.

In addition to this study, many other papers discussed more price strategies in a dual-channel supply chain system and channel selection. Manufacturers and retailers are key members in a product supply chain. Customers have different preferences regarding ways of purchasing. Some people prefer to place an order online, while others are willing to buy products via a retailer. Both the manufacturer and the retailer could cater to various customers by operating in a dual-channel supply chain system. Different sales channels may vary the price policy, which brings competition between different sales channels. Chen, Fang, and Wen focus on the channel competition and brand competition in an environment of one manufacturer with internet and traditional channels, and one retailer who sells two substitutable products (Chen et al., 2013). They built up Nash and Stackelberg game models and applied sensitivity analysis of an equilibrium solution. They believe that brand loyalty is the key to profitability; both the manufacturer and the retailer prefer an appropriate cooperation rather than the Nash strategy. Soleimani built a manufacturer-leader Stackelberg game pricing model with fuzzy cost and customer demand to determine the optimal retail price and the wholesale price (Soleimani,

2016). The result shows that with a decreased fuzzy degree of the parameter, optimal prices also decrease, and the profits of the manufacturer and the entire channel increase, while the retailer's profit drops. Rodriguez and Aydin studied the same system but built an inventory cost and rested-logit customer demand model for pricing and assortment decisions (Rodriguez and Aydin, 2015). They found contradictory results on the preferences of the manufacturer and the retailer. Xiao and Shi developed game-theoretic models to decide pricing and channel priority strategies when the retailer faces insufficient supply from the manufacturer caused by random yield (Xiao and Shi, 2016). They indicated that the dual-channel policy has an advantage in alleviating the retailer's pressure on inadequate supply. Supply priority decision varies in different conditions. Ding, Dong, and Pan assumed that the manufacturer acts as a Stackelberg leader, finding an optimal joint strategy for a wholesale price, retailer price, and direct channel price by hierarchical pricing decision process (Ding et al., 2016). They found that operating a dual-channel system only brings more benefits to the manufacturer and an equal-price policy for both sales channels is not optimal for the manufacturer. Gao, Wang, Yang, and Zhong worked on a similar problem but analyzed customer acceptance of the direct channel on pricing decisions also by building Nash game, manufacturer Stackelberg game, and retailer Stackelberg game models (Gao et al., 2016). They learned that when customer's acceptance of direct channel increases, the wholesale price, retail price, and expected profits of the retailer decrease; the direct sales price and manufacturer's expected earnings in the retailer-Stackelberg game all increase. Besides price decision, Wang, Wang, and Wang opened a way to market new and remanufactured products by identifying the optimal channel strategy (Wang et al., 2016). They believed the manufacturer prefers to open a direct online channel to differentiate new and remanufactured products, while the dual-channel system benefits the customer but sacrifices the retailer.

The papers above only examined the pricing strategy in a forward supply chain. However, some other researchers extended the problem to a closed-loop supply chain (CLSC). Soleimani (2016), Karimabadi, Arshadi-khamseh, and Naderi also examined the optimal pricing decisions by building game theory models in one manufacturer and one retailer environment with fuzzy variables. However, they extended the study to CLSC when the manufacturer operates dual channels selling new products and collecting used parts (Karimabadi et al., 2019). The results show that with a decrease in the fuzzy degree of price sensitivity in the Manufacturer-Stackelberg model, the expected profit for the manufacturer and the retailer increase. Giri, Chakraborty, and Maiti considered a system with one manufacturer, one retailer, and one third-party, where the manufacturer sells new products through retailer and e-retailer. In contrast, third-party and e-retailer are responsible for collecting used parts (Giri et al., 2017). They decided that optimal retail, wholesale, acquisition prices, and return product collection rate in five scenarios—centralized, Nash game, manufacturer-led, retailer-led, and third party-led. They approved that a retailer-led decentralized scenario provides more benefits than other decentralized scenarios.

Apart from the study on one manufacturer and one retailer, another volume of papers focuses on the horizontal competition between two manufacturers or two retailers, and the vertical competition between the manufacturer and the retailer.

Ke and Cai developed price decision models with collection-relevant demand in a closed-loop supply chain, where one manufacturer acts as the Stackelberg leader and two competing retailers are assumed to be the followers (Ke and Cai, 2019). They indicated that the demand is a more significant factor than the cost-saving from remanufacturing for the motivation to recycle. Hsiao and Chen describe the vertical competition and horizontal competition between retailers and e-retailers by analyzing internet channels, the pricing strategies, and the channel structure using the market Nash Equilibrium method (Hsiao and Chen, 2014). They indicated that although the manufacturer's online channel encroaches the retailer's benefit, the manufacturer is willing to give up direct channels and leverage on retailer competition. Other researchers considered the system with two manufacturers and one retailer. Gu and Gao conducted a study on the closed-loop supply chain in an environment with two manufacturers and one retailer (Gu and Gao, 2012). They found optimal prices for all three and believed that the collection of used cores is the best for the retailer. Zhao, Wei, and Li investigated the remanufacturing decision for two substitutable products—new products produced by raw materials and remanufactured products incorporated used parts (Zhao et al., 2017). They examined the research problem in Manufacturer-Stackelberg and Retailer-Stackelberg models with several sub-conditions, such as two manufacturers have equal or different market power with subgame perfect Nash equilibrium solution concept. They found that the whole supply chain receives the highest profit in Manufacturer-Stackelberg when two manufacturers have equal force. Luo, Chen, Chen, and Wang conducted a related study but focused on two differentiated brand products supplied by two manufacturers (Luo et al., 2017). They built up seven-game models with different power structures and found that the competition between the manufacturers benefits the retailer. For the whole supply chain, more profits are received by increased acceptance of the average brand. Aydin, Kwong, and Ji extended the research to the closed-loop supply chain in a multi-objective optimization problem (Aydin et al., 2016). They proposed a case study on tablet PCs in a system with a manufacturer, a remanufacturer, and a chain retailer, aiming to find pricing decisions, product return rates, and product line solutions by Stackelberg game theory and genetic algorithm. Except for the price decision across the competition and the channel power, some researchers were focused on the service's impact. Wang and Wang established the Retailer-Stackelberg, the Manufacturer-Stackelberg, and the Nash Equilibrium models in a monopoly market with one manufacturer, one remanufacturer, and one retailer, indicating that a growth in customer's acceptance level on remanufactured products leads to lower prices and higher service levels and reduce profit in general for the members with less power on price decision (Wang and Wang, 2015).

Most papers we reviewed concentrated on the negotiation between sellers and buyers. They used game theories to discover the optimal prices or supply chain strategies with different bargain powers. However, very few of them took notice of the product differences in both generation and condition. Especially for high-technology products, technology advantage and the quality condition could influence customers' willingness to pay, and bargaining power will change based on the type of products they are selling. In this study, we will use this research opportunity to build knowledge on how to price various kinds of hi-tech products using game theory.

3   PROBLEM STATEMENT

The study analyzed the pricing decision for two types of products. Type 1 product is new commodities belonging to the early generation. Type 2 product is the re-manufactured item belonging to the latest generations. Type 1 products are depreciated in technology and Type 2 products are inferior in quality when compared to new products from the latest generation. The two products' market is separated from the premium market, which sells new products belonging to the latest technology. The manufacturer sells Type 1 product to the retailer at a wholesale price w_1, while the remanufacturer sells Type 2 product at w_2. The retailer sells the product to the customers at different retail prices p_1 and p_2. The cost of manufacturing a new product is c_1, while the cost of a remanufactured product is c_2. Assume $c_1 > c_2$.

Customers are classified into two segments: quality-conscious and technology-savvies. Customers in different segments have different perceived values on Type 1 and Type 2 products. α_1 and α_2 are the discount factors of generation obsolescence in quality for quality-conscious customers and technology-savvies, respectively. β_1 and β_2 are the discount factors of remanufacturing inferiority for quality-conscious and technology-savvy customers. We assume that $\alpha_1 > \alpha_2$, and $\beta_1 < \beta_2$, $\alpha_1 > \beta_1$, and $\alpha_2 < \beta_2$. Supply constraint is not considered. The optimal wholesale prices and retail prices for each product is determined in a Vertical Nash Equilibrium. Manufacturer, remanufacturer, and the retailer make their pricing decisions simultaneously. Manufacturers and re-manufacturers maximize the profits, predicting the retailer's margin profits, while the retailer determines retail prices anticipating the wholesale prices.

4   MODEL

As described in the paper (Luo et al., 2017), the authors considered one product priced at p. The customer perceived value v is uniformly distributed over [0, 1]. Therefore, the demand for the product is $Q = \int_p^1 dv = 1 - p$, for $0 \le p \le 1$. We develop the model into two types of products with different generations and product condition. We assume that v is the perceived value for the new product belonging to the latest generation. Therefore, with discount factors, the perceived value from customers in each segment is as follows (Table 2.1). The conditions when customers in different segments will purchase the products are shown in Table 2.2. We also assume δ is the proportion of the customers in the quality-conscious segment; the proportion of the technology-savvy is $1-\delta$. Therefore, the piecewise demand function can be derived.

Demand Function

When $p_1 < p_2$, all the quality-conscious customers will buy Type 1 products; the technology-savvy customers will choose Type 1 items if $\beta_2 - \alpha_2 < p_2 - p_1$.

When $p_2 < p_1$, all the technology-savvy customers will buy Type 2 products; quality-conscious customers will choose Type 2 items if $\alpha_1 - \beta_1 < p_1 - p_2$.

TABLE 2.1
Perceived Value of Two Types of Products from Customers in Segment

	Customer Segments	
Perceived Value	Quality-Conscious	Technology-Savvy
Type 1 product	$\alpha_1 v$	$\alpha_2 v$
Type 2 product	$\beta_1 v$	$\beta_2 v$

TABLE 2.2
Customers' Selection

Customer Segments	Purchase Option	
	Type 1 Product	Type 2 Product
Quality-conscious	$\alpha_1 v - p_1 >= 0$	$\beta_1 v - p_2 >= 0$
	and	and
	$\alpha_1 v - p_1 >= \beta_1 v - p_2$	$\alpha_1 v - p_1 < \beta_1 v - p_2$
Technology-savvy	$\alpha_2 v - p_1 >= 0$	$\beta_2 v - p_2 >= 0$
	and	an
	$\alpha_2 v - p_1 > \beta_2 v - p_2$	$\alpha_2 v - p_1 <= \beta_2 v - p_2$

Therefore, the demand functions of Type 1 and Type 2 products are as follows through function (1) to (4).

Condition 1: $p_1 < p_2$

Region 1: When $0 < \frac{\beta_2}{\alpha_2} < \frac{p_2}{p_1} < 1$ and $p_1 < \alpha_1$, $p_1 < \alpha_2$, all customers will buy Type 1 products.

$$\begin{cases} D_1(p_1, \ p_2) = \delta(\alpha_1 - p_1) + (1 - \delta)(\alpha_2 - p_1) \\ D_2(p_1, \ p_2) = 0 \end{cases} \tag{2.1}$$

Region 2: When $\frac{\alpha_2}{\beta_2} < \frac{p_1}{p_2} < 1$ and $p_1 < \alpha_1$, $p_1 < \alpha_2$, $p_2 < \beta_2$, all quality-conscious customers will buy Type 1 products, and some technology-savvies will purchase Type 1 products instead of Type 2 products.

$$\begin{cases} D_1(p_1, \ p_2) = \delta(\alpha_1 - p_1) + (1 - \delta)\left(\frac{p_2 - p_1}{\beta_2 - \alpha_2} - \frac{p_1}{\alpha_2} \right) \\ \\ D_2(p_1, \ p_2) = 0 + (1 - \delta)\left(\beta_2 - \frac{p_2 - p_1}{\beta_2 - \alpha_2} \right) \end{cases} \tag{2.2}$$

Condition 2: $p_1 > p_2$,
    Region 3: When $1 < \frac{\alpha_1}{\beta_1} \leq \frac{p_1}{p_2}$ and $p_2 < \beta_1$, $p_2 < \beta_2$, all customers will buy Type 2 products

$$\begin{cases} D_1(p_1, \ p_2) = 0 \\ D_2(p_1, \ p_2) = \delta(\beta_1 - p_2) + (1 - \delta)(\beta_2 - p_2) \end{cases} \quad (2.3)$$

Region 4: When $1 < \frac{p_1}{p_2} < \frac{\alpha_1}{\beta_1}$, some quality-conscious customers will purchase Type 2 products instead of Type 1.

$$\begin{cases} D_1(p_1, \ p_2) = \delta\left(\alpha_1 - \frac{p_1 - p_2}{\alpha_1 - \beta_1}\right) + 0 \\ D_2(p_1, \ p_2) = \delta\left(\frac{p_1 - p_2}{\alpha_1 - \beta_1} - \frac{p_2}{\beta_1}\right) + (1 - \delta)(\beta_2 - p_2) \end{cases} \quad (2.4)$$

The profit models for manufacturer, remanufacturer, and retailer can be generated based on the demand functions.

PROFIT MODEL

Manufacturer 1's profit model:

$$\Pi_{m1} = (w_1 - c_1) * D_1(p_1, \ p_2) \quad (2.5)$$

Manufacturer 2's profit model (Remanufacturer):

$$\Pi_{m2} = (w_2 - c_2) * D_2(p_1, \ p_2) \quad (2.6)$$

Retailer's profit model:

$$\Pi r = (p_1 - w_1) * D_1(p_1, \ p_2) + (p_2 - w_2) * D_2(p_1, \ p_2) \quad (2.7)$$

We denote marginal profits of the retailer from Type 1 and Type 2 products as $m_1 = p_1 - w_1$, and $m_2 = p_2 - w_2$, respectively. The optimal prices can be decided in a vertical Nash model.

Vertical Nash Model
Condition 1, region 1:
Retailer's profit model:

$$\Pi_r = (p_1 - w_1) * \left[\delta(\alpha_1 - p_1) + (1 - \delta)(\alpha_2 - p_1)\right]$$
$$\frac{\partial \Pi r}{\partial p1} = -2p_1 + \delta\alpha_1 + (1 - \delta)\alpha_2 + w_1 \quad (2.8)$$

Then, $\frac{\partial \Pi r^2}{\partial^2 p1} = -2 < 0$, $\Pi r(p_1)$ is concave in p_1.

Manufacturer's profit model:

$$\Pi_{m1} = (w_1 - c_1) * \left[\delta(\alpha_1 - p_1) + (1 - \delta)(\alpha_2 - p_1) \right]$$

$$\frac{\partial \Pi_m(w1)}{\partial w1} = \delta - m_1 - 2w_1 + \delta\alpha_1 + \alpha_2 - \delta\alpha_2 + c_1$$

(2.9)

Then, $\frac{\partial \Pi_m(w1)^2}{\partial^2 w1} = -2 < 0$, $\Pi_{m1}(w_1)$ is concave in w_1.

Remanufacturer's profit model:

$$\Pi_{m2} = (w_2 - c_2) * 0 = 0$$

(2.10)

Let $\frac{\partial \Pi r}{\partial p1} = \frac{\partial \Pi_m(w1)}{\partial w1} = 0$ the optimal wholesale price when the selling price of Type 1 product is obtained. The maximized profits of Manufacturer 1 and retailer can be found accordingly.

$$w_1 * = \frac{\delta\alpha_1 + (1 - \delta)\alpha_2 + 2c_1}{3}$$

$$p_1 * = \frac{2\delta\alpha_1 + 2(1 - \delta)\alpha_2 + c_1}{3}$$

Condition 1, Region 2

Retailer's profit model:

$$\Pi_r = \left(p_1 - w_1 \right) * \left[\delta * \left(\alpha_1 - p_1 \right) + (1 - \delta) * \left(\frac{p_2 - p_1}{\beta_2 - \alpha_2} - \frac{p_1}{\alpha_2} \right) \right]$$

$$+ \left(p_2 - w_2 \right) * \left[(1 - \delta) * \left(\beta_2 - \frac{p_2 - p_1}{\beta_2 - \alpha_2} \right) \right]$$

$$\frac{\partial \Pi r}{\partial p1} = (-2\delta - \frac{2\beta_2(1 - \delta)}{\alpha_2(\beta_2 - \alpha_2)}) * p_1 + \frac{2(1 - \delta)}{\beta_2 - \alpha_2} * p_2 + [\delta + \frac{\beta_2(1 - \delta)}{\alpha_2(\beta_2 - \alpha_2)}] * w_1 - \frac{1 - \delta}{\beta_2 - \alpha_2} * w_2$$

$$+ \delta\alpha_1$$

(2.11)

Then, $\frac{\partial^2 \Pi r}{\partial p_1^2} = -2\delta - \frac{2\beta_2(1 - \delta)}{\alpha_2(\beta_2 - \alpha_2)} < 0$, $\Pi_r(p_1)$ is concave in p_1.

$$\frac{\partial \Pi r}{\partial p_2} = \frac{2(1 - \delta)}{\beta_2 - \alpha_2} * (p_1 - p_2) + \frac{(1 - \delta)}{\beta_2 - \alpha_2} * (w_2 - w_1) + (1 - \delta)\beta_2$$

Then, $\frac{\partial \Pi_r^2}{\partial^2 p_2} = -\frac{2(1 - \delta)}{\beta_2 - \alpha_2} < 0$, $\Pi_r(p_2)$ is concave in p_2.

And $\frac{\partial^2 \Pi_r}{\partial p_2 \partial p_1} = \frac{\partial^2 \Pi_r}{\partial p_1 \partial p_2} = \frac{2(1 - \delta)}{\beta_2 - \alpha_2}$

The value of $\begin{vmatrix} \dfrac{\partial^2 \Pi_r}{\partial p_1^2} & \dfrac{\partial^2 \Pi_r}{\partial p_1 \partial p_2} \\[2mm] \dfrac{\partial^2 \Pi_r}{\partial p_2 \partial p_1} & \dfrac{\partial^2 \Pi_r}{\partial p_2^2} \end{vmatrix} = \dfrac{4\delta(1-\delta)}{\beta_2 - \alpha_2} + \left(\dfrac{\beta_2}{\alpha_2} - 1 \right) * \dfrac{4(1-\delta)}{(\beta_2 - \alpha_2)^2} > 0$

Therefore, $\Pi_r\,(p_1, p_2)$ is a joint concave in p_1 and p_2.
Manufacturer's profit model:

$$\Pi_{m1} = (w_1 - c_1) * \left[\delta * (\alpha_1 - p_1) + (1 - \delta) * \left(\dfrac{p_2 - p_1}{\beta_2 - \alpha_2} - \dfrac{p_1}{\alpha_2} \right) \right] \qquad (2.12)$$

$$\dfrac{\partial \Pi m1\,(w1)}{\partial w1} = \delta * (\alpha_1 - p_1) + (1 - \delta) * \left(\dfrac{p_2 - p_1}{\beta_2 - \alpha_2} - \dfrac{p_1}{\alpha_2} \right)$$

$\dfrac{\partial \Pi^2 m_1\,(w_1)}{\partial^2 w_1} = -\delta - \dfrac{1-\delta}{\beta_2 - \alpha_2} - \dfrac{1-\delta}{\alpha_2} < 0$, $\Pi_{m1}(w_1)$ is concave in w_1

Since $p_1 = m_1 + w_1$, $p_2 = m_2 + w_2$,

$$\dfrac{\partial \Pi m1\,(w1)}{\partial w1} = -\left(\delta + \dfrac{1-\delta}{\beta_2 - \alpha_2} + \dfrac{1-\delta}{\alpha_2} \right) p_1 + \dfrac{1-\delta}{\beta_2 - \alpha_2} p_2 - \left(\dfrac{(1-\delta)}{\alpha_2} + \dfrac{(1-\delta)}{\beta_2 - \alpha_2} + \delta \right)$$

$$w_1 + \left(\dfrac{1-\delta}{\beta_2 - \alpha_2} + \dfrac{1-\delta}{\alpha_2} + \delta \right) c_1 + \delta\alpha_1$$

Remanufacturer's profit model:

$$\Pi_{m2} = (w_2 - c_2) * (1 - \delta) * \left(\beta_2 - \dfrac{P_2 - P_1}{\beta_2 - \alpha_2} \right) \qquad (2.13)$$

$$\dfrac{\partial \Pi m2\,(w2)}{\partial w2} = (1 - \delta) * \left(\beta_2 - \dfrac{P_2 - P_1}{\beta_2 - \alpha_2} \right)$$

And $\dfrac{\partial \Pi^2 m2\,(w2)}{\partial^2 w2} = -\dfrac{1-\delta}{\beta_2 - \alpha_2} < 0$, $\Pi_{m2}(w_2)$ is concave in w_2.

Because $p_1 = m_1 + w_1$, $p_2 = m_2 + w_2$,

$$\dfrac{\partial \Pi m2\,(w2)}{\partial w2} = \dfrac{1-\delta}{\beta_2 - \alpha_2} p_1 - \dfrac{1-\delta}{\beta_2 - \alpha_2} p_2 - \dfrac{1-\delta}{\beta_2 - \alpha_2} w_2 + \dfrac{1-\delta}{\beta_2 - \alpha_2} c_2 + (1 - \delta)\beta_2$$

Let $\dfrac{\partial \Pi r}{\partial p1} = \dfrac{\partial \Pi r}{\partial p2} = \dfrac{\partial \Pi m1\,(w1)}{\partial w1} = \dfrac{\partial \Pi m2\,(w2)}{\partial w2} = 0$, the optimal wholesale and retail prices are obtained.

$$p *_1 = \dfrac{1}{2}w_1 + \dfrac{1}{2} * \dfrac{\alpha_2 \beta_2 (1-\delta) + \delta\alpha_1\alpha_2}{\delta\alpha_2 + 1 - \delta}$$

$$p *_2 = \dfrac{1}{6}w_1 + \dfrac{1}{3}c_2 + \dfrac{1}{2} * \dfrac{\alpha_2 \beta_2 (1-\delta) + \delta\alpha_1\alpha_2}{\delta\alpha_2 + 1 - \delta} + \dfrac{2}{3}\beta_2(\beta_2 - \alpha_2)$$

$$w *_1 = \frac{6 * [(\beta_2 - \alpha_2)(\delta\alpha_2 + 1 - \delta) + \alpha_2(1 - \delta)]}{9 * (\beta_2 - \alpha_2)(\delta\alpha_2 + 1 - \delta) + 8 * \alpha_2(1 - \delta)} * c_1 + \frac{2\alpha_2(1 - \delta)}{9 * (\beta_2 - \alpha_2)(\delta\alpha_2 + 1 - \delta) + 8 * \alpha_2(1 - \delta)} * c_2 +$$

$$\frac{(\beta_2 - \alpha_2)[3\delta\alpha_1\alpha_2 + \alpha_2\beta_2(1 - \delta)]}{9 * (\beta_2 - \alpha_2)(\delta\alpha_2 + 1 - \delta) + 8 * \alpha_2(1 - \delta)}$$

$$w *_2 = \frac{1}{3}w_1 + \frac{2}{3}c_2 + \frac{1}{3}\beta_2(\beta_2 - \alpha_2)$$

Condition 2, region 3:
Retailer's profit model:

$$\Pi_r = (p_2 - w_2) * [\delta * (\beta_1 - p_2) + (1 - \delta) * (\beta_2 - p_2)] \tag{2.14}$$

$$\frac{\partial \Pi r(p_1, p_2)}{\partial p_2} = -2p_2 + \delta\beta_1 + (1 - \delta)\beta_2 + w_2$$

Then ..., $\Pi_r(p_1, p_2)$ is concave in p_2.
Manufacturer's profit model:

$$\Pi_{m1} = (w_1 - c_1) * 0 = 0 \tag{2.15}$$

Remanufacturer's profit model:

$$\Pi_{m2} = (w_2 - c_2) * [\delta * (\beta_1 - p_2) + (1 - \delta) * (\beta_2 - p_2)] \tag{2.16}$$

Because $p_2 = m_2 + w_2$,

$$\frac{\partial \Pi m2(w2)}{\partial w2} = \delta\beta_1 + \beta_2 - \delta\beta_2 - m_2 - 2w_2 + c_2$$

Then, $\frac{\partial \Pi^2 m_2(w_2)}{\partial w_2^2} = -2 < 0$, $\Pi_{m2}(w_2)$ is concave in w_2.
Let $\frac{\partial \Pi r(p_1, p_2)}{\partial p_2} = \frac{\partial \Pi m2(w2)}{\partial w2} = 0$, we get the optimal w_2 and p_2.

$$p *_2 = \frac{2\delta\beta_1 + 2(1 - \delta)\beta_2 + c_2}{3}$$

$$w *_2 = \frac{\delta\beta_1 + (1 - \delta)\beta_2 + 2c_2}{3}$$

Condition 2, region 4:
Retailer's model:

$$\Pi_r = (p_1 - w_1) * \left[\delta\left(\alpha_1 - \frac{p_1 - p_2}{\alpha_1 - \beta_1}\right)\right] + (p_2 - w_2) * \left[\delta\left(\frac{p_1 - p_2}{\alpha_1 - \beta_1} - \frac{p_2}{\beta_1}\right)\right.$$

$$\left. + (1 - \delta)(\beta_2 - p_2)\right] \tag{2.17}$$

$$\frac{\partial \Pi r (p_1, p_2)}{\partial p1} = \frac{2\delta}{\alpha_1 - \beta_1}(p_2 - p_1) + \frac{\delta}{\alpha_1 - \beta_1}(w_1 - w_2) + \delta \alpha_1$$

$$\frac{\partial \Pi r (p_1, p_2)}{\partial p2} = \frac{2\delta}{\alpha_1 - \beta_1}p_1 - \left(\frac{2\delta}{\alpha_1 - \beta_1} + \frac{2\delta}{\beta_1} + 2(1 - \delta)\right)p_2 - \frac{\delta}{\alpha_1 - \beta_1}$$

$$w_1 + \left(\frac{\delta}{\alpha_1 - \beta_1} + \frac{\delta}{\beta_1} + (1 - \delta)\right)w_2 + (1 - \delta)\beta_2$$

Then, $\dfrac{\partial^2 \Pi r}{\partial p_1^2} = -\dfrac{2\delta}{\alpha_1 - \beta_1} < 0,\ \dfrac{\partial^2 \Pi r}{\partial p_2^2} = -\dfrac{2\delta}{\alpha_1 - \beta_1} - \left(\dfrac{2\delta}{\beta_1} + 2 - 2\delta\right) < 0$

Therefore, $\Pi_r (p_1, p_2)$ is concave in p_1 and p_2.

$$\frac{\partial^2 \Pi r}{\partial p_1^2} = -\frac{2\delta}{\alpha_1 - \beta_1} < 0,\ \frac{\partial^2 \Pi r}{\partial p_2^2} = -\frac{2\delta}{\alpha_1 - \beta_1} - \left(\frac{2\delta}{\beta_1} + 2 - 2\delta\right) < 0$$

And $\dfrac{\partial^2 \Pi_r}{\partial p_1 \partial p_2} = \dfrac{\partial^2 \Pi_r}{\partial p_2 \partial p_1} = \dfrac{2\delta}{\alpha_1 - \beta_1}$

Then the value of $\begin{vmatrix} \dfrac{\partial^2 \Pi r}{\partial p_1^2} & \dfrac{\partial^2 \Pi_r}{\partial p_1 \partial p_2} \\ \dfrac{\partial^2 \Pi_r}{\partial p_2 \partial p_1} & \dfrac{\partial^2 \Pi_r}{\partial p_2^2} \end{vmatrix} = \dfrac{4\delta^2 \beta_1 + 4(\alpha_1 - \beta_1)\delta(\delta + \beta_1 - \delta\beta_1)}{\beta_2 (\alpha_1 - \beta_1)^2} > 0$

Therefore, $\Pi_r (p_1, p_2)$ is a joint concave in p_1 and p_2.
Manufacturer's model:

$$\Pi_{m1} = (w_1 - c_1) * \left[\delta\left(\alpha_1 - \frac{p_1 - p_2}{\alpha_1 - \beta_1}\right)\right] \tag{2.18}$$

$$\frac{\partial \Pi m1 (w1)}{\partial w1} = \delta\left(\alpha_1 - \frac{p_1 - p_2}{\alpha_1 - \beta_1}\right)$$

Then, $\dfrac{\partial \Pi^2 m1 (w1)}{\partial w_1^2} = -\dfrac{\delta}{\alpha_1 - \beta_1} < 0$

Therefore, $\Pi_{m1}(w_1)$ is concave in w_1.
Because $p_1 = m_1 + w_1,\ p_2 = m_2 + w_2$

$$\frac{\partial \Pi m1 (w1)}{\partial w1} = \delta\alpha_1 - \frac{\delta}{\alpha_1 - \beta_1}w_1 - \frac{\delta}{\alpha_1 - \beta_1}p_1 + \frac{\delta}{\alpha_1 - \beta_1}p_2 + \frac{\delta}{\alpha_1 - \beta_1}c_1$$

Remanufacturer's model:

$$\Pi_{m2} = (w_2 - c_2) * \left[\delta\left(\frac{p_1 - p_2}{\alpha_1 - \beta_1} - \frac{p_2}{\beta_1}\right) + (1 - \delta)(\beta_2 - p_2)\right] \tag{2.19}$$

$$\frac{\partial \Pi m2(w2)}{\partial w2} = \delta\left(\frac{p_1 - p_2}{\alpha_1 - \beta_1} - \frac{p_2}{\beta_1}\right) + (1 - \delta)(\beta_2 - p_2)$$

Then, $\frac{\partial \Pi^2 m2(w2)}{\partial w_2^2} = -\frac{\delta}{\alpha_1 - \beta_1} - \frac{\delta}{\beta_1} - (1 - \delta) < 0$,

Therefore, $\Pi_{m2}(w_2)$ is concave in w_2.

Because $p_1 = m_1 + w_1$, $p_2 = m_2 + w_2$

$$\frac{\partial \Pi m2(w2)}{\partial w2} = \frac{\delta}{\alpha_1 - \beta_1}p_1 - \left[\frac{\delta}{\alpha_1 - \beta_1} + \frac{\delta}{\beta_1} + (1 - \delta)\right]p_2 - \left[\frac{\delta}{\alpha_1 - \beta_1} + \frac{\delta}{\beta_1} + (1 - \delta)\right]$$
$$w_2 + \left[\frac{\delta}{\alpha_1 - \beta_1} + \frac{\delta}{\beta_1} + (1 - \delta)\right]c_2 + (1 - \delta)\beta_2$$

Let $\frac{\partial \Pi r (p_1, p_2)}{\partial p1} = \frac{\partial \Pi r (p_1, p_2)}{\partial p2} = \frac{\partial \Pi m1 (w1)}{\partial w1} = \frac{\partial \Pi m2(w2)}{\partial w2} = 0$, we can get optimal solutions in this condition.

$$p*_1 = \frac{1}{6}w_2 + \frac{1}{3}c_1 + \frac{2}{3}\alpha_1\left(\alpha_1 - \beta_1\right) + \frac{\delta\alpha_1\beta_1 + (1 - \delta)\beta_1\beta_2}{2\delta + 2\beta_1(1 - \delta)}$$

$$p*_2 = \frac{1}{2}w_2 + \frac{\delta\alpha_1\beta_1 + (1 - \delta)\beta_1\beta_2}{2\delta + 2\beta_1(1 - \delta)}$$

$$w*_1 = \frac{1}{3}w_2 + \frac{2}{3}c_1 + \frac{\alpha_1(\alpha_1 - \beta_1)}{3}$$

$$w*_2 = \frac{2\delta\beta_1}{8\delta\beta_1 + 7\left(\alpha_1 - \beta_1\right)[\delta + (1 - \delta)\beta_1]}c_1$$

$$+ \frac{6\left(\alpha_1 - \beta_1\right)\left[\delta + \beta_1(1 - \delta)\right] + 6\delta\beta_1}{8\delta\beta_1 + 7\left(\alpha_1 - \beta_1\right)[\delta + \beta_1(1 - \delta)]}c_2$$

$$+ \frac{\beta_1\left(\alpha_1 - \beta_1\right)[7\alpha_1\delta + 9\beta_2(1 - \delta)]}{7\left(\alpha_1 - \beta_1\right)\left[\delta + \beta_1(1 - \delta)\right] + 8\delta\beta_1}$$

5   NUMERICAL EXAMPLE

In this section, we will apply a numerical example in the four regions to see how different customer ratios in the two segments vary prices. The impact of perceived value differentiation on the solution is also discussed.

We assume $c_1 = 0.2$, $c_2 = 0.1$. In condition 1, we assure $\alpha_1 = \beta_2 = 0.8$. The difference between α_2 and β_2 ranges from 0.1-0.3. Table 2.3 shows the optimal wholesale price, retail price, and the profit of the retailer, manufacturer, and supply chain in Region 1. Table 2.4 exhibits the result for Region 2.

In region 1, $p_1 < p_2$. The value of p_1 is smaller than α_1 and α_2. The price differentiation between p_1 and p_2 is larger than the difference between the perceived value of Type 2 and Type 1 products for technology-savvies. Therefore, customers will buy Type 1 products only. The result is in line with intuition. The wholesale price and retail price for Type 1 products, and the profits of the retailer and manufacturer, increase as the population proportion of quality-conscious customers increases. The profits of the retailer and the manufacturer remain equally.

In region 2, $p_1 < p_2$, $p_1 < \alpha_1$, $p_1 < \alpha_2$, $p_2 < \beta_2$. The price differentiation between p_1 and p_2 is smaller than the difference between the perceived value of Type 2 and Type 1 products for technology-savvies. Therefore, all quality-conscious customers will buy Type 1 products; some technology savvies will purchase Type 1 products instead of Type 2 products. We learned from Table 2.4 that the wholesale and retail price for both Type 1 and Type 2 products increases with the larger population of quality-conscious customers. However, except for the manufacturer, the margins of the retailer, remanufacturer, and supply chain are dropped. An interesting finding is that the manufacturer begins to earn when the proportion of the quality-conscious segment is large enough. The smaller difference between the perceived value of Type 2 and Type 1 products for technology-savvies requires a more quality-conscious segment proportion for the manufacturer to gain. In general, when the gap between α_2 and β_2 is more substantial, the wholesale prices and retail price for Type 2 products become higher, while Type 2 retail price is getting lower. Also, the prices for Type 1 product is more sensitive to the gap than that of Type 2 product. For the profits of supply chain members, the larger gap makes higher margins for the manufacturer and remanufacturer, but less earning for the retailer and the whole supply chain.

In condition 2, the price for Type 1 is assumed to be larger than the price of Type 2. As the explanation for the model above, all customers will buy Type 2 products in Region 3 while some quality conscious customers will purchase Type 2 products instead of Type 1 in Region 4. The difference between α_1 and β_1 ranges from 0.1-0.3. Tables 2.5 and 2.6 express the results.

In Region 3, the wholesale price, retail price, the profit of the retailer, manufacturer, and the whole supply chain are getting lower with a more substantial proportion of quality-conscious customers and a significant gap between α_1 and β_1. Besides, the wholesale price and retail price are more sensitive to customer proportion with a more significant gap.

In Region 4, when the quality-conscious customer's proportion increases, the optimal wholesale and retail prices for both Type 1 and Type 2 products decrease. The profits of the remanufacturer and supply chain drop and the profit of the manufacturer increases given a larger proportion. There is a peak for the margin of the retailer at an optimal quality-conscious percentage. In the example, the summit appears when the ratio is 0.3.

When the gap between α_1 and β_1 is more significant, the wholesale price and retail price for both products are increased. However, the optimal prices for Type 2

TABLE 2.3

Wholesale Price, Retail Price, and the Profit in Region 1

	Price in Region 1			Profit			
	δ Value	w_1	p_1	Retail	Manufacturer	Remanufacturer	Supply Chain
$\alpha_1=0.8$, $\alpha_2=0.7$, $\beta_1=0.6$, $\beta_2=0.8$	0.1	0.37	0.54	0.0289	0.0289	0	0.0578
	0.3	0.38	0.55	0.0312	0.0312	0	0.0624
	0.5	0.38	0.57	0.0336	0.0336	0	0.0672
	0.7	0.39	0.58	0.0361	0.0361	0	0.0722
	0.9	0.40	0.59	0.0387	0.0387	0	0.0774
$\alpha_1=0.8$, $\alpha_2=0.6$, $\beta_1=0.6$, $\beta_2=0.8$	δ Value	w_1	p_1	Retail	Manufacturer	Remanufacturer	Supply Chain
	0.1	0.34	0.48	0.0196	0.0196	0	0.0392
	0.3	0.35	0.51	0.0235	0.0235	0	0.0470
	0.5	0.37	0.53	0.0278	0.0278	0	0.0556
	0.7	0.38	0.56	0.0324	0.0324	0	0.0648
	0.9	0.39	0.59	0.0374	0.0374	0	0.0748
$\alpha_1=0.8$, $\alpha_2=0.5$, $\beta_1=0.6$, $\beta_2=0.8$	δ Value	w_1	p_1	Retail	Manufacturer	Remanufacturer	Supply Chain
	0.1	0.31	0.42	0.0121	0.0121	0	0.0242
	0.3	0.33	0.46	0.0169	0.0169	0	0.0338
	0.5	0.35	0.5	0.0225	0.0225	0	0.045
	0.7	0.37	0.54	0.0289	0.0289	0	0.0578
	0.9	0.39	0.58	0.0361	0.0361	0	0.0722

TABLE 2.4

Wholesale Price, Retail Price, and the Profit in Region 2.

	Price in Region 2				Profit			
δ Value	w_1	w_2	p_1	p_2	Retail	Manufacturer	Remanufacturer	Supply Chain
$\alpha 1=0.8, \alpha 2=0.7,$ $\beta 1=0.6, \beta 2=0.8$								
0.1	0.1802	0.1534	0.3788	0.4054	0.1211	0.0000	0.0257	0.1468
0.3	0.1867	0.1556	0.4010	0.4255	0.1050	0.0000	0.0216	0.1266
0.5	0.1975	0.1592	0.4281	0.4490	0.0857	0.0000	0.0175	0.1032
0.7	0.2188	0.1663	0.4638	0.4776	0.0809	0.0015	0.0132	0.0956
0.9	0.2813	0.1871	0.5242	0.5171	0.0691	0.0135	0.0076	0.0902
δ Value	w_1	w_2	p_1	p_2	Retail	Manufacturer	Remanufacturer	Supply Chain
$\alpha 1=0.8, \alpha 2=0.6,$ $\beta 1=0.7, \beta 2=0.8$								
0.1	0.1821	0.1807	0.3411	0.4204	0.0870	0.0000	0.0293	0.1164
0.3	0.1927	0.1842	0.3691	0.4448	0.0705	0.0000	0.0248	0.0953
0.5	0.2094	0.1898	0.4047	0.4749	0.0710	0.0003	0.0202	0.0915
0.7	0.2395	0.1998	0.4531	0.5132	0.0697	0.0042	0.0149	0.0889
0.9	0.3103	0.2234	0.5301	0.5667	0.0592	0.0191	0.0076	0.0859
δ Value	w_1	w_2	p_1	p_2	Retail	Manufacturer	Remanufacturer	Supply Chain
$\alpha 1=0.8, \alpha 2=0.5,$ $\beta 1=0.6, \beta 2=0.8$								
0.1	0.1810	0.2070	0.3010	0.4340	0.0729	0.0000	0.0344	0.1072
0.3	0.1939	0.2113	0.3323	0.4609	0.0614	0.0000	0.0289	0.0903
0.5	0.2137	0.2179	0.3735	0.4956	0.0615	0.0006	0.0232	0.0852
0.7	0.2477	0.2292	0.4316	0.5423	0.0606	0.0052	0.0167	0.0826
0.9	0.3204	0.2535	0.5238	0.6104	0.0534	0.0208	0.0079	0.0820

TABLE 2.5

Wholesale Price, Retail Price, and the Profit in Region 3

Price in Region 3			Profit			
δ Value	w_2	p_2	Retail	Manufacturer	Remanufacturer	Supply Chain
$\alpha_1=0.8$, $\alpha_2=0.6$, $\beta_1=0.7$, $\beta_2=0.8$						
0.1	0.33	0.56	0.0529	0.0000	0.0529	0.1058
0.3	0.32	0.55	0.0499	0.0000	0.0499	0.0998
0.5	0.32	0.53	0.0469	0.0000	0.0469	0.0939
0.7	0.31	0.52	0.0441	0.0000	0.0441	0.0882
0.9	0.30	0.51	0.0413	0.0000	0.0413	0.0827
δ Value	w_2	p_2	Retail	Manufacturer	Remanufacturer	Supply Chain
$\alpha_1=0.8$, $\alpha_2=0.6$, $\beta_1=0.6$, $\beta_2=0.8$						
0.1	0.33	0.55	0.0514	0.0000	0.0514	0.1028
0.3	0.31	0.53	0.0455	0.0000	0.0455	0.0910
0.5	0.30	0.50	0.0400	0.0000	0.0400	0.0800
0.7	0.29	0.47	0.0348	0.0000	0.0348	0.0697
0.9	0.27	0.45	0.0300	0.0000	0.0300	0.0601
δ Value	w_2	p_2	Retail	Manufacturer	Remanufacturer	Supply Chain
$\alpha_1=0.8$, $\alpha_2=0.6$, $\beta_1=0.5$, $\beta_2=0.8$						
0.1	0.32	0.55	0.0509	0.0000	0.0509	0.1018
0.3	0.30	0.51	0.0418	0.0000	0.0418	0.0835
0.5	0.28	0.47	0.0336	0.0000	0.0336	0.0672
0.7	0.26	0.43	0.0267	0.0000	0.0267	0.0534
0.9	0.24	0.39	0.0205	0.0000	0.0205	0.0411

TABLE 2.6
Wholesale Price, Retail Price, and the Profit in Region 4

	Price in Region 4				Profit			
δ Value	w_1	w_2	p_1	p_2	Retail	Manufacturer	Remanufacturer	Supply Chain
$\alpha_1=0.8,\ \alpha_2=0.6,$ $\beta_1=0.7,\ \beta_2=0.8$								
0.1	0.2442	0.2527	0.5457	0.5099	0.0710	0.0020	0.0342	0.1071
0.3	0.2185	0.1755	0.5037	0.4422	0.0813	0.0010	0.0185	0.1009
0.5	0.2104	0.1511	0.4746	0.4050	0.0788	0.0005	0.0131	0.0924
0.7	0.2064	0.1392	0.4509	0.3773	0.0740	0.0003	0.0104	0.0846
0.9	0.2040	0.1321	0.4307	0.3547	0.0688	0.0001	0.0087	0.0777
δ Value	w_1	w_2	p_1	p_2	Retail	Manufacturer	Remanufacturer	Supply Chain
$\alpha_1=0.8,\ \alpha_2=0.7,$ $\beta_1=0.6,\ \beta_2=0.8$								
0.1	0.2849	0.2948	0.5975	0.5224	0.0589	0.0036	0.0390	0.1015
0.3	0.2543	0.2029	0.5405	0.4348	0.0689	0.0044	0.0203	0.0936
0.5	0.2423	0.1670	0.5012	0.3835	0.0670	0.0045	0.0123	0.0837
0.7	0.2360	0.1479	0.4707	0.3467	0.0625	0.0045	0.0079	0.0749
0.9	0.2320	0.1360	0.4460	0.3180	0.0576	0.0046	0.0053	0.0675
δ Value	w_1	w_2	p_1	p_2	Retail	Manufacturer	Remanufacturer	Supply Chain
$\alpha_1=0.8,\ \alpha_2=0.6,$ $\beta_1=0.5,\ \beta_2=0.8$								
0.1	0.2906	0.3119	0.6003	0.5309	0.0525	0.0052	0.0337	0.0913
0.3	0.2588	0.2164	0.5427	0.4415	0.0590	0.0082	0.0102	0.0774
0.5	0.2449	0.1748	0.5025	0.3874	0.0559	0.0094	0.0008	0.0660
0.7	0.2372	0.1516	0.4713	0.3485	0.0510	0.0102	0.0000	0.0612
0.9	0.2322	0.1367	0.4461	0.3184	0.0463	0.0109	0.0000	0.0572

products are more sensitive to customer segment ratio and perceived value gap. The profits of the retailer, remanufacturer, and the supply chain decrease with a more substantial perceived value gap, while the manufacturer increases the margin.

6 CONCLUSION

This chapter provided research on high-technology products, considering the product differences in quality and generation. This study fills up the research gap. The demand models for the new product belonging to the early generation and the remanufactured product belonging to the latest generation are based on customers' perceived values. The two types of products have different intrinsic value on quality and technology. Customers are classified into two segments—quality-conscious and technology-savvies. They also have different perceived values on quality and technology. The optimal wholesale price and retail price for each product are obtained in a vertical equilibrium environment. The impact of the customer segment proportion and the perceived value gap between the two segments are analyzed.

The study catches up on research opportunities that only a limited amount of research articles pay attention to—the new and remanufactured products across generations (Zhou et al., 2017; Zhou and Gupta, 2019; Zhou and Gupta, 2019; Zhou and Gupta, 2020). In the future, more exploration can be done to compare the results in the Retailer-Stackelberg and Manufacturer-Stackelberg environment.

REFERENCES

Andrew-Munot M., Ibrahim R. N., and Junaidi E. 2015. "An overview of used-products remanufacturing," *Mechanical Engineering Research*, 5(1), 12–23.

Aydin R., Kwong C., and Ji P. 2016. "Coordination of the closed-loop supply chain for product line design with consideration of remanufactured products," *Journal of Cleaner Production*, 114, 286–298.

Chen Y. C., Fang S.-C., and Wen U.-P. 2013. "Pricing policies for substitutable products in a supply chain with internet and traditional channels," *European Journal of Operational Research*, 224, 542–551.

Council R. I., "What is remanufacturing?" [Online]. Available: http://www.remancouncil. org/educate/remanufacturing-information/what-is-remanufacturing.

Ding Q., Dong C., and Pan Z. 2016. "A hierarchical pricing decision process on a dual-channel problem with one manufacturer and one retailer," *International Journal of Production Economics*, 175, 197–212.

Gao J., Wang X., Yang Q. and Zhong Q. 2016. "Pricing decisions of a dual-channel closed-loop supply chain under uncertain demand of indirect channel," *Mathematical Problems in Engineering*, 2016, 1–13.

Gan S.-S., Pujawan I. N., Suparno, and Widodo B. 2017. "Pricing decision for new and remanufactured product in a closed-loop supply chain with separate sales-channel," *International Journal of Production Economics*, 190, 120–132.

Giri B., Chakraborty A., and Maiti T. 2017. "Pricing and return product collection decisions in a closed-loop supply chain with dual-channel in both forward and reverse logistics," *Journal of Manufacturing Systems*, 42, 104–123.

Gu Q. and Gao T. 2012. "Management for two competitive closed-loop supply chains," *International Journal of Sustainable Engineering*, 5(4), 325–337.

Gungor A. and Gupta S. M. 1999. "Issues in environmentally conscious manufacturing and product recovery: a survey," *Computers & Industrial Engineering*, 36, 811–853.

Hsiao L. and Chen Y.-J. 2014. "Strategic motive for introducing internet channels in a supply chain," *Production and Operations Management*, 23(1), 36–47.

Ilgin M. A. and Gupta S. M. 2010. "Environmentally conscious manufacturing and product recovery (ECMPRO): a review of the state of the art," *Environmental Management*, 91, 563–591.

Karimabadi K., Arshadi-khamseh A., and Naderi B. 2019. "Optimal pricing and re-manufacturing decisions for a fuzzy dual-channel supply chain," *International Journal of Systems Science: Operations & Logistics*, 02/2019, 1–14.

Ke H. and Cai M. 2019. "Pricing decisions in closed-loop supply chain with collection-relevant demand," *Journal of Remanufacturing*, 9, 75–87.

Kuo C.-W. and Huang K.-L. 2012. "Dynamic pricing of limited inventories for multi-generation products," *European Journal of Operational Research*, 217, 394–403.

Liu C., Guo C., Guo L., and Lei R. 2016. "In dual-channel closed-loop supply chain, re-manufactured product pricing strategy based on the heterogeneity of consumers," *Revista de la Facultad de Ingeniería*, 31, 100–113.

Luo Z., Chen X., Chen J., and Wang X. 2017. "Optimal pricing policies for differentiated brands under different supply chain power structures," *European Journal of Operational Research*, 259(2)437–451.

Rodriguez B. and Aydin G. 2015. "Pricing and assortment decision for a manufacturer selling through dual channels," *European Journal of Operational Research*, 242, 901–909.

Soleimani F. 2016. "Optimal pricing decisions in a fuzzy dual-channel supply chain," *Soft Computing*, 20(2), 689–696.

Wang B. and Wang J. 2015. "Price and service competition between new and remanufactured products," *Mathematical Problems in Engineering*, 2015, 1–18.

Wang Z.-B., Wang Y.-Y. and Wang J.-C. 2016. "Optimal distribution channel strategy for new and remanufactured products," *Electronic Commerce Research*, 16(2), 269–295.

Xiao T. and Shi J. 2016. "Pricing and supply priority in a dual-channel supply chain," *European Journal of Operational Research*, 254, 813–823.

Zhao J., Wei J., and Li Y. 2017. "Pricing and remanufacturing decisions for two substitutable products with a common retailer," *Journal of Industrial and Management Optimization*, 13(2), 1125–1147.

Zhou L., Gupta S. M., Kinoshita Y. and Yamada T. 2017. "Pricing decision models for remanufactured short-life cycle technology products with generation consideration," *Procedia CIRP*, 61, 195–200.

Zhu X. and Yu L. 2018. "Differential pricing decision and coordination of green electronic products from the perspective of service heterogeneity," *Applied Sciences*, 8(7), 1207–1216.

Zhou L. and Gupta S. M. 2019. "Marketing research and life cycle pricing strategies for new and remanufactured products," *Journal of Remanufacturing*, 9(1), 29–50.

Zhou L. and Gupta S. M. 2019. "A pricing and acquisition strategy for new and re-manufactured high-technology products," *Logistics*, 3, 1(8), 1–26.

Zhou L. and Gupta S. M. 2020. "Value depreciation factors for new and remanufactured high-technology products: a case study on iPhones and iPads," *International Journal of Production Research*, 58(23), 7218–7249.

3 Disassembly Line Balancing of an End-of-Life Product: A Case of Wind Turbine

İbrahim Miraç Eligüzel, Eren Özceylan, and Süleyman Mete
Department of Industrial Engineering, Gaziantep University, Gaziantep, 27100, Turkey

1 INTRODUCTION

Renewable energy sources gain importance rapidly to handle increasing electricity demand, releasing the dependency on fossil fuels and other economies for countries. On the other hand, increasing environmental awareness also contributes to the usage of renewable energy sources. Therefore, the establishment of a system to utilize renewable energy sources drastically increases. One of the promising renewable electricity sources is wind power. For instance, Germany has installed an important wind power capacity in 2018, which corresponds to 29% of the total wind power capacity in Europe. In addition, cumulative installed wind capacity for onshore wind turbines was demonstrated as 17 GW in 2000, while it reached 542 GW globally in 2018. Another aspect of wind power usage is the number of materials that are utilized to produce wind turbines. On average, 2 MW wind turbine contains 71% steel, 12% fiberglass, 13.3% iron, and 1.6% aluminum (Moné et al. 2015). Preferability and usage of wind power are increasing worldwide. A report proposed by "Wind Europe" includes the construction and financing activity in European onshore and offshore wind farms and has given the total number of wind turbines and installed capacity for the European countries during 2019. According to the findings, the UK has an outstanding total wind capacity of all installations in Europe (75% offshore). Spain is second with 15%, followed by Germany (14%), Sweden (10%), and France (0.8%) (Komusanac et al. 2019).

Installed capacity occupies a key place and most of the countries prefer to build onshore wind turbines except UK, Germany, Belgium, and Denmark. In addition, the UK and Germany are the leading countries with offshore wind power installation among other European countries (Komusanac et al. 2019). Also, it can be expressed that most of the wind turbines in Europe are in the possession of Germany (30%), followed by Spain (13%) (Komusanac et al. 2019). When the general view of the investments for wind energy in Europe is considered, new asset financing and project acquisitions are the leading parts of the total investments (Brindley 2019). Less than half of the total investment—which is €26.7B—was utilized for new wind farms construction and, because of the aim to reduce cost in new project financing, 2018 can be considered as an outstanding year for new capacity financed (16.7 GW) (Brindley 2019). Overall, there is an increasing trend for the usage of wind turbines and there is a significant number of investments in Europe. Besides the European region, the US is also a leading figure when it comes to the wind energy industry. The US wind industry is rapidly growing and investments in it occupy a significant amount. According to US annual and cumulative capacity, their annual installation is fluctuating for 19 periods from 2000 to 2019 (AWEA 2019). However, the annual installation reaches an elevated level, except for 2013, and peaked in 2012. All in all, there is a growing tendency for the use of wind power which leads to an increase in investments in wind turbines. Wind turbines consist of several parts and require a significant amount of raw materials. Also, when we consider the average lifespan of wind turbines (20-25 years) and assume the construction date of MW wind turbines in the 2000s, it can be said that most of them are about to fulfill their lifespan. This fact raises another concern—recycling of established systems classified under the recycling operation (R&O). R&O can be thought of as a significant part of cost reduction because most of the parts of wind tribune are gathered from other countries like China. In addition, there is a massive amount of raw materials required to manufacture a wind turbine, depending on its capacity. In Table 3.1, some of the instances for wind turbines are given to demonstrate mass relation with the capacity of wind turbines (Moné et al. 2015).

To reduce dependency on other economies in producing energy by utilizing wind turbines, it is required to have a well-organized recycling operation for current assets. Currently, dismantle to recycle is done by undetailed processes and onside. Cracked rotor blades, separation of tower segments, and cutting nacelle into smaller pieces are the dismantling processes that lead to waste of time, inefficiency, and both economic and ecological risks (Westbomke et al. 2018). The current trend in the global energy market gives renewable energy systems a raise and wind energy grows accordingly. Also, wind turbine farms can be expressed as an efficient and effective way to generate electricity for most countries. Moreover, most system parts are produced in different countries, and end of life wind turbines can provide an enormous source of steel, iron, fiberglass, etc. Another aspect is the total amount of raw materials gathered from R&O. From Table 3.1, it can be expressed that raw materials of wind turbine have a high potential to be recycled and a high volume of the recycled materials (steel, fiberglass, copper, iron) can be obtained to reuse in the industry. By taking the aforementioned facts into account, it is also required to consider the number of established wind turbines and trends globally. Therefore, R&O gains importance, and wind turbines becomes an inevitable issue to consider due to the age and total number of wind turbines built so far. Disassembly is one of the most important phases for remanufacturing and product recovery. It is a systematic separation of used products. Hence, the disassembly line has attracted much more attention to increasing the productivity and flexibility of disassembly automation. Therefore, in this study, disassembly of wind turbines is considered, and the

TABLE 3.1
Condensed Bill of Materials for Wind Turbines Instances

	OEM						
Turbine make	Micon	Nordex	Micon	Vestas	Vestas	Vestas	Vestas
Turbine model	NM52	N-62	NM72	V82 1.65	V90 2.0	V100 2.0	V110 2.0
Nameplate capacity	0.9 MW	1.3 MW	1.5 MW	1.65 MW	2.0 MW	2.0 MW	2.0 MW
Hub height	60.7 m	69 m	80 m	78 m	80 m	80 m	80 m
Rotor diameter	52.2 m	62 m	72 m	82 m	90 m	100 m	110 m
						Mass (kg per kW)	
Steel	111.2	104.5	110.1	96.3	82.2	83.9	92.2
Fiberglass/resin/plastic	18.8	23.9	20.9	18.2	16.0	14.1	14.2
Iron/cast iron	7.2	17.3	8.7	17.8	20.5	13.3	13.3
Copper	1.6	1.5	1.2	1.8	0.9	0.6	0.7
Aluminum	N/A	N/A	N/A	1.9	2.1	1.7	1.9
Total	139.9	148.2	141.7	138.9	124.0	115.0	124.0
						% of total turbine mass	
Steel	79%	71%	78%	69%	66%	73%	74%
Fiberglass/resin/plastic	13%	16%	15%	13%	13%	12%	11%
Iron/cast iron	5%	12%	6%	13%	17%	12%	11%
Copper	1%	1%	1%	1%	1%	1%	1%
Aluminum	N/A	N/A	N/A	1%	2%	1%	2%
Total	99.2%	99.4%	99.4%	97.8%	98.0%	98.7%	98.5%

rotor part is taken for a case study. The mathematical model, proposed by Koc et al. (2009), is used for solving the disassemble rotor parts. The case of rotor parts is applied with GAMS- CPLEX solver. The rest of the study is given as follows: in the second part, the literature review is analyzed, which includes the disassembly line balancing problems (DLB) and DLB application on renewable energy systems. In the third part, a mathematical model for the DLB problem is given. The application of DLB on the rotor is given in the fourth part. Lastly, the conclusion is made on the findings of the study.

2 LITERATURE REVIEW

The disassembly is a significant step in the recycling process due to the extraction of desired parts from the end of the life products and circular economy. Disassembly is the process that comprehends the systematic extraction of desired parts or materials by dividing it into sub-assemblies or other groupings (Kalayci et al. 2016). In addition, disassembly line balancing is the allocating of tasks to workstations with varying objectives, such as minimizing the number of workstations, ensuring similar idle time at each workstation, etc., and its importance gradually inclines with environmental awareness (Özceylan et al., 2019). Therefore, effective and efficient disassembly processes have prime importance for remanufacture or reuse of the materials. Exact, heuristic, and meta-heuristic approaches are developed to solve the problem. The mathematical models are developed to solve the problem by Altekin et al. (2008), Koç et al. (2009), and the linear physical programming approach has also been developed for the problem by Ilgin et al. (2017). Mete et al. (2018) proposed a new line design for assembly and DLB problem and proposed a mathematical model. Recently, Li et al. (2019) proposed a fast branch bound and remember algorithm to solve the DLB problem. Since the DLBs are considered as NP-Hard problem (McGovern and Gupta 2007), and due to the importance of DLB planning, several approaches such as greedy algorithm (McGovern and Gupta 2003), hybrid genetic algorithm (Kalayci et al. 2016), and multi-criteria optimization model (Igarashi et al. 2016) are applied. One of the approaches proposes the mathematical model for multi-objective fuzzy DLB; the model is solved by a Pareto-improved artificial fish swarm algorithm (Zhang et al. 2017). There are several types of performance measures for the DLB which consider the cycle time, number of workstations, profit (Altekin et al. 2016), and the number of resources (Mete et al. 2016a), etc. In literature, there are several algorithms to solve DLB besides the aforementioned algorithms such as the Pareto firefly algorithm (Zhu et al. 2018), variable neighborhood search (Kalayci et al. 2015), hybrid migrating birds (Zhu et al. 2020), and artificial bee colony (Liu and Wang 2017). Moreover, Mete et al. (2019) proposed a novel model for the rebalancing situations of DLB problem.

Another factor that gives rise to the importance of DLB is environmental awareness. By utilizing the DLB approach in the industry, most environmental hazards can be avoided because of a decrease in waste, reuse of the raw materials instead of mining, and the elimination of hazardous materials. In this aspect, renewable energy systems can be taken into consideration due to its sustainability and environmental friendly regard. However, there is concern that requires coping with the recycling of end of life systems. Recycling of renewable energy systems comprehends the battery, cable, steel, fiberglass, etc. To achieve efficient and effective energy production in both economic and environmental objectives via renewable energy sources, it is required to have a well-designed recycling process. Therefore, DLB has become a widely used process and a critical issue in the renewable energy industry. From this point of view, critical raw materials (CRM) are fundamental for both high-tech applications and renewable energy systems (Işıldar et al. 2019), including most CRM, supplied from the waste of electrical and electronic equipment (WEEE). Therefore, the study is conducted to demonstrate the steady-state frontiers in CRMs recycle from WEEE by utilizing biotechnology and present the current research and development (R&D) activities (Işıldar et al. 2019). When renewable energy systems are considered, it can be said that energy storage through batteries occupies an important regard and their recycling processes gain importance. Therefore, there are several studies conducted in this manner. The focus of these studies can be summarized as a novel utility-scale flywheel energy storage system that provides higher energy density and easy recycling (Li et al. 2018), analysis of the life cycles of batteries that are classified as lead-acid, lithium-ion, and nickel-metal hydride, and safe recycling or separation process for the end of the life batteries (Dutt 2013), proposition for recycling process of lithium-ion batteries retrieved from used mobile-phones to reuse as energy storage for solar lightening (Diouf et al. 2015), presentation of policies and guidelines for recycling of solar PV modules (Sharma et al. 2019), and comparison of dedicated full recovery facility and laminated glass recycling facility for recycling of crystalline silicon solar panel (Faircloth et al. 2019). Also, it is crucial to monitor the current situation of wind turbines to facilitate the provision of predictive maintenance and increase the performance of end-of-life operations, in which the study proposes embedding sensors into wind turbines (Dulman and Gupta, 2018). Comparison is done between the regular and embedded sensors systems in wind turbine via utilization of discrete-event simulation and it is concluded that embedded sensor in wind turbines shows significant benefits (Dulman and Gupta, 2018). As a result, there are a lot of research papers that aim to utilize renewable energy systems more efficiently and effectively via recycling. However, to the best knowledge of the authors, there is no comprehensive study that proposes a full disassembly of the renewable energy systems. Therefore, in this chapter, disassemble of rotor parts of wind turbines are considered as a case study. Furthermore, the task precedence diagram proposed by Koç et al. (2009) is considered. A transformed AND/OR graph (TAOG) is used for determining precedence relations between disassembly tasks. The reason behind choosing the rotor is the retrieval of more detailed information from it.

3   MATHEMATICAL MODEL

The objective function of the model is to minimize the number of disassembly workstations under a pre-known cycle time. Assumptions for the model are also given as follows (Mete et al. 2016b):

- A single product type (rotor part of a wind turbine) is to be disassembled on a disassembly line.
- The supply of the end-of-life product is infinite.
- A disassembly task cannot be divided between two workstations.
- Part removal times are deterministic, constant, and discrete.
- Each product undergoes complete disassembly.
- All products contain all parts with no additions, deletions, modifications, or physical defects.
- Each task is assigned to one and only one workstation.
- Transformed And/Or Graph of each product is known and should be satisfied.

In this study, the integer programming formulation proposed by Koç et al. (2009) is used. The mathematical model of the DLB problem is as follows:

Objective function

$$\text{Min } \sum_{j=1}^{J} j \cdot F_j \tag{3.1}$$

Objective function (3.1) minimizes the total number of open workstations. j is the workstation index, $j = 1, 2, \ldots, J$. If workstation j is opened, F_j is 1; otherwise, it is 0.

Subject to

$$\sum_{i:B_i \in S(A_k)} Z_i = 1 \forall_{k=0} \tag{3.2}$$

$$\sum_{i:B_i \in S(A_k)} Z_i = \sum_{i:B_i \in P(A_k)} Z_i \forall_{k=1,2,\ldots,h} \tag{3.3}$$

Constraints (3.2) and (3.3) assure that exactly one of the OR-successors is selected. A_k is the artificial nodes in *transformed AOG*, $k = 0, 1, 2, \ldots, K$; B_i is the normal nodes in *transformed AOG*, $i = 1, 2, \ldots, I$. If task B_i is performed, Z_i is 1; otherwise, it is 0.

$$\sum_{j=1}^{J} X_{ij} = Z_i \forall_{i=1,2,\ldots,I} \tag{3.4}$$

Constraint (3.4) makes sure that if the task is selected it is assigned to one of the workstations. If task B_i is assigned to workstation j, X_{ij} is 1; otherwise, it is 0.

$$\sum_{i:B_i \in P(A_k)} \sum_{j=1}^{v} X_{ij} \le \sum_{i:B_i \in S(A_k)} X_{iv} \forall_{k=1,2,\ldots,K} \tag{3.5}$$

Constraint (3.5) handles the precedence relations between the normal nodes since exactly one of the OR-predecessors and one of the OR-successors of an artificial node will be selected. Constraint (3.5) makes sure that the successor chosen among the OR-successors will be assigned to the higher-indexed workstation than the predecessor chosen among the OR-predecessors. d_{Bi} is the task time of B_i (normal node); $P(A_k)$, $P(B_i)$ is the immediate predecessor set of artificial node A_k, B_i, respectively; $S(A_k)$, $S(B_i)$ is the immediate successor set of artificial node A_k, B_i, respectively. C is the cycle time.

$$\sum_{i=1}^{I} X_{ij} \cdot d_{Bi} \le C \cdot F_j \forall_{j=1,2,\ldots,J} \tag{3.6}$$

Constraint (3.6) is the cycle time constraint that forces the total workload of a workstation to be less than the cycle time if the workstation is opened.

$$X_{ij}, F_j, Z_i \in \{0, 1\} \forall_{i=1,2,\ldots,I; j=1,2,\ldots,J} \tag{3.7}$$

Constraint (3.7) is the binary constraints.

4   A CASE OF A WIND TURBINE

Wind turbines consist of several parts that can show variance depending on the brand and load of a wind turbine. However, wind turbine parts can be expressed under three main parts—rotor, nacelle, and tower. In the case of a 2 MW wind turbine, the tower, blades, and gearbox correspond to 61% of the total wind turbine cost and current disassembly methods for wind turbine parts deal with open recycling (Ortegon et al., 2013). Therefore, it can be said that savings in materials and lower disposal rates could be accomplished by the aforementioned three components of the wind turbines. In addition, there are main subassemblies under the mentioned parts: blades, pitch, and hub assembly for rotor; nacelle structural, drivetrain, nacelle electrical, and yaw assembly for nacelle; tower module assembly (Moné et al. 2015). In this study, the rotor is disassembled because of retrieving more detailed information on its parts and demonstrated for full disassembly. Basic illustration for disassembly parts for wind turbines given in Figure 3.1.

To have a better understanding of the wind turbine disassembly, it is required to have general knowledge about their parts. As it is explained before, whole parts in the wind turbine can be gathered under three main parts. The rotor is the part that transmits wind power to the nacelle to produce electricity. Rotor deduces the right blade angle for maximum efficiency of wind via the win wane data. In addition, the yaw system takes a place between the tower and rotor, which provides mobility for the rotor with respect to the analysis of data gathered from win vane with the help of electric or hydraulic motors (AWEA 2011). Nacelle is the part that contains shafts, electronic systems, generator, and brake systems, which take a place on the tower and next to the rotor. Therefore, it can be said that the right angle of blades arranged through information received from the nacelle and the yaw system provides the right position for rotor and nacelle altogether. Besides all points expressed before, precedence relation between parts and the task time for each operation is a significant issue in DLB. To achieve a well-designed DLB, it is required to generate AOG and TAOG. Therefore, precedence relations, tasks to manage, and overall view of the DLB can be comprehended easily and accurately. An ordinary AOG demonstrates all the possible subassemblies, except that precedence relations between the tasks and TAOG show a more detailed version of an AOG, which comprehends whole information about the disassembly trees including the precedence relations between tasks (Koc et al. 2009).

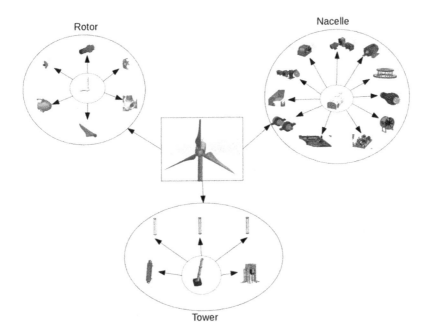

FIGURE 3.1 Basic Illustration for Disassembly of Wind Turbine Parts.

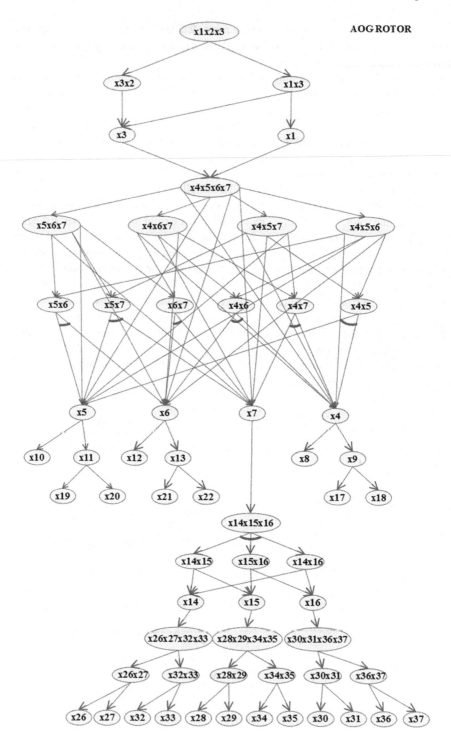

FIGURE 3.2   AOG for Disassembly of Wind Turbine Rotor.

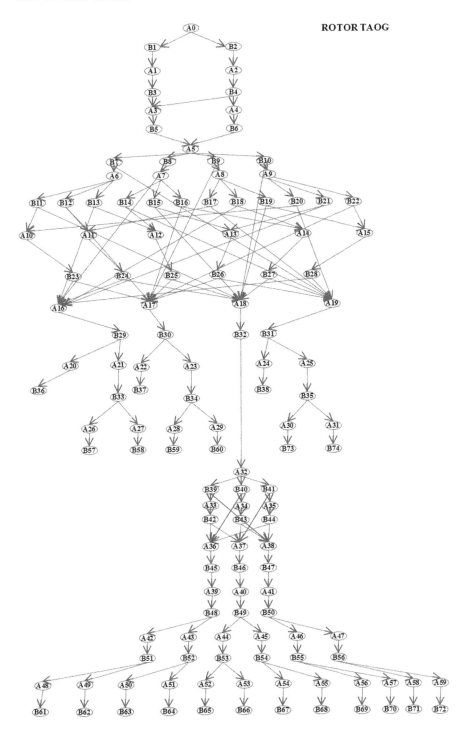

FIGURE 3.3 TAOG for Disassembly of Wind Turbine Rotor.

TABLE 3.2
Wind Turbine Rotor Parts and Codes

Part	Part Code
Plane	X1
Spinner nose cone	X2
Nose cone	X3
Blade mechanism	X4, X5, X6
Rotor hub	X7
Blade	X8, X10, X12
Blade bearing	X9, X11, X13
Blade pitch	X14, X15, X16
Outer ring	X17, X19, X21
Inner ring	X18, X20, X22
Ball-bearing (deep grove)	X23, X24, X25, X32, X34, X36
Drive gear	X26, X28, X30
Shaft	X27, X29, X31
Planet mechanism	X33, X35, X37

TABLE 3.3
Tasks and Task Times

Level No	Level Time Range	Level Tasks	Task Times
1	$30 < d_{Bi} < 50$	1, 2, 3, 4, 5, 6	44, 43, 37, 41, 44, 39
2	$25 < d_{Bi} < 45$	7, 8, 9, 10, 11, 12, 13, 14, 15, 16, 17, 18, 19, 20, 21, 22	30, 37, 40, 35, 25, 31,37, 26, 28, 39, 36, 29, 39, 32, 29, 35
3	$20 < d_{Bi} < 40$	23, 24, 25, 26, 27, 28, 29, 30, 31, 32	22, 26, 22, 34, 24, 31, 28, 37, 26, 27
4	$15 < d_{Bi} < 35$	33, 34, 35, 36, 37, 38	22, 35, 29, 31, 34, 11
5	$10 < d_{Bi} < 30$	39, 40, 41, 42, 43, 44, 45, 46, 47, 48, 49, 50, 51, 52, 53, 54, 55, 56, 57, 58, 59, 60, 61, 62, 63, 64, 65, 66, 67, 68, 69, 70, 71, 72, 73, 74	11, 29, 25, 19, 15, 17,30, 11, 12, 28, 21, 13, 26, 28, 18, 11, 10, 17, 19, 24, 29, 10, 30, 16, 28, 27,12, 19, 28, 26, 26, 16, 12, 26, 18

Figure 3.2 demonstrates the AOG for the rotor which includes the disassembly trees for the rotor. Figure 3.3 also shows the TAOG which covers the 74 disassembly tasks; it can deduce the precedence relations between the tasks. Therefore,

TABLE 3.4

Generated Scenarios

Scenario	CPU Time (s)	Number of Opened Workstations	Cycle Time (s)
1	1.89	15	80
2	12.69	17	75
3	16.22	18	70
4	23.36	19	65
5	–	Infeasible	60

TABLE 3.5

Results of Scenario 1

Workstation	1	2	3	4	5	6	7	8
Tasks	1, 3	5, 9	17, 31	24, 32, 40, 43	54, 49, 46	55, 56, 67, 50, 47	30, 35, 71	53, 69, 73

Workstation	9	10	11	12	13	14	15
Tasks	29, 37, 74	34, 45, 48, 66, 72	33, 52, 70	38, 57, 60	36, 58, 64	51, 59, 61, 65	62, 63, 68

TABLE 3.6

Results of Scenario 2

Workstation	1	2	3	4	5	6	7	8	9
Tasks	1	3	5, 8	15, 32, 39, 47	26, 30, 42	29, 33, 45	31, 50, 56	46, 49	48, 53, 54

Workstation	10	11	12	13	14	15	16	17
Tasks	38, 51, 55, 57	67, 68, 69	36, 61, 62	34, 65, 70	35, 59, 71	37, 60, 66	52, 58, 74	63, 64, 73

the case can be applied after generating TAOG. The part codes in AOG and corresponding parts are given in Table 3.2.

The next step is deciding the task times. Task time for each disassembly tasks is determined randomly between the intervals according to part size; the interval is decreased by five for each level of size for our case. Task times are given in Table 3.3.

TABLE 3.7

Results of Scenario 3

Workstation	1	2	3	4	5	6	7	8	9
Tasks	1	3	5	8, 29	15, 32, 40	26, 36, 43	45, 47, 48, 52	50, 55, 56, 70	46, 51, 69
Workstation	**10**	**11**	**12**	**13**	**14**	**15**	**16**	**17**	**18**
Tasks	31, 49	30, 34, 54	33, 64, 72	53, 6, 66	35, 71, 74	38, 57, 58	37, 60, 61	63, 68, 73	59, 65, 67

TABLE 3.8

Results of Scenario 4

Workstation	1	2	3	4	5	6	7	8	9	10
Tasks	1	3	5	7, 12	24, 32, 40	30, 31	37, 38	35, 45, 48	29, 43, 47	46, 50, 51
Workstation	**11**	**12**	**13**	**14**	**15**	**16**	**17**	**18**	**19**	**11**
Tasks	34, 63, 52	49, 54, 74	33, 56, 64	55, 53, 73	71, 62, 58	36, 59, 72	57, 60, 67	61, 65, 68	66, 69, 70	34, 63, 52

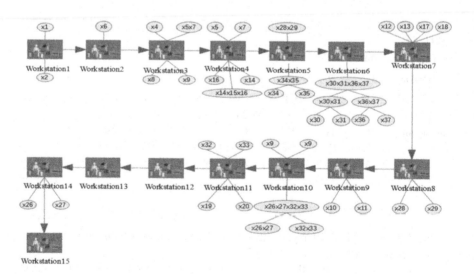

FIGURE 3.4   Demonstration of Scenario 1.

After deciding on the task times, the rotor case with 74 determined tasks is solved for five different cycle-time scenarios. 80, 75, 70, 65, and 60 is given as the cycle-times for Scenario 1, 2, 3, 4 and 5, respectively. All scenarios are solved by GAMS-CPLEX solver win64 23.8.2 in a computer that has Intel® Core™ i7 - CPU at 2.60 GHz 8.00 RAM. Results from each scenario are given in Table 3.4.

In Table 3.4, it can be expressed that cycle time has an inverse proportion with both "Number of Opened Stations" and "CPU Time". While cycle time is increased, the number of open workstations decreases with CPU time.

Tables 3.5 to 3.8 demonstrate the results from each scenario except Scenario 5 due to gathered infeasible solutions. As can be seen from the results, a reduction in cycle time increases the workload on each workstation. For instance, three tasks on average are included and first-level tasks, which have slightly high task time, are mostly together. However, workload is decreased with the incline in cycle time. Therefore, the number of open stations is increased accordingly.

Figure 3.4 is an illustration of Scenario 1 and it shows the output of each task on the related workstation. Each part extracted from the main body is demonstrated within ellipses in Figure 3.4. In addition, some of the workstations do not have any part to be illustrated, which means end-tasks (exit tasks from the system) in TAOG are allocated in a given workstation.

5   CONCLUSION

In this study, a mathematical model is applied to the full disassembly of the renewable energy systems. A real end-of-life wind turbine is analyzed. Different scenarios are also performed by altering cycle times to observe its effect on the number of open workstations. To the best knowledge of the authors, this study is the first adaptation of a mathematical model to balance the wind turbine disassembly line. The rotor part of the wind turbine is considered to retrieve more detailed information on it. The case of a rotor with 74 determined tasks is solved for five different cycle-time scenarios. The objective function is to minimize the total number of workstations under determined cycle time scenarios. As a result, 74 tasks are assigned to 15 workstations optimally with a cycle time of 80s.

In the future, other parts of the wind turbine can be considered and solved. Heuristic and meta-heuristic approaches can be developed for the problem. In addition to the straight line in this study, different line layouts can be considered—parallel, two-sided, and U-type.

REFERENCES

Altekin, F. T., Kandiller, L., and Ozdemirel, N. E. (2008). "Profit-oriented disassembly-line balancing." *International Journal of Production Research* 46, 2675–2693.

Altekin, F. T., Bayındır, Z. P., and Gümüşkaya, V. (2016). "Remedial actions for disassembly lines with stochastic task times." *Computers & Industrial Engineering* 99, 78–96.

AWEA. (2019). *Quarterly Market Report First Quarter 2019 Table of Contents*, 1–37. https://www.awea.org/Awea/media/Resources/PublicationsandReports/MarketReports/3Q-2019-AWEA-Market-Report-Public-Version.pdf. Access date: 15/02/2020.

AWEA. (2011). *Wind Energy Industry Manufacturing Supplier Handbook*, 41. http://www. awea.org/Issues/Content.aspx?ItemNumber=4611. Access date: 15/02/2020.

Brindley, Guy. (2019). "Financing and investment trends: the European wind industry in 2018." *WindEurope*, 32. https://windeurope.org/wp-content/uploads/files/about-wind/reports/Financing-and-Investment-Trends-2018.pdf. Access date: 15/02/2020.

Diouf, Boucar, Ramchandra Pode, and Rita Osei. (2015). "Recycling mobile phone batteries for lighting." *Renewable Energy* 78, 509–515.

Dulman, M. T., and Gupta, S. M. (2018). "Maintenance and remanufacturing strategy: using sensors to predict the status of wind turbines." *Journal of Remanufacturing*, 8(3), 131–152.

Dutt, Devraj. (2013). "Life cycle analysis and recycling techniques of batteries used in renewable energy applications." Conference and Exhibition – 2013 International Conference on New Concepts in Smart Cities: Fostering Public and Private Alliances, SmartMILE 2013, 1–7.

Faircloth, C. C., Wagner, K. H., Woodward, K. E., Rakkwamsuk, P., and Gheewala, S. H. (2019). "The environmental and economic impacts of photovoltaic waste management in Thailand." *Resources, Conservation and Recycling*, 143 (December 2018), 260–272.

Igarashi, K., Yamada, T., Gupta, S. M., Inoue, M., and Itsubo, N. (2016). "Disassembly system modeling and design with parts selection for cost, recycling and CO2 saving rates using multi criteria optimization." *Journal of Manufacturing Systems*, 38, 151–164.

Ilgin, M. A., Akçay, H., Araz, C. (2017). "Disassembly line balancing using linear physical programming." *International Journal of Production Research* 55, 6108–6119.

Işıldar, A., van Hullebusch, E. D., Lenz, M., Du Laing, G., Marra, A., Cesaro, A., ... Kuchta, K. (2019). "Biotechnological strategies for the recovery of valuable and critical raw materials from waste electrical and electronic equipment (WEEE) – A review." *Journal of Hazardous Materials*, 362 (August 2018), 467–481.

Kalayci, C. B., Polat, O., and Gupta, S. M. (2015). "A variable neighbourhood search algorithm for disassembly lines." *Journal of Manufacturing Technology Management* 26(2), 182–194.

Kalayci, C. B., Polat, O., and Gupta, S. M. (2016). "A hybrid genetic algorithm for sequence-dependent disassembly line balancing problem." *Annals of Operations Research* 242(2), 321–354.

Koc, Ali, Ihsan Sabuncuoglu, and Erdal Erel. (2009). "Two exact formulations for disassembly line balancing problems with task precedence diagram construction using an AND/OR graph." *IIE Transactions (Institute of Industrial Engineers)* 41(10), 866–881.

Komusanac, Ivan, Guy Brindley, and Daniel Fraile. (2019). *Wind Energy in Europe*. https://windeurope.org/wp-content/uploads/files/about-wind/statistics/WindEurope-Annual-Statistics-2019.pdf. Access date: 15/02/2020.

Li, X., Anvari, B., Palazzolo, A., Wang, Z., and Toliyat, H. (2018). "A utility-scale flywheel energy storage system." *IEEE Transactions on Industrial Electronics*, 65(8), 6667–6675.

Li, Z., Çil, Z. A., Mete, S., and Kucukkoc, I. (2019). "A fast branch, bound and remember algorithm for disassembly line balancing problem." *International Journal of Production Research*, 58(11), 3220–3234.

Liu, Jia, and Shuwei Wang. (2017). "Balancing disassembly line in product recovery to promote the coordinated development of economy and environment." *Sustainability (Switzerland)* 9(2), 309.

McGovern, S. M., and Gupta S. M. (2003). "Greedy algorithm for disassembly line scheduling." Proceedings of the IEEE International Conference on Systems, Man and Cybernetics 2, 1737–1744.

McGovern, S. M., and Gupta S. M. (2007). "Combinatorial optimization analysis of the unary NP-complete disassembly line balancing problem." *International Journal of Production Research*, 45(18–19), 4485–4511.

Mete, S., Cil, Z. A., Ağpak, K., Özceylan, E., and Dolgui, A. (2016a). "A solution approach based on beam search algorithm for disassembly line balancing problem." *Journal of Manufacturing Systems*, 41, 188–200.

Mete, S., Abidin Çil, Z., Özceylan, E., Ağpak, K., (2016b). Resource constrained disassembly line balancing problem. IFAC-PapersOnLine, 8th IFAC Conference on Manufacturing Modelling, Management and Control MIM 2016Troyes, France, 49, 921–925.

Mete, S., Çil, Z. A., Özceylan, E., Ağpak, K., Battaïa, O., (2018). "An optimisation support for the design of hybrid production lines including assembly and disassembly tasks." *International Journal of Production Research*, 56(24), 7375–7389.

Mete, S., Çil, Z. A., Celik, E., and Ozceylan, E. (2019). "Supply-driven rebalancing of disassembly lines: A novel mathematical model approach." *Journal of Cleaner Production*, 213, 1157–1164.

Moné, C., Hand, M., Bolinger, M., Rand, J., Heimiller, D., and Ho, J. (2015). Nrel *2015 Cost of Wind Energy Review NREL/TP-6A20-66861.* https://www.nrel.gov/docs/fy17osti/66861.pdf. Access date: 23/11/2020.

Ortegon, K., Nies, L. F., and Sutherland, J. W. (2013). "Preparing for end of service life of wind turbines." *Journal of Cleaner Production*, *39*, 191–199.

Özceylan, E., Kalayci, C. B., Güngör, A., and Gupta, S. M. (2019). "Disassembly line balancing problem: A review of the state of the art and future directions." *International Journal of Production Research*, 57(15–16), 4805–4827.

Sharma, Arvind, Suneel Pandey, and Mohan Kolhe. (2019). "Global review of policies & guidelines for recycling of solar Pv modules." *International Journal of Smart Grid and Clean Energy,* 8(5), 597–610.

Westbomke, M., Piel, J.-H., Breitner, M. H., Nyhius, P., and Stonis, M. (2018). "Operations research." In *Physics Today*, eds.Natalia Kliewer, Jan Fabian Ehmke, and Ralf Borndörfer. Springer, 239–244.

Zhang, Zeqiang, Kaipu Wang, Lixia Zhu, and Yi Wang. (2017). "A pareto improved artificial fish swarm algorithm for solving a multi-objective fuzzy disassembly line balancing problem." *Expert Systems with Applications,* 86, 1339–1351.

Zhu, Lixia, Zeqiang Zhang, and Yi Wang. (2018). "A pareto firefly algorithm for multi-objective disassembly line balancing problems with hazard evaluation." *International Journal of Production Research,* 56(24), 7354–7374.

Zhu, Lixia, Zeqiang Zhang, Yi Wang, and Ning Cai. (2019). "On the end-of-life state oriented multi-objective disassembly line balancing problem." *Journal of Intelligent Manufacturing*, 31, 1403–1428.

4 A Differential Evolution Algorithm for Balancing Mixed-Model Disassembly Lines

Ibrahim Kucukkoc[1], Gulsen Aydin Keskin[1], and Zixiang Li[2]
[1]Balikesir University, Department of Industrial Engineering, Balikesir, Turkey
[2]Wuhan University of Science and Technology, Engineering Research Centre for Metallurgical Automation and Measurement Technology of Ministry of Education, Wuhan, China

1 INTRODUCTION

Sustainable development and green production concepts constantly increase their popularity in the manufacturing industry. Due to the rapid advancement of technology and customer expectations, the use of many products has decreased, and the average product life is constantly getting shorter. As a result, these products waste environmental resources and harm the environment. Therefore, disassembly has drawn attention as it contributes to the recycling of materials by reusing valuable parts or components (Ren et al. 2018). The most critical process in the product recovery system requires disassembly at a certain level for product recovery options such as recycling and reproduction (Edis, Ilgin, and Edis 2019).

As proposed by Zhu et al. (2019), disassembly is a key to recycle, reproduce, and/or reuse products due for disposal. Performing a product recovery involves several steps. The first and essential stage of the recycling of the manufactured products is the disassembly process. At this procedure, valuable components and subassemblies are methodically separated from the discarded products (Avikal, Jain, and Mishra 2014, Ding et al. 2010, Mete et al. 2016).

The disassembly line is a key for the automatic removal of returned products. The disassembly line balancing problem (DLBP) quests the best sequence with one or several optimization criteria, minimizes the number of workstations, and provides similar idle times between workstations, as well as other end-of-life (EOL) specific concerns (McGovern and Gupta 2003, Li, Cil, et al. 2020).

After the disassembly operation, reusable parts are cleaned, refurbished, tested, and directed to the sub-assembly inventory for remanufacturing. After this process, the recyclable materials can be sold to raw material suppliers, while the remaining can be sent to landfills (Avikal, Jain, and Mishra 2014, McGovern, and Gupta 2007a).

Since DLBPs belong to the NP-hard problem group, many heuristic and meta-heuristic algorithms were developed to deal with the complexity of DLBPs to achieve acceptable solutions in a reasonable time, focusing on the workstation and its workload (Mete et al. 2016, Wang, Gao, and Li 2020).

Disassembly has drawn attention in literature thanks to its significant role in product recovery over the past decade (Zhang et al. 2019). As it is a significant issue in this field, an extensive amount of research has been conducted on disassembly line balancing, which focuses on minimizing the use of valuable resources and maximizing the level of process capability in disassembly operation (Ding et al. 2010).

Due to the importance of the subject, we aim to contribute by demonstrating one of the preliminary applications of the differential evolution (DE) algorithm on DLBPs. Moreover, based on the fact that it is not economically convenient to construct a different line for each disassembled model, the mixed-model disassembly line balancing problem (MDLBP) is addressed in this research. The contributions of our study to the literature are summarized as follows:

- To the best of our knowledge, this is the first DE algorithm developed to solve any type of DLBP (MDLBP is addressed in this research).
- While most studies address single-model disassembly lines, this research aims to minimize the workload smoothness index in mixed-model disassembly lines for the first time. Thus, a more measurable objective is sought rather than just the number of workstations utilized.
- The proposed DE algorithm applies a priority-based decoding scheme (used by Pitakaso (2015) to solve a simple assembly line balancing problem—ALBP) for an MDLBP. So, no repairing mechanism is needed, which enhances the running efficiency of the proposed approach.

The remainder of this chapter is organized as follows: in Section 2, a detailed literature review has been realized for DE and disassembly line balancing. The MDLBP is described in Section 3 and the implementation of the proposed DE algorithm is presented in Section 4 with implicit examples of decoding and reproduction procedures. A numerical example is provided in Section 5 to exhibit the solution steps of the proposed DE algorithm; computational tests were conducted in Section 6 to test the performance of DE. Finally, concluding remarks and future research directions are outlined in Section 7.

2   LITERATURE REVIEW

This section is devoted to a detailed literature review about the DE algorithm and disassembly line balancing that is the subject of this chapter.

2.1 Literature Review on DE Algorithm

DE algorithm is a relatively new member of population-based stochastic optimization algorithms. The studies using the DE algorithm for line balancing (both assembly and disassembly) have been presented below.

Nearchou (2005) expressed the application of the DE algorithm to type-1 simple ALBP to minimize the number of workstations within a fixed cycle time given. Nearchou (2007) considered type-2 simple ALBPs by optimally partitioning the operations of the tasks in an assembly line among the workstations with the minimization of the cycle time, and proposing an approach based on the DE method. Nearchou (2008) proposed a new population heuristic to solve the multi-objective ALBP based on the general DE method. Nourmohammadi and Zandieh (2011) proposed a multi-objective DE algorithm to solve the multi-objective type-2 simple ALBP.

Vincent et al. (2014) used variants of a DE-based algorithm to schedule flexible assembly lines. Pitakaso (2015) proposed a DE algorithm for solving type-1 simple ALBPs. Pitakaso and Sethanan (2016) proposed a DE and modified DE algorithm to solve type-1 simple ALBPs when the maximum number of machine types in a workstation is considered as a constraint. Nearchou and Omirou (2017) presented a DE algorithm to solve the simple ALBP. Sun and Wang (2019) proposed a decomposition-based matheuristic based on the DE algorithm by combining metaheuristic and model-based approaches to solve the ALBPs.

When the studies above were examined in detail, the DE algorithm has yielded very successful results, but it has only been used in ALBPs.

2.2 Literature Review on DLBP

The DLBP was introduced by Gungor and Gupta (1999). Since the concepts of sustainability and recycling have become crucial in recent years, the disassembly balancing lines have gained immense importance. Özceylan, Kalaycı, Güngör, and Gupta (2019) presented a detailed review of DLBPs in one of their recent research.

Mcgovern and Gupta (2007a) developed a hybrid approach by integrating genetic algorithm and ant colony optimization metaheuristics, a greedy algorithm, a greedy/hill-climbing, and greedy/2-optimal heuristics to the DLBP. McGovern and Gupta (2007b) proposed a genetic algorithm to obtain optimal or near-optimal solutions for a single-product DLBPs. Ding et al. (2010) proposed a novel multi-objective ant colony optimization algorithm for the single-model DLBP. Paksoy et al. (2013) proposed a mixed-integer programming model for an MDLBP considering cycle time, the number of workstations, and balanced workload per station by binary fuzzy goal programming and the fuzzy multi-objective programming. Avikal et al. (2014) proposed a Kano model, fuzzy-AHP, and M-Topsis techniques to find the optimal order of component removal using precedence relation for single-product disassembly line balancing. Duta et al. (2016) proposed a tree search algorithm to find the optimal schedule for the single-product disassembly line. Kalaycılar et al. (2016) presented various bounding procedures, assigning the tasks to the workstations to maximize the total net revenue for the single-product DLBP. Mete et al. (2016) proposed a beam search algorithm for the single-product DLBP—they specified that the minimization number of workstations is a performance measure. Zhang et al. (2017) addressed the multi-objective DLBP under fuzzy disassembly times for a single product. They used a Pareto-improved artificial fish swarm algorithm to solve the problem. Ren et al. (2017) presented a mathematical model for a profit-oriented partial DLBP and proposed an approach based on a gravitational search algorithm to solve the problem. Moreover, particle swarm optimization, DE, and artificial bee colony algorithms were presented to demonstrate the proposed approach. Wang, Xinyu, et al. (2019a) established a mathematical model of the stochastic two-sided partial disassembly line balancing with multi-objective, multi-constraint, and uncertainty factors. For the optimization of the problem, they used a multi-objective discrete flower pollination algorithm based on Pareto dominance relations. Wang, Xinyu, et al. (2019b) proposed a new multi-objective genetic simulated annealing algorithm to construct an evaluation system for partial disassembly line to balance environmental impacts and economic benefits. Li, Cil et al. (2019)(2020) developed a fast branch and bound and remember algorithm for a simple DLBP, using an AND/OR graph. Li, Kucukkoc et al. (2019) developed a mixed-integer programming model to solve the sequence-dependent U-shaped DLBP with multiple objectives. Li, Chen et al. (2019) proposed a branch, bound, and remember algorithm for DLBP under AND/OR precedence conditions. Kucukkoc et al. (2020) proposed linear and nonlinear models to solve the type-E multi-manned DLBP considering both cycle time and number of workstation objectives. Fang et al. (2019) proposed a mathematical programming model to minimize cycle time, total energy consumption, peak workstation energy consumption, and number of robots used for mixed-model disassembly line balancing. An evolutionary algorithm consisting of an encoding/decoding scheme, initialization approach, and variation operators was also developed in the same research. Kucukkoc (2020) introduced the two-sided disassembly line balancing problem and proposed a MILP model and a genetic algorithm approach (called 2-GA) to solve it. Wang et al. (2020) introduced a multi-objective disassembly line balancing model conceiving partial destructive mode and U-shaped layout considering the number of workstations, the workload balancing, energy consumptions, and the disassembly profit. They also proposed a new multi-objective discrete flower pollination algorithm to solve the problem. Cevikcan et al. (2020) developed a mixed-linear integer programming model and two heuristic algorithms (namely, multi-manned disassembly line balancing heuristic-I and multi-manned disassembly line balancing heuristic-II) to multi-manned stations according to the objectives of minimizing the workstations and the number of workers.

When the papers summarized above have been analyzed from a broad perspective, it was seen that single and partial disassembly line models are the most common. Although heuristic and meta-heuristic methods have been proposed for DLBP, the DE algorithm has not been used in any of those studies.

3 PROBLEM STATEMENT

MDLBP involves assigning a set of disassembly tasks ($i \in I$) to workstations ($k = 1, 2, ..., NS$) with the aim of optimizing an objective function while satisfying the precedence relationship and capacity constraints. More than one product model ($m \in M$) is disassembled on the line and some disassembly tasks may require different processing times for models. The term t_{im} denotes the processing time of task i for model m. The workload time of workstation k for model m is simply calculated as $WT_{mk} = \Sigma_i \, t_{im}$. As this is a mixed-model line, the workload of a workstation at a time point depends on the model type being disassembled in that workstation. Therefore, the proportion of each model (d_m) on the line needs to be used to calculate the weighted workload of workstation k, i.e., $\Sigma_{m \in M} \, d_m WT_{mk}$.

In literature, the frequently encountered objectives are minimization of cycle time (C), minimization of the number of workstations (NS), and removal of hazardous parts and highly demanded parts at earlier workstations. In this study, we aim to maximize the weighted smoothness index which can be calculated using equation (4.1).

$$FV = \sqrt{\sum_{k \in K} \left(CT - \sum_{m \in M} d_m WT_{mk} \right)^2} \tag{4.1}$$

Minimizing FV results in maximizing the so-called objective line efficiency — this objective ensures a smooth workload distribution across workstations. FV gives a better idea than NS in terms of the quality of the obtained line balance. This is because two solutions may require the same NS but different FVs.

One of the main differences of a DLBP from an ALBP is the consideration of OR precedence relationship as well as AND precedence relationship constraint. Figure 4.1 represents a sample precedence relationship diagram for ten tasks and processing times belonging to two models (A and B). The nodes represent tasks and the arrows between nodes represent the AND precedence relationships. The arcs between the arrows denote OR precedence relationship. Note that there is a dummy task (A1) to keep simplicity in the representation.

To have a feasible solution when assigning tasks to the workstations, it can become available if and only if all its AND predecessors and at least one of its OR

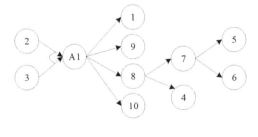

Task	A	B
1	14	11
2	10	19
3	12	12
4	18	22
5	23	23
6	16	16
7	20	20
8	36	21
9	14	14
10	10	10

FIGURE 4.1 Data for Mixed-Model P10-OR Problem.

predecessors have been assigned (if any). Assume that AND_i is the set of AND predecessors of task i and OR_i is the set of OR predecessors of task i. In Figure 4.1, $AND_9 = \varnothing$ and $OR_9 = \{2, 3\}$. As task-9 has no AND precedence relationship, it can be assigned to a workstation once task-2 or task-3 is assigned (Kucukkoc, Li, and Li 2019).

When the problem given above is solved under $C = 39$ time-unit constraint with equal model proportions $(d_A = d_B = 0.5)$, the solution minimizing the FV could be obtained in Figure 4.2a. As seen, five workstations have been utilized, which gives an FV of 12.67 and a fairly smooth workload distribution. If the same problem were solved "without" considering OR relationship (so all precedence relations in Figure 4.1 are assumed to be AND precedence relationship), the solution given in Figure 4.2b could be obtained with the FV of 27.17. As seen from this particular example, solving the problem under OR precedence constraint increases the probability of having more available tasks and increased FV.

The mixed-model disassembly process prevents the repeated construction of lines to disassemble similar product models (Wang et al. 2019). In this chapter, a DE algorithm is developed to solve the MDLBPs to minimize FV. The assumptions are given as follows:

- The line is paced and linear.
- Disassembly times of all components are known and deterministic.
- The precedence relationships are known.
- Cycle time is known and deterministic.
- Model mix on the line is known and deterministic.
- Tasks cannot be split.
- No setup is needed between model changes.
- Failures or breakdowns are not considered.
- Models are completely disassembled, not partly.
- Work in process does not allow (no buffers) between workstations.

In the next section, we present the proposed DE algorithm and exemplify its decoding and reproduction procedures. A numerical example will also be provided following the next section.

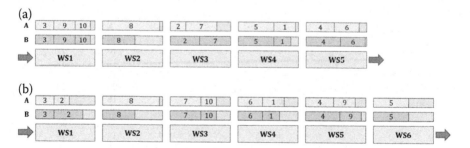

FIGURE 4.2 The Assignment Configuration of Tasks (a) with OR Precedence Relationships and (b) without OR Precedence Relationships for P10-OR Problem.

4 PROPOSED APPROACH

DE is an evolutionary algorithm introduced by Storn and Price (1997). The method proposed for continuous optimization problems initially based its basic principles on two components — the design of a simple mutation operator and a crossover operation (Ponsich and Coello Coello 2013). DE was inspired by Darwin's theory of evolution, mimicking the biological model of evolution and natural selection. The advantage of DE has been derived from its fast and simple structure that performs well in a variety of continuous optimization problems (Ali, Essam, and Kasmarik 2020). Several researchers have investigated applications of DE algorithm in areas such as scheduling (Nguyen et al. 2019, Wisittipanich and Kachitvichyanukul 2012), clustering (Ertuğrul 2020), engineering design (Gong, Cai, and Zhu 2009), power system planning and operational studies (Varadarajan and Swarup 2008), continuous optimization (Jitkongchuen and Thammano 2014), time series forecasting (Donate et al. 2013), and optimization of economic load dispatch in power plant management (Nascimento et al. 2017).

Like other evolutionary algorithms, DE iterates through a population consisting of a predetermined number of vectors. The general structure is very similar to a real-coded genetic algorithm. Candidate solutions are to be improved iteratively via mutation, crossover, and selection operators. One of the most distinctive features of the DE is that vectors are mutated by adding weighted random vector differentials to them. The following subsection presents the general outline of the proposed DE algorithm.

4.1 GENERAL OUTLINE

The pseudo-code which shows the general outline of the proposed approach is given in Algorithm 1. As seen from the pseudo-code given, the DE algorithm starts with the generation of the initial population. The population consists of a set of D-dimensional vectors, where D is the number of disassembly operations (i.e., $I = \{1, 2, ..., D\}$) in the problem. A vector is represented by X_{eG}; where e denotes the vector number ($e = 1, 2, ..., N$) and G denotes the iteration (generation) number. So, the population is made of N vectors generated randomly. Figure 4.3 represents a vector and a population sample with randomly determined parameter values between [0,1] for a 10-task problem.

The parameter values of the vectors (determined randomly within the bounds) correspond to the relative selection probability (*rsp*) of tasks. For example, considering X_{1G}, the *rsp* values of task-1 and task-2 are 0.336 and 0.742, respectively. These probabilities will be used to decode and calculate the objective function values of the vectors. The initial population is generated randomly to uniformly represent the search space, such that $S = \{X_{1G}, X_{2G}, ..., X_{NG}\}$, and $G = 0$. Note that the population size (N) must be equal to or larger than four (i.e. $N \geq 4$, since three different vectors (except the target vector) are needed to apply mutation and crossover).

Algorithm 1   The General Outline of the DE Algorithm.

1 Start the algorithm and initialize parameters
2 Randomly generate initial population, S
3 Calculate objective function values (FVs) of vectors in the population
4 For iter=1 to MaxIter
    4.1 Select a random target vector (X_{eG}) and three different vectors $(X_{aG}, X_{bG}, \text{ and } X_{cG})$
    4.2 Obtain a trial vector (U_{eG}) applying mutation and crossover
    4.3 Decode and calculate the objective function value of the trial vector, $FV(U_{eG})$
    4.4 Accept trial vector if better than the target vector, $FV(U_{eG}) < FV(X_{eG})$
5 Endfor
6 Report best solution

A parameter value

A vector

X_{1G}	0.336	0.742	0.784	0.812	0.560	0.874	0.527	0.623	0.883	0.644
X_{2G}	0.579	0.873	0.551	0.916	0.877	0.479	0.320	0.169	0.422	0.809
X_{3G}	0.699	0.460	0.266	0.78	0.429	0.425	0.720	0.296	0.161	0.272
X_{4G}	0.582	0.266	0.298	0.333	0.431	0.479	0.864	0.271	0.620	0.831

FIGURE 4.3   Representation of a Vector ($D = 10$) and a Sample Population with $N = 4$ Vectors.

4.2   DECODING PROCEDURE

The proposed algorithm utilizes a priority-based decoding scheme (Pitakaso 2015) when allocating tasks to workstations (see Algorithm 2). This scheme ensures that the precedence relationships are satisfied so that infeasibility is prevented. When selecting a task to assign to a workstation, the selection is made among *available tasks* based on their *rsp* values (stored on the vector). A roulette wheel selection is applied—the higher the *rsp* value of a task, the higher the chance that the task to be selected for the assignment. When determining available tasks for a current workstation (k), all unassigned task $(i \in U)$ carrying the following characteristics are considered to be available:

- The model-based idle times of the current workstation are large enough to perform the task (i.e. $CT - cur_{mj} \geq t_{im}$).
- The task has no unassigned AND-predecessor and at least one of its OR-predecessors has been assigned (if any).

Let us exemplify the utilized decoding scheme with roulette wheel selection. Consider the vector [0.336, 0.742, 0.784, 0.812, 0.560, 0.874, 0.527, 0.623, 0.883,

---

Algorithm 2   Priority-based Decoding Scheme

1 Add all tasks to unassigned tasks list, U
2 Open the first workstation, $k \leftarrow 1$
3 While $U \neq \varnothing$
4 Build available tasks list, AT
   4.1 While $AT \neq \varnothing$
   4.2 Select a task (i) using the roulette wheel
   4.3 Assign selected task to current workstation
   4.4 Increase model-based workloads of current workstation, $cur_m \leftarrow cur_m + t_{im}$
5 Update AND-OR precedence conditions
6 Remove task i from U
Update AT
Endwhile
Open a new workstation, $k \leftarrow k + 1$
Reset model-based workstation workloads, $cur_m \leftarrow 0$
Endwhile
Calculate objective function value

0.644] given in Figure 4.3. Assume the 10-task P10-OR problem (given in Figure 4.1) with two models.

Based on the precedence relationships of the P10-OR problem, the tasks available for assignment are task-2 and task-3. Assume that the idle time in the workstation is large enough for both tasks (the workstation is empty as of now). The sum of rsp values is 0.336+0.742=1.078. So, the selection probabilities of task-2 and task-3 are calculated as 0.312 (=0.336/1.078) and 0.688 (=0.742/1.078). The tasks are ranked in decreasing order based on their selection probabilities (i.e. 0.688 for task-3 and 0.312 for task-2) and a random number between [0,1] is generated. If the random number is equal to or smaller than 0.688, task-3 is selected and assigned to the workstation (otherwise, task-2 is selected). Available tasks list is updated every time a new task is assigned and the roulette wheel selection is applied. When there is no available task, a new workstation is opened and this procedure continues until all tasks are assigned. When the decoding process is over, the assignment configuration of tasks to workstations becomes known. The objective function value (FV) is calculated using the equation (4.1) given in Section 3, considering the weighted idle times of workstations. Thus, a smooth workload distribution across the workstations is obtained and the number of workstations is minimized.

4.3   REPRODUCTION AND ACCEPTANCE

Different from the common tendency in the literature (e.g., see Nearchou and Omirou (2017) and Ali et al. (2020)), reproduction (mutation and crossover) is not performed for *each* target vector in each iteration. Instead, the target vector is selected randomly. Thus, a randomly selected vector is targeted at each iteration, and competitions are held to replace the target vector. Mutation and crossover operators are performed consequently to obtain a mutant and a trial vector eventually. Acceptance operator is applied based on the function value of the trial vector obtained (Nearchou and Omirou 2017).

Reproduction is applied until a predetermined iteration number, *MaxIter* is exceeded. First, a mutant vector V_{eG} is obtained in the mutation stage (see equation (4.2)) using three vectors chosen randomly (i.e. X_{aG}, X_{bG}, and X_{cG}). These three vectors are not equal to each other or the target vector (Nearchou and Omirou 2017).

$$V_{eG}^r = X_{aG}^r + F|X_{bG}^r - X_{cG}^r| \text{ where } a,\ b,\ c \in \{1, 2, \ldots, N\} \text{ and } F \geq 0 \quad (4.2)$$

F is a parameter called weighting factor, usually between [0–2] used to control the amplification of the differential variation $(X_{bG}^r - X_{cG}^r)$,\; the superscript r represents the r-th component of the corresponding vector. Notice the use of absolute operation to avoid negative parameter values.

Second, the trial vector (U_{eG}) is obtained via a crossover operator using the target and mutant vectors (i.e. X_{eG} and V_{eG}) as in equation (4.3). This is called binomial crossover operation (Ali et al., 2020).

$$U_{eG}^r = \begin{cases} V_{eG}^r & \text{if } R^r \leq CR \text{ or } k = I_e \\ X_{eG}^r & \text{if } R^r > CR \text{ or } k \neq I_e \end{cases} \quad (4.3)$$

where R^r is a random number $(0,1)$ determined for each r-th component of the corresponding vector and I_e is a randomly chosen integer (i.e. $I_e \in I$).

In the acceptance phase, equation (4.4) is utilized to update the target vector comparing the *FVs* of trial vector and target vector (i.e. $FV(U_{eG})$ and $FV(X_{eG})$, respectively).

$$X_{eG+1} = \begin{cases} U_{eG} & \text{if } FV(U_{eG}) \leq FV(X_{eG}) \\ X_{eG} & \text{otherwise} \end{cases} \quad (4.4)$$

It is ensured that the competing individual is conveyed to the next generation and reproduction is continued until the maximum iteration limit, *MaxIter*, is exceeded. The algorithm is terminated and the best individual in the population is considered as the solution to the problem.

5   NUMERICAL EXAMPLE

This section presents the solution steps of a numerical example to exhibit the running mechanism of the proposed DE algorithm. P10-OR problem is considered for this. Remember that the precedence relationship diagram and task times of the P10-OR were given in Figure 4.1. Assume that the cycle time is 59 time-units. The two models are disassembled on the line and the model proportions on the line are the same (i.e. $d_A = d_B = 0.5$).

With the initialization of the DE algorithm, a population consisting of four vectors is created with parameter values generated randomly between [0,1]. Task assignments of each vector are obtained using the decoding procedure presented in Section 4.2 and the function values are calculated. The initial population with four-vectors is presented in Figure 4.4.

The mutation phase takes place after the FV values of the vectors are calculated. In this stage, three vectors ($X_{a,0}$, $X_{b,0}$, and $X_{c,0}$) selected randomly from the population are combined using equation (4.2) to get the mutant vector ($V_{e,0}$). For example, the first parameter value of $V_{e,0}$ is obtained as $V_{e,0}^1 = 0.747 + 2|0.649 - 0.615| = 1.013$. All the parameter values of the mutant vector are obtained this way for each $r = 1, 2, \ldots D$. Following the mutation, crossover is applied using equation (4.3). A trial vector, $U_{e,0}$ in Figure 4.5, is obtained from the target and mutant vectors.

When the crossover is completed, the trial vector is decoded using the procedure given in Section 4.2 and the FV is calculated. When the trial vector obtained in Figure 4.5 is decoded, the assignment configuration is obtained as in Table 4.1 and the FV is calculated as 4.5 using equation (4.1) given in Section 3.

As seen in Table 4.1, the FV of the trial vector is better than that of the target vector (i.e. $4.5 < 40.954$). Hence, acceptance is applied and the target vector in the population is replaced with the trial vector presented in Figure 4.6.

The population is formed, and the first iteration is completed in this way. In the next iteration, new target and random vectors are selected and reproduction is applied. This procedure is repeated until a predefined maximum iteration number is exceeded. In this example, the algorithm was run for 100 iterations and the best solution is taken. It was observed that the best solution has been found in the first

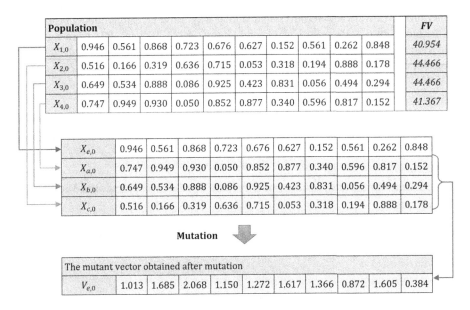

FIGURE 4.4 The Initial Population and Mutation in DE.

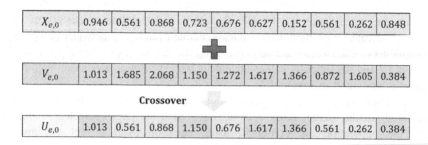

FIGURE 4.5 The Crossover Operation in DE.

iteration with the *FV* value of 4.5 (see the solution given in Table 4.1). This shows the promising decoding and reproduction procedures of the proposed approach.

6   COMPUTATIONAL TESTS

A comprehensive set of experiments has been conducted to test the performance of the proposed DE algorithm. Test problems on single-model DLBPs have been derived from literature and modified to get mixed-model DLBPs. For this, new product models have been generated while considering the minimum and maximum of task times in the original problem. For example, P10-OR_3m is generated using the 10-task single-model DLBP — P10-OR by Avikal and Mishra (2012). Task times in the original P10-OR have been assumed to belong to model A, and two new models (B and C) were generated with task processing times between 10 and 36. For P10-OR_2m, the generated P10-OR_3m is solved considering only two models, A and B. Other problems were also retrieved in a similar way using the single-model problems in the literature (i.e. P8 from Kalayci and Gupta (2014), P10 and P25 from Kalayci and Gupta (2013), P10-OR from Avikal and Mishra (2012), P22-OR and P47-OR from Li, Cil et al. (2020) and P47A, P47B, and P47C from Kalayci et al. (2015). For two-model problems, model proportions were assumed to be the same ($d_A = d_B = 0.5$), while for three-model problems they were assumed to be $d_A = 0.5$, $d_B = 0.25$, and $d_C = 0.25$.

TABLE 4.1
Task Assignment Configuration of the Trial Vector and Calculated Weighted Workloads

Workstation	1	2	3
Task assignments	2, 8, 3	9, 7, 5	6, 4, 1, 10
Workload for model A	58	57	58
Workload for model B	52	57	59
Weighted workload	55	57	58.5

Population										FV
$X_{1,1}$	1.013	0.561	0.868	1.150	0.676	1.617	1.366	0.561	0.262	0.384	4.500
$X_{2,1}$	0.516	0.166	0.319	0.636	0.715	0.053	0.318	0.194	0.888	0.178	44.466
$X_{3,1}$	0.649	0.534	0.888	0.086	0.925	0.423	0.831	0.056	0.494	0.294	44.466
$X_{4,1}$	0.747	0.949	0.930	0.050	0.852	0.877	0.340	0.596	0.817	0.152	41.367

FIGURE 4.6 The Population after the Acceptance.

Each test problem was solved under different cycle time constraints (see column C) and the results are presented in Tables 4.2 and 4.3. The algorithm was run for a predefined number of iterations (see column *MaxIter*) based on the size and complexity of each test problem. The statistical results for five runs are presented in terms of the number of stations utilized and the function values of the solutions. The columns *minFV* and *maxFV* represent the function values for the best and worst solutions of five runs. As additional information, relative percentage deviation is also presented in the column *RpdFV*, calculated as $RpdFV = (maxFV-minFV)/minFV$.

However, to the best of the authors' knowledge, there is no result in the literature on the minimization of the smoothness index for the problem addressed in this research. Therefore, the number of stations of the solutions obtained are also presented to enable a comparison with the theoretical lower bound for the number of workstations (K_{LB}). The columns *minNS* and *maxNS* present the number of stations required for the best and the worst solutions obtained in five runs. *RpdNS* is the relative percentage deviation, calculated as $RpdNS=(maxNS-minNS)/minNS$. The average CPU time and its standard deviation in five runs are presented in the columns *avCPU* and *stdCPU*, respectively. Note that the time is given in milliseconds (ms).

The problems in Tables 4.2 and 4.3 are organized in an increasing order based on their size and complexity. It is known that the increase in the number of tasks corresponds to the increase in problem size. In DLBPs, OR constraints also increase the problem complexity.

From the results presented, it is seen that the deviation between the FVs of the solutions are pretty reasonable. While this is not a complimentary indication about the performance of the algorithm itself, it gives an idea regarding the robustness of the search mechanism of the proposed DE algorithm. While the *RpdFV* is quite small (or even zero) for small size problems (see for example P8_2m, P8_3m, P8-OR_2m, and P8-OR_3m), the gap between *minFV* and *maxFV* relatively enlarges with the increase in the problem size; so does the complexity.

When the results are compared to K_{min}, it is seen that the algorithm performs quite well. As seen in Table 4.2, the optimal solutions have been found for 41 problems over 42, in terms of number of stations. The *minNS* is obtained 5 for P8-OR_2m under 38 time-units cycle time constraint while $K_{LB} = 4$. Therefore, it is not known whether the solution is optimal for this particular example. Similarly, 39 over 40 solutions found by the proposed DE algorithm for problems in Table 4.3 are optimal in terms of the total number of workstations required.

TABLE 4.2

Computational Test Results-1

Test Problem	MaxIter	m	C	K_{LB}	minNS	maxNS	RpdNS	DE Algorithm				
								minFV	maxFV	RpdFV	avCPU	stdCPU
P8_2m	1000	2	39	4	4	4	0	5.24	5.24	0	34	10.70
		2	41	4	4	4	0	8.91	8.91	0	11	2.89
		2	43	4	4	4	0	12.78	12.78	0	8	1.04
		2	45	4	4	4	0	16.71	16.71	0	12	3.60
P8_3m		3	46	4	4	4	0	15.89	15.89	0	11	2.09
		3	48	4	4	4	0	19.84	19.84	0	10	2.72
		3	50	4	4	4	0	23.80	23.80	0	9	1.61
		3	52	4	4	4	0	27.77	27.77	0	13	3.01
P8-OR_2m	1000	2	38	4	5	5	0	23.98	23.98	0	12	3.46
		2	46	4	4	4	0	22.79	22.79	0	12	5.59
		2	49	4	4	4	0	28.97	28.97	0	14	2.98
		2	56	3	3	3	0	14.64	15.31	0.04	10	3.66
P8-OR_3m		3	38	5	5	5	0	21.92	21.92	0	30	9.17
		3	40	5	5	5	0	25.69	25.69	0	18	3.27
		3	46	4	4	4	0	18.53	18.53	0	17	5.19
		3	48	4	4	4	0	22.34	22.34	0	18	3.57
P10_2m	1000	2	36	5	5	5	0	11.05	11.05	0	14	2.77
		2	47	4	4	4	0	13.68	14.32	0.04	15	3.56
		2	49	4	4	4	0	17.58	18.09	0.02	17	2.72
		2	58	3	3	3	0	8.90	8.90	0	12	2.62
P10_3m		3	36	5	5	5	0	10.41	10.41	0	15	2.19

(Continued)

TABLE 4.2 (Continued)

Test Problem	MaxIter	m	C	K_{LB}	minNS	maxNS	RpdNS	DE Algorithm				
								minFV	maxFV	RpdFV	avCPU	stdCPU
		3	38	5	5	5	0	14.08	14.53	0.03	22	3.61
		3	40	5	5	5	0	17.45	18.25	0.04	23	8.82
		3	46	4	4	5	0.25	11.98	39.15	2.26	13	1.44
P10-OR_2m	1500	2	39	5	5	5	0	12.67	12.79	0.01	27	5.53
		2	47	4	4	4	0	8.78	9.75	0.11	24	5.85
		2	49	4	4	4	0	12.77	13.46	0.05	33	15.44
		2	59	3	3	3	0	4.50	6.02	0.33	18	4.09
P10-OR_3m		3	39	5	5	5	0	13.08	13.08	0	25	3.59
		3	46	4	4	4	0	7.42	7.42	0	20	1.09
		3	49	4	4	4	0	13.34	13.46	0.01	19	5.45
		3	60	3	3	3	0	6.37	6.37	0	19	2.12
P25_2m	2000	2	18	10	10	11	0.1	7.84	13.61	0.73	167	47.23
		2	20	9	9	9	0	7.77	9.40	0.20	138	11.77
		2	22	8	8	9	0.12	7.24	14.15	0.95	128	9.09
		2	25	7	7	8	0.14	6.28	15.21	1.42	135	22.88
		2	29	6	6	7	0.16	6.96	17.73	1.54	138	12.62
P25_3m		3	18	11	11	11	0	12.34	13.10	0.06	173	23.59
		3	20	10	10	11	0.1	13.11	18.93	0.44	152	9.96
		3	23	9	9	9	0	15.19	15.33	0.01	150	14.76
		3	26	8	8	8	0	15.88	17.25	0.08	145	13.49
		3	28	7	7	8	0.14	12.87	22.46	0.74	183	6.57

TABLE 4.3

Computational Test Results-2

Test Problem	MaxIter	m	C	K_{LB}	DE Algorithm								
					minNS	maxNS	RpdNS	minFV	maxFV	RpdFV	avCPU	stdCPU
P22-OR_2m	2500	2	24	11	11	12	0.09	16.59	22.72	0.36	160	9.59
		2	26	10	10	11	0.1	14.18	23.15	0.59	152	12.48
		2	28	9	9	10	0.11	10.71	21.56	1.01	152	7.85
		2	29	9	9	9	0	13.84	16.13	0.16	145	13.05
		2	33	8	8	8	0	15.83	16.84	0.06	153	26.99
		2	37	7	7	7	0	15.20	15.74	0.03	135	11.10
P22-OR_3m		3	24	11	12	12	0	20.50	21.66	0.05	247	29.51
		3	28	9	9	10	0.11	10.76	19.28	0.79	172	21.85
		3	30	9	9	9	0	15.83	16.25	0.02	166	10.62
		3	32	8	8	9	0.125	12.39	21.76	0.75	156	7.57
		3	34	8	8	8	0	17.06	17.63	0.03	153	11.36
		3	36	7	7	8	0.14	11.28	22.93	1.03	154	13.08
P47_2m	4000	2	104	9	9	9	0	54.70	55.72	0.01	970	66.63
		2	110	8	8	8	0	34.23	34.45	0.01	971	85.47
		2	115	8	8	8	0	48.74	49.12	0.01	905	29.74
		2	120	8	8	8	0	63.69	64.70	0.01	888	26.74
		2	125	7	7	7	0	34.56	35.22	0.01	907	42.40
		2	129	7	7	7	0	45.59	47.07	0.03	949	107.71
		2	150	6	6	6	0	47.42	48.14	0.01	859	24.16
P47_3m		3	110	10	10	10	0	88.30	89.27	0.01	1047	64.88
		3	118	9	9	10	0.11	77.91	116.69	0.49	1072	61.02

(Continued)

TABLE 4.3 (Continued)

Test Problem	MaxIter	m	C	K_{LB}	minNS	maxNS	RpdNS	DE Algorithm				
								minFV	maxFV	RpdFV	avCPU	stdCPU
		3	120	9	9	9	0	83.84	85.29	0.01	1026	63.79
		3	124	9	9	9	0	95.79	96.64	0.01	1052	79.37
		3	132	8	8	9	0.12	80.08	122.34	0.52	1037	63.79
		3	140	8	8	8	0	102.29	103.52	0.01	959	22.38
		3	150	7	7	8	0.14	83.36	136.87	0.64	974	30.95
P47-OR_2m	5000	2	104	9	9	9	0	53.72	57.30	0.06	1247	71.50
		2	110	8	8	8	0	34.60	34.95	0.01	1399	243.13
		2	114	8	8	8	0	46.31	47.49	0.02	1170	38.15
		2	117	8	8	8	0	56.52	58.99	0.04	1237	69.83
		2	121	8	8	8	0	69.64	72.00	0.03	1243	96.59
		2	130	7	7	7	0	48.47	49.93	0.03	1172	73.36
		2	136	7	7	7	0	65.96	73.70	0.11	1206	81.66
P47-OR_3m		3	110	10	10	10	0	87.55	87.73	0.01	1405	99.10
		3	115	10	10	10	0	103.27	104.06	0.01	1455	112.14
		3	118	9	9	9	0	78.44	79.19	0.01	1364	53.35
		3	120	9	9	9	0	84.29	85.13	0.01	1414	107.72
		3	132	8	8	8	0	80.22	81.02	0.01	1344	39.18
		3	135	8	8	8	0	88.45	88.75	0.01	1318	25.28
		3	150	7	7	8	0.14	83.41	134.51	0.61	1290	30.36

Notice that *minFV* and *maxFV* values can be different, while *minNS* and *maxNS* values are the same. See, for example, the solution for P47_2m under the 104 time-units cycle time: while *minNS* and *maxNS* are the same, *minFV* and *maxFV* values are 54.70 and 55.72, respectively. This indicates that the use of the proposed objective function value gives a better measurement in terms of solution quality. It is also worthy to note that the proposed DE algorithm consumes a very reasonable CPU time even for large-sized problems. The largest computational time required to solve the P47-OR_3m problem under 120 time-units cycle time is 1414 ms. Consequently, the computational results exhibit the efficient solution building capacity of the proposed DE algorithm.

7 CONCLUSIONS AND FUTURE RESEARCH

Recycling of EOL products is now getting even more popular with the increase in environmental considerations. According to a recent report published by Platform for Accelerating the Circular Economy - PACE, Europe ranks second behind Asia in terms of waste in the electrical-electronic equipment (PACE, 2019). In earlier reports, the role of recycling in the green economy has been explained and the interaction between recycling and job creation has been represented by the European Environment Agency (EEA, 2011).

Disassembly lines are used to take apart EOL products and recycle their valuable components. The disassembly process is the most critical stage in this context. This chapter contributes by demonstrating one of the preliminary applications of the DE algorithm on disassembly line balancing problems.

A DE algorithm was proposed in balancing the disassembly lines where more than one model of a product can be disassembled concurrently. The main objective of the proposed algorithm is to minimize the weighted smoothness index considering the ratio of models on the line while satisfying the AND/OR precedence relationships together. Thus, it is ensured that the number of work-stations is also minimized, and the line efficiency is maximized. It is worthy to note that the proposed DE algorithm applies a priority-based decoding scheme to ensure the feasibility of the obtained solutions and prevent any unnecessary (and time-consuming) repairing mechanisms. A numerical example is provided to represent the decoding and reproduction mechanisms of the proposed algorithm. Test problems have been generated to modify existing ones on single-model DLBP in literature and the performance of the proposed algorithm is tested through a comprehensive set of computational tests. It was observed that efficient solutions have been obtained within less than 1 s (1000 ms) for many of the test problems (including some complex large-size problems). Also, except in two cases, the results are optimum in terms of the number of workstations required.

In future research, the proposed DE algorithm can be extended or modified to solve more complex problems. The authors' ongoing work aims to use the proposed DE algorithm to solve partial DLBPs, where parts are not disassembled completely. Also, other considerations (i.e. hazardous parts and part demands) can be considered in solving MDLBP within a certain range of cycle time.

APPENDIX
DATA ON TEST PROBLEMS

P8

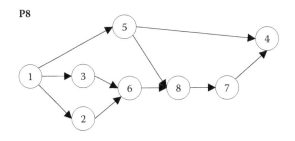

Task	A	B	C
1	14	22	14
2	10	0	10
3	12	17	22
4	18	18	18
5	23	10	14
6	16	21	28
7	20	21	26
8	36	36	36

P8-OR

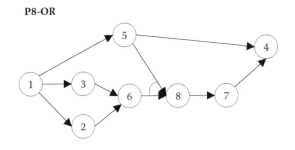

Task	A	B	C
1	14	18	18
2	10	10	12
3	12	0	12
4	18	18	18
5	23	28	23
6	16	20	20

(*Continued*)

Task	A	B	C
7	20	22	22
8	36	21	36

P10

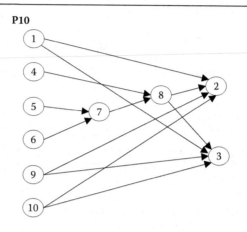

Task	A	B	C
1	14	17	13
2	10	10	10
3	12	12	12
4	17	11	26
5	23	23	23
6	14	14	0
7	19	17	14
8	36	36	36
9	14	14	14
10	10	0	10

P10-OR

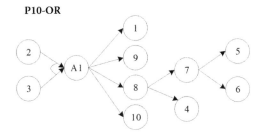

Task	A	B	C
1	14	11	14
2	10	19	10
3	12	12	12
4	18	22	22
5	23	23	23
6	16	16	16
7	20	20	20
8	36	21	23
9	14	14	14
10	10	10	10

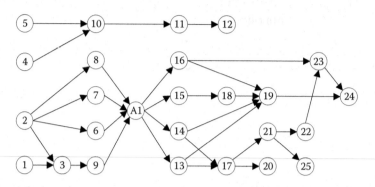

Task	A	B	C
1	3	3	3
2	2	8	8
3	3	3	0
4	10	9	7
5	10	10	10
6	15	15	15
7	15	11	18
8	15	9	18
9	15	15	15
10	2	0	2
11	2	2	2
12	2	4	6
13	2	2	2
14	2	0	5
15	2	2	2
16	2	8	7
17	2	2	2
18	3	4	3
19	18	18	16
20	5	5	5
21	1	1	1
22	5	5	6
23	15	15	15
24	2	2	2
25	2	12	17

P22OR

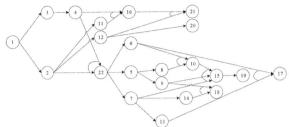

Task	A	B	C
1	10	10	10
2	6	0	0
3	12	14	18
4	17	0	17
5	9	10	4
6	19	13	19
7	9	9	9
8	15	15	15
9	6	0	10
10	11	11	11
11	1	1	1
12	19	4	10
13	12	14	2
14	6	14	15
15	4	15	4
16	17	12	12
17	10	10	10
18	6	8	10
19	16	9	8
20	7	7	7
21	14	14	0
22	19	8	16

P47

Part	Time A	Time B	Time C	AND Predecessor(s)
1	14	16	18	-
2	28	32	36	1
3	3	4	9	2
4	2	4	6	2
5	3	4	9	-
6	4	6	8	5
7	8	10	12	-
8	12	16	20	-
9	4	5	6	8
10	28	32	36	8
11	3	4	9	9,10
12	4	6	8	10
13	6	8	10	-
14	1	2	3	13
15	20	24	28	-
16	5	7	9	15
17	28	32	36	16
18	4	6	8	17
19	3	4	9	14
20	12	16	20	18
21	3	4	9	20
22	3	4	9	18
23	28	32	36	22
24	28	32	36	22
25	12	16	20	23,24
26	76	88	100	19
27	6	8	10	26
28	28	32	36	27
29	3	6	11	28
30	6	8	10	29
31	3	4	9	30
32	98	104	110	1,5,9,18
33	14	18	22	32
34	2	4	6	33
35	6	8	10	33
36	7	8	12	34,35
37	60	72	84	36
38	6	8	10	37
39	60	72	84	38

(Continued)

Part	Time A	Time B	Time C	AND Predecessor(s)
40	8	10	12	39
41	12	16	20	40
42	3	4	9	41
43	12	16	20	36
44	3	4	9	43
45	28	32	36	44
46	2	4	6	45
47	3	4	9	18

REFERENCES

Ali, Ismail M., Daryl Essam, and Kathryn Kasmarik. 2020. "A novel design of differential evolution for solving discrete traveling salesman problems." *Swarm and Evolutionary Computation* 52:100607. doi: https://doi.org/10.1016/j.swevo.2019.100607.

Avikal, S., and P. Mishra. 2012. "A new U-shaped heuristic for disassembly line balancing problems." *International Journal of Science* 1 (1):21–27.

Avikal, S., Jain, R., and P. K. Mishra. 2014. "A Kano model, AHP and M-TOPSIS method-based technique for disassembly line balancing under fuzzy environment." *Applied Soft Computing* 25:519–529. doi: https://doi.org/10.1016/j.asoc.2014.08.002.

Cevikcan, Emre, Dicle Aslan, and Fatma Betul Yeni. 2020. "Disassembly line design with multi-manned workstations: a novel heuristic optimisation approach." *International Journal of Production Research* 58 (3):649–670. doi: 10.1080/00207543.2019.1587190.

Ding, Li-Ping, Yi-Xiong Feng, Jian-Rong Tan, and Yi-Cong Gao. 2010. "A new multi-objective ant colony algorithm for solving the disassembly line balancing problem." *The International Journal of Advanced Manufacturing Technology* 48 (5):761–771. doi: 10.1007/s00170-009-2303-5.

Donate, Juan Peralta, Xiaodong Li, Germán Gutiérrez Sánchez, and Araceli Sanchis de Miguel. 2013. "Time series forecasting by evolving artificial neural networks with genetic algorithms, differential evolution and estimation of distribution algorithm." *Neural Computing and Applications* 22 (1):11–20. doi: 10.1007/s00521-011-0741-0.

Duta, L., I. Caciula, and P. C. Patic. 2016. "Column generation approach for disassembly line balancing." *IFAC-PapersOnLine* 49 (12):916–920. doi: https://doi.org/10.1016/j.ifacol.2016.07.892.

Edis, Emrah B., Mehmet Ali Ilgin, and Rahime Sancar Edis. 2019. "Disassembly line balancing with sequencing decisions: A mixed integer linear programming model and extensions." *Journal of Cleaner Production* 238:117826. doi: https://doi.org/10.1016/j.jclepro.2019.117826.

EEA. 2011. Earnings, jobs and innovation: the role of recycling in a green economy. European Environment Agency. https://www.eea.europa.eu/publications/earnings-jobs-and-innovation-the.

Ertuğrul, Ömer Faruk. 2020. "A novel clustering method built on random weight artificial neural networks and differential evolution." *Soft Computing* 24:12067–12078. doi:10.1007/s00500-019-04647-324.

Fang, Yilin, Quan Liu, Miqing Li, Yuanjun Laili, and Duc Truong Pham. 2019.

"Evolutionary many-objective optimization for mixed-model disassembly line balancing with multi-robotic workstations." *European Journal of Operational Research* 276 (1):160–174. doi: https://doi.org/10.1016/j.ejor.2018.12.035.

Gong, Wenyin, Zhihua Cai, and Li Zhu. 2009. "An efficient multiobjective differential evolution algorithm for engineering design." *Structural and Multidisciplinary Optimization* 38 (2):137–157. doi: 10.1007/s00158-008-0269-9.

Gungor, A., and S. M. Gupta. 1999. "Disassembly line balancing." Proceedings of the 1999 Annual Meeting of the Northeast Decision Sciences Institute:193–195.

Jitkongchuen, Duangjai, and Arit Thammano. 2014. "A self-adaptive differential evolution algorithm for continuous optimization problems." *Artificial Life and Robotics* 19 (2):201–208. doi: 10.1007/s10015-014-0155-z.

Kalayci, Can B., and Surendra M. Gupta. 2013. "Artificial bee colony algorithm for solving sequence-dependent disassembly line balancing problem." *Expert Systems with Applications* 40 (18):7231–7241. doi: https://doi.org/10.1016/j.eswa.2013.06.067.

Kalayci, Can B., and Surendra M. Gupta. 2014. "A tabu search algorithm for balancing a sequence-dependent disassembly line." *Production Planning & Control* 25 (2):149–160. doi: 10.1080/09537287.2013.782949.

Kalayci, Can B., Arif Hancilar, Askiner Gungor, and Surendra M. Gupta. 2015. "Multiobjective fuzzy disassembly line balancing using a hybrid discrete artificial bee colony algorithm." *Journal of Manufacturing Systems* 37:672–682. doi: https://doi.org/10.1016/j.jmsy.2014.11.015.

Kalaycılar, Eda Göksoy, Meral Azizoğlu, and Sencer Yeralan. 2016. "A disassembly line balancing problem with fixed number of workstations." *European Journal of Operational Research* 249 (2):592–604. doi: https://doi.org/10.1016/j.ejor.2015.09.004.

Kucukkoc, Ibrahim. 2020. "Balancing of Two-sided Disassembly Lines: Problem Definition, MILP Model and Genetic Algorithm Approach." *Computers & Operations Research*. doi: 10.1016/j.cor.2020.105064.

Kucukkoc, Ibrahim, Zixiang Li, and Yuchen Li. 2019. "Type-E disassembly line balancing problem with multi-manned workstations." *Optimization and Engineering* 21:611–630. doi: 10.1007/s11081-019-09465-y.

Li, Jinlin, Xiaohong Chen, Zhanguo Zhu, Caijun Yang, and Chengbin Chu. 2019. "A branch, bound, and remember algorithm for the simple disassembly line balancing problem." *Computers & Operations Research* 105:47–57. doi: https://doi.org/10.1016/j.cor.2019.01.003.

Li, Zixiang, Ibrahim Kucukkoc, and Zikai Zhang. 2019. "Iterated local search method and mathematical model for sequence-dependent U-shaped disassembly line balancing problem." *Computers & Industrial Engineering* 137:106056. doi: https://doi.org/10.1016/j.cie.2019.106056.

Li, Zixiang, Zeynel Abidin Cil, Süleyman Mete, and Ibrahim Kucukkoc. 2020. "A fast branch, bound and remember algorithm for disassembly line balancing problem." *International Journal of Production Research* 25 (11):3220–3234. doi: 10.1080/00207543.2019.1630774.

McGovern, S. M., and S. M. Gupta. 2007a. "Combinatorial optimization analysis of the unary NP-complete disassembly line balancing problem." *International Journal of Production Research* 45 (18–19):4485–4511. doi: 10.1080/00207540701476281.

McGovern, SM, and SM Gupta. 2003. "2-opt heuristic for the disassembly line balancing problem." Proceedings of the SPIE International Conference on Environmentally Conscious Manufacturing III. Providence, RI2003.

McGovern, Seamus M., and Surendra M. Gupta. 2007b. "A balancing method and genetic algorithm for disassembly line balancing." *European Journal of Operational Research* 179 (3):692–708. doi: https://doi.org/10.1016/j.ejor.2005.03.055.

Mete, Süleyman, Zeynel Abidin Cil, Kürşad Ağpak, Eren Özceylan, and Alexandre Dolgui. 2016. "A solution approach based on beam search algorithm for disassembly line balancing problem." *Journal of Manufacturing Systems* 41:188–200. doi: https://doi.org/10.1016/j.jmsy.2016.09.002.

Nascimento, Manoel Henrique Reis, Marcus Vinícius Alves Nunes, Jorge Laureano Moya Rodríguez, and Jandecy Cabral Leite. 2017. "A new solution to the economical load dispatch of power plants and optimization using differential evolution." *Electrical Engineering* 99 (2):561–571. doi: 10.1007/s00202-016-0385-2.

Nearchou, A. C. 2008. "Multi-objective balancing of assembly lines by population heuristics." *International Journal of Production Research* 46 (8):2275–2297. doi: 10.1080/00207540600988089.

Nearchou, Andreas C. 2005. "A Differential Evolution Algorithm for Simple Assembly Line Balancing." *IFAC Proceedings Volumes* 38 (1):247–252. doi: https://doi.org/10.3182/20050703-6-CZ-1902.01463.

Nearchou, Andreas C. 2007. "Balancing large assembly lines by a new heuristic based on differential evolution method." *The International Journal of Advanced Manufacturing Technology* 34 (9):1016–1029. doi: 10.1007/s00170-006-0655-7.

Nearchou, Andreas C., and Sotiris L. Omirou. 2017. "Assembly Line Balancing Using Differential Evolution Models." *Cybernetics and Systems* 48 (5):436–458. doi: 10.1080/01969722.2017.1319238.

Nguyen, Su, Dhananjay Thiruvady, Andreas T. Ernst, and Damminda Alahakoon. 2019. "A hybrid differential evolution algorithm with column generation for resource constrained job scheduling." *Computers & Operations Research* 109:273–287. doi: https://doi.org/10.1016/j.cor.2019.05.009.

Nourmohammadi, A., and M. Zandieh. 2011. "Assembly line balancing by a new multiobjective differential evolution algorithm based on TOPSIS." *International Journal of Production Research* 49 (10):2833–2855. doi: 10.1080/00207540903473367.

Özceylan, Eren, Can B. Kalayci, Aşkıner Güngör, and Surendra M. Gupta. 2019. "Disassembly line balancing problem: a review of the state of the art and future directions." *International Journal of Production Research* 57 (15–16):4805–4827. doi: 10.1080/00207543.2018.1428775.

PACE. 2019. "A New Circular Vision for Electronics: Time for a Global Reboot", The Platform for Accelerating the Circular Economy (PACE). http://www3.weforum.org/docs/WEF_A_New_Circular_Vision_for_Electronics.pdf.

Paksoy, Turan, Aşkıner Güngör, Eren Özceylan, and Arif Hancilar. 2013. "Mixed model disassembly line balancing problem with fuzzy goals." *International Journal of Production Research* 51 (20):6082–6096. doi: 10.1080/00207543.2013.795251.

Pitakaso, Rapeepan. 2015. "Differential evolution algorithm for simple assembly line balancing type 1 (SALBP-1)." *Journal of Industrial and Production Engineering* 32 (2):104–114. doi: 10.1080/21681015.2015.1007094.

Pitakaso, Rapeepan, and Kanchana Sethanan. 2016. "Modified differential evolution algorithm for simple assembly line balancing with a limit on the number of machine types." *Engineering Optimization* 48 (2):253–271. doi: 10.1080/0305215X.2015.1005082.

Ponsich, Antonin, and Carlos A. Coello Coello. 2013. "A hybrid Differential Evolution—Tabu Search algorithm for the solution of Job-Shop Scheduling Problems." *Applied Soft Computing* 13 (1):462–474. doi: https://doi.org/10.1016/j.asoc.2012.07.034.

Ren, Yaping, Daoyuan Yu, Chaoyong Zhang, Guangdong Tian, Leilei Meng, and Xiaoqiang Zhou. 2017. "An improved gravitational search algorithm for profit-oriented partial disassembly line balancing problem." *International Journal of Production Research* 55 (24):7302–7316. doi: 10.1080/00207543.2017.1341066.

Ren, Yaping, Chaoyong Zhang, Fu Zhao, Guangdong Tian, Wenwen Lin, Leilei Meng, and Hongliang Li. 2018. "Disassembly line balancing problem using interdependent weights-based multi-criteria decision making and 2-Optimal algorithm." *Journal of Cleaner Production* 174:1475–1486. doi: https://doi.org/10.1016/j.jclepro.2017. 10.308.

Storn, Rainer, and Kenneth Price. 1997. "Differential Evolution – A Simple and Efficient Heuristic for global Optimization over Continuous Spaces." *Journal of Global Optimization* 11 (4):341–359. doi: 10.1023/A:1008202821328.

Sun, Bin-qi, and Ling Wang. 2019. "A decomposition-based matheuristic for supply chain network design with assembly line balancing." *Computers & Industrial Engineering* 131:408–417. doi: https://doi.org/10.1016/j.cie.2019.03.009.

Varadarajan, M., and K. S. Swarup. 2008. "Volt-Var Optimization Using Differential Evolution." *Electric Power Components and Systems* 36 (4):387–408. doi: 10.1080/ 15325000701658524.

Vincent, Lui Wen Han, S. G. Ponnambalam, and G. Kanagaraj. 2014. "Differential evolution variants to schedule flexible assembly lines." *Journal of Intelligent Manufacturing* 25 (4):739–753. doi: 10.1007/s10845-012-0716-8.

Wang, Kaipu, Xinyu Li, Liang Gao, and Akhil Garg. 2019. "Partial disassembly line balancing for energy consumption and profit under uncertainty." *Robotics and Computer-Integrated Manufacturing* 59:235–251. doi: https://doi.org/10.1016/j.rcim.2019. 04.014.

Wang, Kaipu, Xinyu Li, and Liang Gao. 2019a. "A multi-objective discrete flower pollination algorithm for stochastic two-sided partial disassembly line balancing problem." *Computers & Industrial Engineering* 130:634–649. doi: https://doi.org/10.1016/j.cie. 2019.03.017.

Wang, Kaipu, Xinyu Li, and Liang Gao. 2019b. "Modeling and optimization of multi-objective partial disassembly line balancing problem considering hazard and profit." *Journal of Cleaner Production* 211:115–133. doi: https://doi.org/10.1016/j.jclepro. 2018.11.114.

Wang, Kaipu, Liang Gao, and Xinyu Li. 2020. "A multi-objective algorithm for U-shaped disassembly line balancing with partial destructive mode." *Neural Computing and Applications* 32:12715–12736. doi: 10.1007/s00521-020-04721-0.

Wisittipanich, W., and V. Kachitvichyanukul. 2012. "Two enhanced differential evolution algorithms for job shop scheduling problems." *International Journal of Production Research* 50 (10):2757–2773. doi: 10.1080/00207543.2011.588972.

Zhang, Lei, Xikun Zhao, Qingdi Ke, Wanfu Dong, and Yanjiu Zhong. 2019. "Disassembly Line Balancing Optimization Method for High Efficiency and Low Carbon Emission." *International Journal of Precision Engineering and Manufacturing-Green Technology*. doi: 10.1007/s40684-019-00140-2.

Zhang, Zeqiang, Kaipu Wang, Lixia Zhu, and Yi Wang. 2017. "A Pareto improved artificial fish swarm algorithm for solving a multi-objective fuzzy disassembly line balancing problem." *Expert Systems with Applications* 86:165–176. doi: https://doi.org/10.1016/j. eswa.2017.05.053.

Zhu, Lixia, Zeqiang Zhang, Yi Wang, and Ning Cai. 2019. "On the end-of-life state oriented multi-objective disassembly line balancing problem." *Journal of Intelligent Manufacturing* 31:1403–1428. doi: 10.1007/s10845-019-01519-3.

5 Integrated Collection-Disassembly-Distribution Problem

Zülal Diri Kenger[1], Çağrı Koç[2], and Eren Özceylan[1]
[1]Department of Industrial Engineering, Gaziantep University, Gaziantep, 27470, Turkey
[2]Department of Business Administration, Social Sciences University of Ankara, Ankara, 06830, Turkey

1 INTRODUCTION

Increased consumption and depletion of natural resources are becoming a rapidly growing danger affecting the environment. Enterprises aware of the dangers around them adopt recycling and remanufacturing as an important business strategy; they provide economic advantages and environmentally conscious manufacturing. Therefore, successful reverse logistics (RL) management is crucial for a sustainable clean environment, lower cost, and shorter response time. Reverse supply chain (RSC) network optimization is a subgroup of RL management where all used and discarded products, components, and materials are efficiently managed (Özceylan and Paksoy, 2013). One of the crucial steps of RSC processes is disassembly. The disassembly line is the best option for automated systems. It is also a part of a disassembly system where end-of-life (EOL) products are taken apart for recycling and remanufacturing (Güngör and Gupta, 2001).

The disassembly process became an active study field in the last two decades due to its role in product recovery (Mete et al. 2016). The returned products save cost and energy as a result of providing cheaper components. From the environmental point of view, it decreases the need for new resources and prevents waste burial (Habibi et al. 2017b). The disassembly line balancing (DLB) problem can be defined as assigning of disassembly tasks of a product to sequential workstations to provide certain conditions (Güngör and Gupta, 2001; Altekin and Akkan, 2012). Collecting EOL products is considered a compulsory operation before the disassembly process (Habibi et al. 2017a). The DLB problem has been considered for factors such as line types (straight, U-type, two-sided, etc.), performance measures (removing hazardous parts early, maximization profit, minimization workstations, etc.), and product types (single, mixed, multi) (Özceylan et al. 2019). However, a few studies consider the integration of DLB with other problems (Özceylan and Paksoy, 2013; Özceylan and Paksoy 2014; Özceylan et al. 2014; Habibi et al. 2017a; Habibi et al. 2017b; Kannan et al. 2017; Koç 2017; Habibi et al. 2018). To the best of our knowledge, the vehicle routing (VR) problem has not been considered for both collections of EOL products and the distribution of recycled components.

Diri Kenger et al. (2020) introduce an optimization problem called the integrated disassembly line balancing and routing problem (I-DLB-RP). The I-DLB-RP simultaneously optimizes two well-known problems. The former one balances the disassembly lines in disassembly centers, whereas the latter constructs a routing plan to distribute usable components, generated by the disassembly process, from the disassembly center to the remanufacturing centers. While I-DLB-RP does not consider the collection of EOL products from collection centers, I-CDDP does. In this study, we make three main scientific contributions to the literature: (i) we introduce I-CDDP; (ii) we develop two mixed-integer mathematical formulations for a RSC network cooperating with DLB and VR problems. The first model is presented for cases where one type component is demanded, and the second model is for a multi-component case. Lastly, (iii) we conduct computational experiments on a realistic case study including several scenario analyses on models.

The rest of the study is structured as follows: in Section 2, we provide a brief literature review on the DLB and the VR problem. In Section 3, the current problem is defined and our assumptions and the mixed-integer mathematical formulations are provided. Section 4 presents the experimental results and scenario analysis. Finally, the conclusions are summarized in Section 5.

2   LITERATURE REVIEW

The VR problem introduced about 60 years ago has been intensively studied in the literature. For further details on the VR problem and its rich variants, we refer the reader to the book of Toth and Vigo (2014), the surveys of Koç et al. (2016), Koç, and Laporte (2018), and Vidal et al. (2019).

In the last two decades, the classical DLB problem and its many variations have been studied due to the importance of recycling. Güngör and Gupta (2001) introduced the DLB, analyzed task failures, and described a formula to decrease their effects on the line. Özceylan et al. (2019) and Deniz and Ozcelik (2019) provide a detailed review of the DLB and its variants.

Due to resource scarcity and pollution, recycling and remanufacturing have attracted attention in recent years. Although the DLB is an essential step in RSC, the integration with RSC is rarely studied in literature. Özceylan and Paksoy (2013) considered an RSC network consisting of collection/disassembly centers and plants, and proposed a mathematical formulation to optimize the problem. Özceylan and Paksoy (2014) studied the integration of RSC and DLB problem and proposed a fuzzy programming approach. Habibi et al. (2017a) introduced the collection-disassembly problem in RSC and proposed several formulations. Habibi et al. (2017b) proposed a two-phase iterative heuristic for the collection-disassembly problem. Kannan et al. (2017) integrated DLB and RSC in terms of a third-party provider for multiple types of products. Habibi et al. (2018) considered the collection-disassembly problem under uncertainty and proposed sample average approximation to solve the problem. Diri Kenger et al. (2020) introduced the I-DLB-RP and simultaneously optimized the DLB and VR problems. In the I-DLB-RP, a single disassembly center first disassembles EOL products. A vehicle with fixed capacity, then, dispatches usable disassembled components to re-manufacturers. While I-DLB-RP does not consider the collection of EOL products from collection centers, I-CDDP does.

3 PROBLEM DEFINITION AND MODELING

We developed two mathematical formulas for different cases of the current problem. In model #1, all remanufacturing centers demand one type of component. In model #2, all remanufacturing centers demand more than one type of component.

We consider a two-phase RSC network. In the first phase, EOL products are transported from collection centers to disassembly center. In the second phase, recycled components via line balancing in the disassembly center are distributed to remanufacturing centers. The current problem is a two-echelon VR problem — one for the collection of EOL products and the other for distributing recycled components after disassembly operation. The I-CDDP is illustrated in Figure 5.1.

3.1 MATHEMATICAL MODEL #1

The first problem considers a single disassembly center, a single type of product, and a single component. Mathematical formulation is developed by extending the

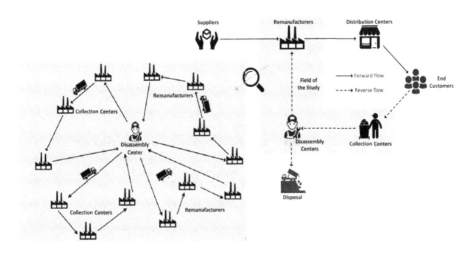

FIGURE 5.1 The Illustration of the I-CDDP.

classical capacitated VR problem, integrating the formula of Koc et al. (2009). The assumptions of model #1 are as follows:

- A single type of product is completely disassembled.
- A task cannot be split among two or more workstations.
- All workstations can process any of the tasks and all have the same associated costs.
- Each disassembly takes apart the product or sub-assembly into exactly two or more new sub-assemblies.
- Capacities of vehicles are known.
- Vehicles are ready for service in the disassembly center.
- The demand of each remanufacturing center for a component is known and must be fully satisfied.
- All remanufacturing centers demand only one type and the same component.
- The total demand for a remanufacturing center cannot exceed the vehicle capacity.
- Collection centers supply one type of EOL product to the disassembly center.

Indices

c, g remanufacturing center, $c, g = \{1,2, \dots, N\}$
m, t collection center, $m, t = \{1,2, \dots, E\}$
i Task
j, h workstation
a artificial task

Sets and Parameters

A_a artificial nodes of a
B_i normal nodes of i
$P(A_a)$ immediate predecessor set of A_a
$P(B_i)$ immediate predecessor set of B_i
$S(A_a)$ immediate successor set of artificial node A_a
$S(B_i)$ immediate successor set of artificial node B_i
d_{B_i} task time (normal node) of B_i (s)
J number of workstations (upper bound)
W_{time} working time (s)
De_g demand of remanufacturing center g (unit)
Ct fixed cost of shipping from disassembly center to remanufacturing center g or collection center t (cents)
D_{cg} distance between disassembly center and remanufacturing center or between remanufacturing centers (km)
O fixed cost to open a workstation in the disassembly line ($/workstation)
Q vehicle capacity (unit)
Cyc cycle time (s)
S_t supply of collection center t (unit)

DC_{mt} distance between disassembly center and collection or between collection centers (km)

Integer Variables

U_g load of vehicle after visiting remanufacturing center c
UU_t load of vehicle after visiting collection center m

Binary Variables

x_{ij} 1, if task B_i is assigned to workstation j; 0 otherwise
y_{cg} 1, if a vehicle travels from disassembly center to remanufacturing center or between remanufacturing centers; 0 otherwise
yy_{mt} 1, if a vehicle travels from disassembly center to collection center or between collection centers; 0 otherwise
z_i 1, if task B_i is performed; 0 otherwise
F_j 1, if workstation j is opened; 0 otherwise

The mathematical model #1:

$$Minimize\ O * \sum_{j}^{J} F_j + Ct * \sum_{c}^{N} \sum_{g,g \neq c}^{N} D_{cg} * y_{cg} + Ct * \sum_{m}^{E} \sum_{t,m \neq t}^{E} DC_{mt} * yy_{mt} \quad (5.1)$$

subject to

$$\sum_{i:B_i \in S(A_0)} z_i = 1 \quad (5.2)$$

$$\sum_{i:B_i \in S(A_a)} z_i = \sum_{i:B_i \in P(A_a)} z_i \quad \forall\ a, a \neq 0 \quad (5.3)$$

$$\sum_{j=1}^{J} x_{ij} = z_i \quad \forall\ i \quad (5.4)$$

$$\sum_{i:B_i \in S(A_a)} x_{ih} \leq \sum_{i:B_i \in P(A_a)} \sum_{j=1}^{h} x_{ij} \quad \forall\ a,\ a \neq 0, h \in J \quad (5.5)$$

$$\sum_{i=1}^{I} x_{ij} * d_{Bi} \leq Cyc * F_j \quad \forall\ j, j = 1, ..., J \quad (5.6)$$

$$Cyc = (W_{time}/(\sum_{g=2}^{N} De_g)) \tag{5.7}$$

$$F_{j+1} \leq F_j \quad \forall j, j = 1, \ldots, J - 1 \tag{5.8}$$

$$\sum_{t=1, t \neq m}^{E} yy_{mt} \leq 1 \quad \forall m, m = 2, \ldots, E \tag{5.9}$$

$$\sum_{m=1, m \neq t}^{E} yy_{mt} - \sum_{m=1, m \neq t}^{E} yy_{tm} = 0 \quad \forall t, t = 1, \ldots, E \tag{5.10}$$

$$UU_m - UU_t + Q * yy_{mt} \leq Q - S_t * yy_{mt} \quad \forall m, t, m, t = 2, \ldots, E, m \neq t \tag{5.11}$$

$$S_m * yy_{1m} \leq UU_m \quad \forall m, \ m = 2, \ldots, E \tag{5.12}$$

$$S_m * \sum_{t=1}^{E} yy_{mt} \leq UU_m \quad \forall m, \ m = 2, \ldots, E \tag{5.13}$$

$$\sum_{g=2}^{N} De_g \leq \sum_{m=2}^{E} \sum_{t=1}^{E} S_t * yy_{mt} \tag{5.14}$$

$$\sum_{g=1, g \neq c}^{N} y_{cg} = 1 \quad \forall c, c = 2, \ldots, N \tag{5.15}$$

$$\sum_{c=1, c \neq g}^{N} y_{cg} - \sum_{c=1, c \neq g}^{N} y_{gc} = 0 \quad \forall g, g = 1, \ldots, N \tag{5.16}$$

$$U_c - U_g + Q * y_{cg} \leq Q - De_g \quad \forall c, g, c, g = 2, \ldots, N, c \neq g \tag{5.17}$$

$$De_g \leq U_g \leq Q \quad \forall g \tag{5.18}$$

$$0 \leq UU_t \leq Q \quad \forall t \tag{5.19}$$

$$x_{ij}, \ z_i, \ y_{cg}, \ yy_{mt} F_j \in \{0, 1\} \quad \forall i, j, c, g. \tag{5.20}$$

The objective function (5.1) minimizes the total traveling cost of vehicles and the total fixed cost of operating disassembly workstations.

Constraints (5.2) and (5.3) guarantee that exactly one of the OR-successors is selected in the disassembly center. Constraint (5.4) ensures that each task is assigned to exactly one workstation in the disassembly center. Constraint (5.5) ensures the precedence relationships. Constraint (5.6) prevents the cycle time from being exceeded by a disassembly workstation. Constraint (5.7) shows the cycle time calculated by dividing the total working time by the total demand of re-manufacturing centers. Constraint (5.8) ensures that workstations open sequentially. Constraint (5.9) states that all collection centers do not need to be visited. Constraint (5.10) defines the flow for collection centers. Constraints (5.11)-(5.13) are capacity constraints. Constraint (5.14) ensures that the total supply of collection centers must not be less than the demand for remanufacturing centers. Constraint (5.15) guarantees that each remanufacturing center must be visited once. Constraint (5.16) defines the flow for remanufacturing centers. Constraint (5.17) is a capacity constraint. Constraints (5.18) and (5.19) enforce the non-negativity restriction on the decision variables. Finally, constraint (5.20) represents the binary variables.

3.2 MATHEMATICAL MODEL #2

We now present model #2 for the I-CDDP. In this case, the disassembly line is balanced according to maximum demand among remanufacturing centers to satisfy all remanufacturing centers. Assumptions and additional notations are as follows:

- All valid assumptions of model #1 are valid except, "all remanufacturing centers demand only one type and same component".
- All remanufacturing centers demand more than one type of component.

Indices

k component

Parameters

De_{gk} demand of remanufacturing center g for component k

The model #2:
Objective function (5.1),
subject to

$$Max_{Demand} = \max \left\{ \sum_{g=1}^{N} De_{gk} | k = 1, \ldots, K \right\} \qquad (5.21)$$

$$Cyc = \frac{W_{time}}{Max_{Demand}} \qquad (5.22)$$

$$U_c - U_g + Q * y_{cg} \leq Q - \sum_{k=1}^{K} De_{gk} \quad \forall\, c,\, g,\quad c,\, g = 2,\, ...,N,\quad c \neq g \quad (5.23)$$

$$Max_{Demand} \leq \sum_{m=2}^{E} \sum_{t=1}^{E} S_t * yy_{mt} \qquad\qquad (5.24)$$

$$\sum_{k=1}^{K} De_{gk} \leq U_g \leq Q \quad \forall\, g \qquad\qquad (5.25)$$

Constraints (5.2)-(5.6), (5.8), (5.9)-(5.13), (5.15), (5.16), (5.19), and (5.20).

Constraints (5.21) and (5.22) calculate the cycle time since the demand of all remanufacturing centers will be provided. Constraint (5.23) is a capacity constraint. Constraint (5.24) ensures that the total supply of collection centers must not be less than the maximum demand for remanufacturing centers. Constraint (5.25) states the non-negativity restriction.

4   COMPUTATIONAL EXPERIMENTS

This section provides the results of computational experiments for models #1 and #2 on test instances. All experiments were conducted on a server with Intel Core i5-8250U 1.6 GHz dual processor and 8.00 GB memory. We used GAMS/CPLEX to solve models #1 and #2.

We considered the realistic product, a toy car sample of Mete et al. (2018), for the DLB with 97 tasks and 12 components. Figure 5.2 shows the toy car and its components. Figure 5.3 illustrates the Transformed AND/OR Graph (TAOG) for the toy car.

For the VR problem phase, we considered the classical VR problem instances of Augerat et al. (1995) and modified them to generate our test instances. We arranged the smallest data set (P_n16_k8) from Set P and divided the nodes as collection centers and remanufacturing center. The first node is considered a disassembly center and the next eight nodes are considered as remanufacturing centers without changing the coordinate of nodes. The remaining seven nodes are considered as collection centers. Demand and supply characteristics are generated using a discrete uniform distribution. Demand for single-component and multi-component cases are generated within the range [1–5] and [1–6], respectively. Table 5.1 presents the demands where C1, C2, C3, C4, C4, C5, C6, C7, C8, C9, C10, C11, and C12 denote components 1, 2, 3, 4, 5, 6, 7, 8, 9, 10, 11, and 12, respectively. RC denotes the remanufacturing center. Table 5.2 presents the supply for both cases that are generated from the range [5–10] (Tables 5.1 and 5.2).

Ct is fixed to 5.23 cents per ton-km for a general freight truck; O in the disassembly line is fixed to 100 $ per workstation; $Wtime$ is set to 1250 seconds (s), and the number of maximum workstations (J) is set to 10. For model #1, the vehicle capacity is 35 units for both collection and distribution processes. For model #2, the vehicle capacity is 35 units for collection and 70 units for the distribution process.

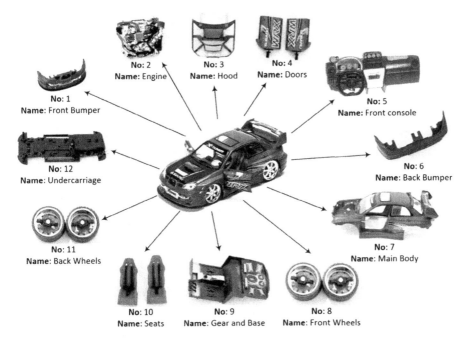

FIGURE 5.2 Toy Car and Its Components (Mete et al. 2018).

4.1 RESULTS FOR THE NUMERICAL EXAMPLE

This section presents the results for models #1 and #2. Table 5.3 presents the details of optimal results — Obj denotes the objective function value, NS denotes the number of workstations, CT denotes the cycle time in seconds, and CPU denotes the run time in seconds.

Figures 5.4 and 5.5 show the optimal distribution plans for models #1 and #2, respectively. "C" denotes the collection center, "DC" denotes the disassembly center, and "R" denotes the remanufacturing center. Since the total products in the collection centers are much higher than the demand for remanufacturing centers, not all collection centers are visited. Therefore, collection centers 2, 3, 6, and 7 are only visited for single-component cases, and 3, 4, 5, and 8 are only visited for multi-component cases. The demand of each remanufacturing center must be fully satisfied, so all remanufacturing centers are visited.

4.2 SCENARIO ANALYSES

We now present the managerial insights for models #1 and #2. The effects of changing working time on performance measures are considered and results are analyzed. For this scenario, working time is increased by 500 s at every turn. Since the supply and demand quantities are not changed for scenario analysis, the results of routing are the same for each time.

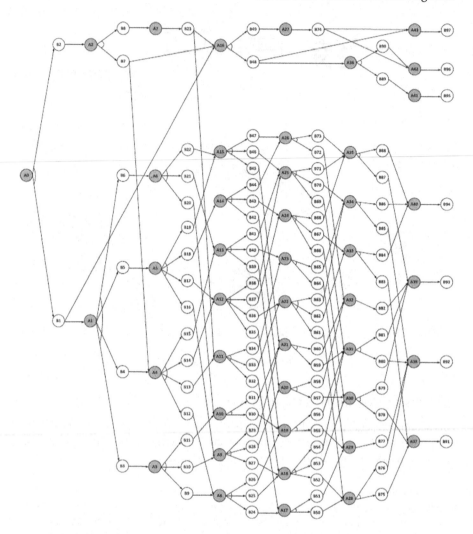

FIGURE 5.3   The Transferred AND/OR Graph of the Toy Car (Mete et al. 2018).

Table 5.4 and Figure 5.6 present the obtained results for model #1. It is clear that increased working time decreases objective function value and number of work-stations. However, the cycle time systematically increases. A new disassembly workstation is opened to satisfy the demand for remanufacturing centers under decreased working time.

Table 5.5 and Figure 5.7 present the obtained results for model #2. When we increase the working time, the objective function value, and the number of work-stations decreases. However, cycle time increases. Since the size of the test problem is not changed, a significant change in CPU time is not observed.

TABLE 5.1

Demands for Single Component and Multi-component Cases

RC	One Component Case Demand	Multi-component Case											
		C1	C2	C3	C4	C5	C6	C7	C8	C9	C10	C11	C12
2	1	3	1	4	3	2	1	2	2	3	5	4	2
3	5	2	4	3	2	3	1	2	1	1	3	2	6
4	4	1	1	6	2	2	1	3	2	2	1	1	4
5	2	2	1	2	3	4	3	4	1	1	2	3	1
6	3	1	1	5	2	1	2	1	3	2	1	2	3
7	5	5	2	1	1	3	3	4	2	1	4	1	2
8	5	2	1	3	3	4	2	1	1	3	2	3	1
9	4	1	3	2	1	3	3	2	4	2	1	1	3

TABLE 5.2

Supply Quantities for Single-Component and Multi-component Cases

Collection Centers	2	3	4	5	6	7	8
Supply	6	8	7	8	10	9	5

TABLE 5.3

Optimal Results for Models #1 and #2

	Obj ($)	NS	CT (s)	CPU (s)
Model #1	410.40	4	43.10	0.312
Model #2	415.30	4	48.07	0.297

5   CONCLUSIONS

Due to economic, legislative, and environmental reasons, the need for RSC has increased in recent years — a great potential for tackling waste. Disassembly is one of the most essential elements of RSC, and DLB is used in disassembly operations due to its high productivity. The VR problem plays a crucial role in the collection of EOL products and the distribution of recycled components in product recovery.

We have studied the optimization of the RSC network through collection, disassembly, and remanufacturing centers; we have introduced the I-CDDP. We have

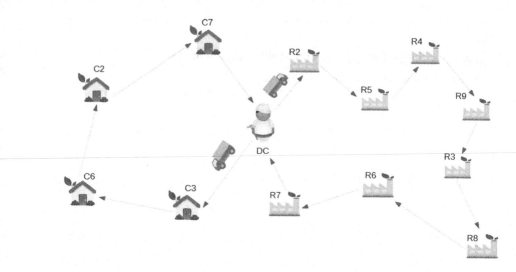

FIGURE 5.4   Optimal Distribution Plan for Model #1.

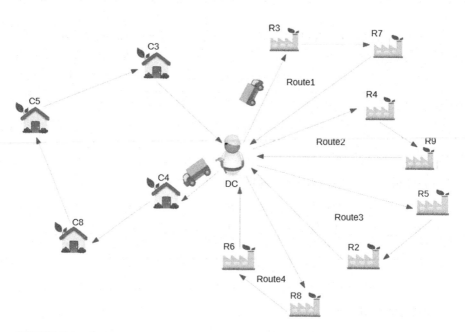

FIGURE 5.5   Optimal Distribution Plan for Model #2.

considered two different scenarios for the I-CDDP and have presented two mixed-integer mathematical formulations —the first one is for a single-component case and the second is for multi-component case. We have considered a case study and have conducted scenario analyses using mathematical models to see the impact of changing working time. Results have shown that both models tend to open new

TABLE 5.4
Results of Model #1 on Changing Working Time

WT	Obj ($)	NS	CT (s)	CPU (s)
1250	410.40	4	43.10	0.312
1750	310.40	3	60.30	0.297
2250	310.40	3	77.50	0.312
2750	210.40	2	94.80	0.297
3250	210.40	2	112.06	0.297
3750	210.40	2	129.30	0.453

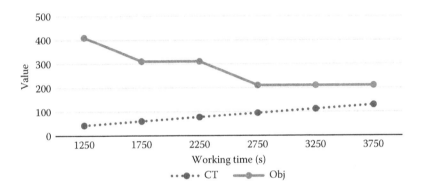

FIGURE 5.6   Relationship between Cycle Time, Objective Function, and Working Time for Model #1.

TABLE 5.5
Results of Model #2 on Changing Working Time

WT	Obj ($)	NS	CT (s)	CPU (s)
1250	415.30	4	48.07	0.297
1750	315.30	3	67.30	0.297
2250	215.30	2	86.50	0.297
2750	215.30	2	105.70	0.297
3250	215.30	2	125.00	0.282
3750	215.30	2	144.20	0.297

disassembly workstations to satisfy the demand of remanufacturing centers under decreasing working time. Furthermore, as working time increases, the objective function value tends to decrease.

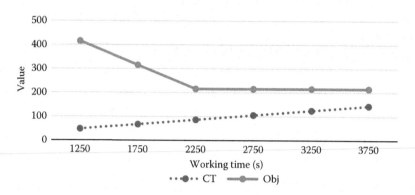

FIGURE 5.7   Relationship between Cycle Time, Objective Function, and Working Time for Model #2.

For future work, different scenario cases — inventory, multi-period, uncertainty — may be considered for the novel problem. Heuristic algorithms can be developed for large-size benchmark instances.

ACKNOWLEDGMENTS

The authors thank the Higher Education Council of Turkey for the 100/2000 PhD scholarship and TÜBİTAK (The Scientific and Technological Research Council of Turkey) for the 2211/C PhD research scholarship.

REFERENCES

Altekin, F. T., and Akkan, C. (2012). Task-failure-driven rebalancing of disassembly lines. *International Journal of Production Research* 50(18), 4955–4976.
Augerat, P., Belenguer, J. M., Benavent, E., Corberan, A., Naddef, D., and Rinaldi, G. (1995). Computational results with a branch and cut code for the capacitated vehicle routing problem. Research report RR949-M. ARTEMIS-IMAG, France.
Deniz, N., and Ozcelik, F. (2019). An extended review on disassembly line balancing with bibliometric and social network and future study realization analysis. *Journal of Cleaner Production* 225, 697–715.
Diri Kenger, Z., Koç, Ç., and Özceylan, E. (2020). Integrated disassembly line balancing and routing problem. *International Journal of Production Research* 58(23), 7250-7268.
Güngör, A., and Gupta, S. M. (2001). A solution approach to the disassembly line balancing problem in the presence of task failures. *International Journal of Production Research* 39(7), 1427–1467.
Habibi, M. K. K., Battaïa, O., Cung, V. D., and Dolgui, A. (2017a). Collection-disassembly problem in reverse supply chain. *International Journal of Production Economics* 183, 334–344.
Habibi, M. K. K., Battaïa, O., Cung, V. D., and Dolgui, A. (2017b). An efficient two-phase iterative heuristic for collection-disassembly problem. *Computers & Industrial Engineering* 110, 505–514.

Habibi, M. K. K., Battaïa, O., Cung, V. D., Dolgui, A., and Tiwari, M. K. (2018). Sample average approximation for multi-vehicle collection–disassembly problem under uncertainty. *International Journal of Production Research* 57(8), 1–20.

Kannan, D., Garg, K., Jha, P. C., and Diabat, A. (2017). Integrating disassembly line balancing in the planning of a reverse logistics network from the perspective of a third party provider. *Annals of Operations Research* 253(1), 353–376.

Koc, A., Sabuncuoglu, I., and Erel, E. (2009). Two exact formulations for disassembly line balancing problems with task precedence diagram construction using an and/or graph. *IIE Transactions* 41(10), 866–881.

Koç, Ç., Bektaş, T., Jabali, O., and Laporte, G. (2016). 30 years of heterogeneous vehicle routing. *European Journal of Operational Research* 249(1), 1–21.

Koç, Ç. (2017). An evolutionary algorithm for supply chain network design with assembly line balancing. *Neural Computing and Applications* 28(11), 3183–3195.

Koç, Ç., and Laporte, G. (2018). Vehicle routing with backhauls: Review and research perspectives. *Computers and Operations Research* 91, 79–91.

Mete, S., Çil, Z. A., Ağpak, K., Özceylan, E., and Dolgui, A. (2016). A solution approach based on beam search algorithm for disassembly line balancing problem. *Journal of Manufacturing Systems* 41, 188–200.

Mete, S., Çil, Z. A., Özceylan, E., Ağpak, K., and Battaïa, O. (2018). An optimisation support for the design of hybrid production lines including assembly and disassembly tasks. *International Journal of Production Research* 56(24), 7375–7389.

Özceylan, E., and Paksoy, T. (2013). Reverse supply chain optimisation with disassembly line balancing. *International Journal of Production Research* 51(20), 5985–6001.

Özceylan, E., and Paksoy, T. (2014). Fuzzy mathematical programming approaches for reverse supply chain optimization with disassembly line balancing problem. *Journal of Intelligent and Fuzzy Systems* 26, 1969–1985.

Özceylan, E., Paksoy, T., and Bektaş, T. (2014). Modeling and optimizing the integrated problem of closed-loop supply chain network design and disassembly line balancing. *Transportation Research* Part E 61, 142–164.

Özceylan, E., Kalayci, C. B., Güngör, A., and Gupta, S. M. (2019). Disassembly line balancing problem: a review of the state of the art and future directions. *International Journal of Production Research* 57(15-16), 4805–4807.

Toth, P. and Vigo, D., eds. (2014). *Vehicle Routing: Problems, Methods, and Applications.* MOS-SIAM Series on Optimization, Philadelphia.

Vidal, T., Laporte, G., and Matl, P. (2019). A concise guide to existing and emerging vehicle routing problem variants. *European Journal of Operational Research* 286(2), 401–416.

6 Trade-in-to-Upgrade Policymaking in a Digital Twin Disassembly-to-Order System for the Consumer Electronics Industry

Özden Tozanlı[1]*, Elif Kongar[2], and Surendra M. Gupta[3]

[1]Center for Transportation and Logistics, Massachusetts Institute of Technology, Cambridge, MA 02142, USA

[2]Departments of Mechanical Engineering and Technology Management, School of Engineering, University of Bridgeport, Bridgeport, CT 06604, USA

[3]Department of Mechanical and Industrial Engineering, Northeastern University, Boston, MA 02115, USA

1 INTRODUCTION

In line with technological developments, supply chain operations today evolved into a highly complex structure due to expanding customer demand and expectations. One viable way original equipment manufacturers (OEMs) can adjust to these purchasing habits is to redesign their strategic and operational activities to build far-reaching information and resource avenues through groundbreaking IT developments. These newly implemented operations, however, need to comply with extended producer responsibility (EPR) guidelines that also align with rising concerns towards the shortening product life cycles and increasing need for natural resources (Aizawa, Yoshida, & Sakai, 2008; Johnson & McCarthy, 2014; Torrente-Velásquez, Ripa, Chifari, Bukkens, & Giampietro, 2020). To achieve this, manufacturers must create efficient end-of-life product (EOLP) return structures and ensure value creation through product recovery operations to dwindle the cascading waste of discarded products. These constraints dictate two structural challenges: how fast and efficiently OEMs can respond to the changing market and capital needs while preserving their sustainability levels, and how manufacturers and consumers can become active participants in EOLP treatment activities (Ilgin & Gupta, 2010; Tozanli, Duman, Kongar, & Gupta, 2017).

Motivated by these challenges, product trade-in programs today become a compelling marketing strategy for many companies, specifically in oversaturated industries such as electronics and automotive (Ray, Boyaci, & Aras, 2005). In addition to its ability to alleviate environmentally benign manufacturing and supply chain operations, trade-in incentives help OEMs remarkably stimulate additional revenue channels through product remanufacturing, refurbishing, or repair operations. Moreover, offering exclusive discounts or credits in place of old devices fuels new product sales.

This chapter builds on the previous work published by Tozanlı, Kongar, and Gupta (2020) and introduces a predictive analysis for trade-in-to-upgrade policy-making in a digital twin product recovery model. The model aims to enable an autonomous, sensor-embedded, and decentralized EOLP management system. The study investigates the feasibility of cost- and resource-effective electronic product recovery strategy in a smart reverse logistics network, where quality-dependent trade-in-to-upgrade decisions take place in an autonomous ecosystem. The proposed model can be employed in manufacturing industries for precise assessment of value creation amid digital advancements in a future-oriented platform.

2 TRADE-IN-TO-UPGRADE POLICYMAKING IN SMART PRODUCT RECOVERY

2.1 END-OF-LIFE PRODUCT RECOVERY OPERATIONS

End-of-life product (EOLP) management is typically appended as the last stage of the value chain as a reverse logistics flow in closed-loop supply chain operations (Tozanli et al., 2017), with producers permanently engaged in product recovery operations (Fiksel, 2009; Wu & Wu, 2016). Product recovery relies on the collection of used devices, followed by the disassembly, inspection and sorting, remanufacturing, recycling, and/or reuse operations. Retrieving the value-added in EOLPs by converting them into a "like-new" condition via remanufacturing operations and recycling for material recovery has been proven to convey a significant environmental impact (Alqahtani & Gupta, 2017a; Ray et al., 2005).

Achieving an effective EOLP recovery necessitates well-established dismantling operations for selective separation of valuable parts and components (Alqahtani & Gupta, 2017a; Lambert & Gupta, 2005; Soh, Ong, & Nee, 2015). Disassembly problem handles a variety of approaches in literature, including scheduling, design for disassembly, disassembly line balancing, disassembly sequencing, and disassembly-to-order systems. Disassembly scheduling pinpoints the timing and quantity of disassembly to meet the demand for components and materials over a period (Ehm, 2018; Gupta & Taleb, 1994; Liu & Zhang, 2018), while the design for disassembly model concentrates on designing products for easier disassembly to attain value creation from EOLPs (Battaïa, Dolgui, Heragu, Meerkov, & Tiwari, 2018; Joshi & Gupta, 2019; Rios, Chong, & Grau, 2015). Additionally, disassembly line balancing research involves the disassembly task assignments to an optimal set of workstations by minimizing the number of stations and idle times between them (Kalayci, Hancilar, Gungor, & Gupta, 2015; McGovern & Gupta, 2007; Özceylan, Kalayci, Güngör, & Gupta, 2019). Disassembly sequencing enables a systematic dismantling method of valuable parts and components in the used product architectures (Alshibli, El Sayed, Kongar, Sobh, & Gupta, 2016; Gungor & Gupta, 1998). Disassembly-to-order systems, on the other hand, form a generic model of product recovery, considering the optimum number of EOL products to be disassembled to meet the demand for components and materials ordered from multiple origins (Kongar & Gupta, 2002; Ondemir & Gupta, 2014b). Recently, autonomous architectures in disassembly gained increased attention due to its high impact on decreasing manual labor costs (Alshibli et al., 2018; ElSayed, Kongar, & Gupta, 2012; Zheng et al., 2017).

Disassembly-to-order (DTO) systems, the focus of this chapter, deal with finding optimum levels of disassembly to meet the demand for materials and components derived from EOLPs (Lambert & Gupta, 2002). Kongar and Gupta (2002) presented a multi-objective optimization model of a DTO system to find the best combination of end-of-life product take-backs to fulfill the demand for materials and items retrieved through disassembly to maximize overall profit, minimize the number of items sent to landfills, minimize the inventory cost, and minimize the total disassembly cost.

Yet, along with the exponential growth in the number of item returns, possible modifications in the returned product structures, and the lack of information about the remaining useful life of components, the disassembly yields of EOLPs demonstrate a highly unpredictable pattern (Alqahtani & Gupta, 2018; Dulman & Gupta, 2018a). To overcome this uncertainty, producers perform disassembly operations regardless of components' functionality prior to inspection and sorting. Even though this process gives manufacturers the ability to sort products and components more efficiently, the upshot is that numerous unnecessary disassembly steps result from this conventional inspection process.

Remanufacturing is considered as another key foundation of product regaining processes, which is initiated by disassembly operations (Soh et al., 2015; Tian & Zhang, 2019; Zhou et al., 2019). In accordance with the unpredictable nature of disassembly yield due to the possible changes in the product structures during and after their use, remanufacturing applications are also acknowledged in a highly variable and uncertain framework. Even though remanufactured electronics are characterized by low cost and high technology, such ambiguity culminates inescapable financial damping to OEMs, leading to a laborious and time-consuming procedure in overall recovery operations (Bumblauskas, Gemmill, Igou, & Anzengruber, 2017; Dulman & Gupta, 2018a).

Digital twin technology, on the other hand, rises as a viable solution to eliminate the ambiguity related to the remaining life span of EOLPs. The following section discusses the utilization of digital twins in product recovery operations to decrease the unpredictability in dismantling and increase the quality of remanufacturing yields.

2.1 Digital Twin as a Disruptive Technology

Disruptive technologies today are signified as Cyber-Physical Systems (CPS), Internet-of-Things (IoT), cloud computing, big data analytics, artificial intelligence, blockchains, advanced robotics, and autonomous systems, augmented reality, and additive manufacturing. The utilization of one or more of these innovative concepts in business processes can be characterized as a paradigm shift since it carries conventional industrial operations to the digital era. Building smart infrastructures, companies also become capable of eradicating several extraneous operational costs across their supply chain layers. Among these digital tools, CPS and IoT here can be remarked as the most potent core enablers that form the basis of digital-to-physical and physical-to-digital platform cycles to maintain communication between intelligent structures and humans.

Digital twins are characterized as part of CP that demonstrates the virtual representation of a physical asset by transforming its properties and behavior into simulation models, information, and data (Rodič, 2017; Stark, Kind, & Neumeyer, 2017). Digital twins are enabled through IoT devices (Madni, Madni, & Lucero, 2019). IoT infrastructure consists of widely distributed and virtually connected ubiquitous sensors embedded in individual products to leverage data exchange and traceability along their life spans (Tozanlı & Kongar, 2020). A series of embedded processors—high-quality sensors, actuators, RFID tags, and microprocessors—can capture external data such as temperature, vibration, humidity, magnetism, or chemicals that products are exposed to aside from transferring this data to the cloud wirelessly. Through this instrument, OEMs become capable of collecting and storing product health and operational status data continuously (Brettel, Friederichsen, Keller, & Rosenberg, 2014).

Embedding IoT devices in products facilitates a high degree of traceability of individual goods (Brettel et al., 2014; Hehenberger et al., 2016). Capturing the product data via a cloud, digital twins generate virtual instances of products and processes with the help of 3D models and discrete-event simulations, respectively (Bottani, Cammardella, Murino, & Vespoli, 2017; Karanjkar et al., 2018). These abstract models help OEMs build a fully automated network with a high degree of surveillance and, therefore, monitor the omnipresent condition of products (Negri, Fumagalli, & Macchi, 2017; Subic, Xiang, Pai, & de La Serve, 2017; Zhang, Zhong, Farooque, Kang, & Venkatesh, 2020).

Through data analytics, large volumes of data transmitted to digital twins from IoT processors can be transformed into valuable information. Bartodziej (2016) delineated this structure as a self-optimizing intelligent system that stimulates the reduction of lead time and energy consumption while increasing quality. Such continuous physical-to-digital-to-physical chain also allows manufacturers to envision possible outcomes and to remarkably reduce error rates at operational levels (Subic et al., 2017). This provides an unprecedented degree of end-to-end control, transparency, and efficiency in the value chain by facilitating a real-time optimization and analytical capability (Hofmann & Rüsch, 2017).

In this setting, blockchain can play a crucial role in performing data transmission to digital twins in an immutable ecosystem and ensure security and perpetuity

(Teslya & Ryabchikov, 2017; Wang, Singgih, Wang, & Rit, 2019; Yadav & Singh, 2020). Blockchain technology can be defined as a distributed peer-to-peer ledger technology that can log transactions in a shared database between parties by eradicating many trusted intermediaries securely and permanently (Christidis & Devetsikiotis, 2016; Mandolla, Petruzzelli, Percoco, & Urbinati, 2019; Zhang et al., 2020). Adopting blockchain-powered compositions brings distinctive potential advantages to corporations due to blockchain's ability to accommodate a decentralized, knowledge-based, modular, and self-organized digital structure. Digital twins can achieve a high degree of interconnection in a blockchain-enabled environment (Smetana, Seebold, & Heinz, 2018).

Utilizing IoT-powered digital twins in a blockchain platform is allied with the vision of prospective EOLP management practices. Manufacturers have the privilege to fetch and permanently store dynamic product usage data in the digital ledger in post-sale periods, including the end-of-life stage. This allows the tracking of end-to-end real-time product information and material flow with manufacturing and product recovery processes. By embedding product life cycle data in a simulated scenario, firms can proactively take appropriate actions to mitigate the uncertainty in disassembly yield (Alqahtani & Gupta, 2017a; Dulman & Gupta, 2018a; Ondemir & Gupta, 2014a). Such value-creating capability helps not only eliminate unnecessary EOLP operation time and costs, but also boost remanufactured item sales by ensuring the correct quantity, quality, cost, time, and place requirements (Alqahtani, Kongar, Pochampally, & Gupta, 2019; Bartodziej, 2016). This novel capacity also paves the way for a remarkable level of resource- and eco-efficiency in terms of sustainability.

2.3 TRADE-IN POLICYMAKING IN A DIGITAL TWIN DISASSEMBLY-TO-ORDER SYSTEM

In EOLP processing operations, the entire process is reliant on the customers' willingness to return their discarded equipment, making the collection step a crucial one in the overall system. OEMs in the electronics industry design environmentally and economically benign product take-back strategies that would spark the volume of product returns as well as achieve sustainable price advantage against independent remanufacturers.

Trade-up promotions in the electronics industry are set forth as part of long-term marketing strategies that aim to entice current and potential customers to trade-in their used products with newer generations at a discounted price (Fudenberg & Tirole, 1998; Heese, Cattani, Ferrer, Gilland, & Roth, 2005). Several OEMs such as Apple, HP, Xerox, Mercedes, and Amazon launch trade-in programs as their marketing strategy to encourage existing and future consumers to substitute their low generation products with successive versions (Agrawal, Ferguson, & Souza, 2016; Cao, Xu, Bian, & Sun, 2019; Yin, Li, & Tang, 2015). Product take-back plans convey unique environmental and financial benefits to firms. By aggregating EOLPs through product acquisitions, manufacturers align their operations with EPR principles as well as facilitate the accessibility of used goods. This process influences an unprecedented degree of EOLP treatment in the CLSCs (Johnson & McCarthy, 2014). This catalyst allows firms to obtain a unique profit-generating venue (Zhang & Zhang, 2015).

Trade-in policies can be examined within the fields of economics and closed-loop supply chain management (Agrawal et al., 2016; Chen & Hsu, 2015). Determining a fitting replacement trade-off decision elicits strategic leverage for companies since it seamlessly absorbs numerous cost factors. These trade-in rebates can be studied from various perspectives such as the competition between OEMs and remanufacturers (Agrawal et al., 2016; Oraiopoulos, Ferguson, & Toktay, 2012), the comparison online or offline platforms (Cao, Wang, Dou, & Zhang, 2018; Cao et al., 2019), the analysis of buyback and discounts programs (Cole, Mahapatra, & Webster, 2017), and the optimal pricing decision in the business-to-consumer ecosystem (Han, Yang, Shang, & Pu, 2017; Kim, Rao, Kim, & Rao, 2011; Zhang & Zhang, 2015; Zhou & Gupta, 2019). Engaging policy-making can be implemented through a quality-dependent approach, where rebate decisions vary based on the operational conditions of returned products (Guide, Teunter, & van Wassenhove, 2003). Ray et al. (2005) analyzed three pricing schemes taking the continuous age of returned devices into account.

For purchasing behavior, trade-in programs positively impact customers' buying decisions by granting buyers the ability to claim the scrap value of their used devices. Many companies offer business-to-consumer trade-in practices today and implement a quality-dependent plan with processes that last for at least a month (Hahler & Fleischmann, 2017). Such a product acquisition plan comprises a multi-dimensional structure that relies on prolonged bargaining processes between customers and OEMs.

A typical product acquisition system on a business-to-consumer basis succinctly begins with customers returning their outmoded products to OEMs (Cao et al., 2019). Upon the delivery of eligible items, OEMs provide buyers with trade-in discounts, instant credits, or gift cards to be redeemed in their future purchases. This offer is usually exclusively applicable for upgraded device purchases. Following this, compiled appliances are incorporated into the end-of-life recovery process and are first sent to the disassembly field to be inspected for their valuable components. Hinging on this condition analysis, manufacturers conduct one of the proper handling methods for EOLPs, viz., remanufacturing for secondary markets, reusing high-quality components, or recycling mostly the precious metals and other materials in demand. Figure 6.1 depicts the workflow of a traditional product acquisition plan.

This lengthy process emerges due to the unpredictability surrounding the actual quality status of returned products. The conventional intransigent trade-in schemes fail to address this uncertainty, leading to various unnecessary inspection and sorting, disassembly, and shipping steps, eventually resulting in increased complexity and product recovery cost.

Yet, in today's fast-changing market dynamics, achieving a competent trade-in scheme requires agile, responsive, and customer-centered solutions that traditional manufacturing and supply chain technologies are incapable of offering by design. Such a challenging task inevitably needs strategic initiatives by utilizing innovative

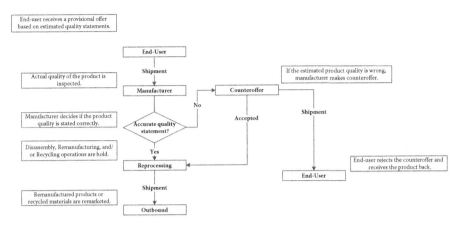

FIGURE 6.1 A General Workflow of Traditional Trade-in Programs.

information technologies for rapid response to customer needs and reducing complexity across all operational layers (Tozanlı et al., 2020). Considering the unprecedented capability of digital twins in eliminating the uncertainty in EOLP management, trade-in rebate decisions can be transformed into an autonomous platform. This critical talent provides manufacturers a noteworthy opportunity to determine the proper trade-in-to-upgrade policy based on the remaining useful life of components within a product. A well-established acquisition policy also leads to a critical upsurge in sales based on the changing customer behavior, making EOLP returns more appealing, eventually increasing the companies' overall profitability.

Motivated by this, a manufacturer is assumed to acquire real-time data streams regarding product usage patterns and quality conditions with the help of IoT sensors. Each product is distinguished via a unique identification code—a serial number—sealed in blockchains. Product serialization designates a unique code to each product carrying critical information—the product's origin, bill of materials, remaining lifetime, assembly, and disassembly instructions—in addition to the user's manual. Blockchain technology allows firms to permanently store such data in addition to timestamp product usage via a decentralized network. Inquiring the serial number of a product in online servers, the product information and data usage of this product stored in the blockchain is retrieved and simultaneously transmitted to product and process twins via the cloud. Performing recovery operations in a digital twin-supported platform, the OEM implements a real-time performance analysis and cost optimization for the product recovery cost of an individual product via online simulation models. Accessing the actual quality status of products and the expected total cost of the DTO system authorizes the producer to virtually predict trade-in prices for discarded products without relying on subjective perceptions. Therefore, manufacturers can determine more realistic trade-in pricing policies for customers with different purchasing preferences for products with varying quality levels.

Predictive analysis in determining the optimal trade-up pricing helps foresee the patterns of large datasets and their results (Duman, Kongar, & Gupta, 2019). This chapter builds on the recent work published by Tozanlı et al. (2020)—finding the optimal trade-in-to-upgrade incentives for returned products with discrete sets of quality standards in an IoT-embedded blockchain-powered DTO platform.

The following section elaborates on the proposed methodology for trade-in-to-upgrade policymaking in a digital twin disassembly-to-order system for the consumer electronics industry.

3   SYSTEM DESCRIPTION

This section presents a conceptual model of a discrete event digital twin of a DTO system. Discrete event simulation (DES) is deployed as an illustration to pinpoint the implementation of a conceptual digital twin of the product recovery line.

A DES model is developed from the OEM viewpoint to mimic the performance degradation pattern of smart products and their components throughout their in-use periods. DES captures the stored data to evaluate real-time behavior and performance of the overall processes in a simulated scheme. Retrieving the characteristics of returned products, along with their quality status, the simulation creates a series of possible events that can occur over time. This gives companies the opportunity to observe and analyze various what-if scenarios. Performing alternative scenarios, the DES model chooses appropriate disassembly processes for returned products and decides on the right combination of recycling, remanufacturing reuse, or disposal operations (Alqahtani & Gupta, 2017b; Dulman & Gupta, 2018b). The simulation model returns the expected cost of product recovery operations—disassembly, re-manufacturing, inventory, backorder, and transportation costs.

Such a smart product recovery system is analyzed using sensor-embedded end-of-life (EOL) game consoles (GCs) as returned products. Customers return their old generation GCs in exchange for newer counterparts. Initially, EOL GCs arrive at the system for key information retrieval using RFID sensors stored in the blockchain. Following this, they are processed in a six-station disassembly and a six-station remanufacturing line.

The cost of information retrieval is assumed to be lower than the cost of inspection and sorting. Returned GCs are examined for their age classes where devices with less than one year of handling are considered as high-quality products, devices with one to two years of usage are classified as medium-quality products. Low-quality products are products older than two years. It should be noted that in case a returned product with less usage time has lost its functionality, the quality is downgraded. Devices older than three years are not accepted in trade-in events. Poisson distribution is used to replicate interarrival rates of EOLP game consoles and demand for each component; disassembly and assembly times at each station follows an exponential distribution.

In some rare situations, customers return their products after less than 30 days of handling. The system investigates whether the returned products are fully functional or in need of repair. In case of full functionality, the EOLPs are directly sent to high-quality product inventory to resell in the secondary market. Otherwise, product recovery processes are successively performed. The flowchart of the product recovery system is depicted in Figure 6.2.

Game consoles consist of six components: a hard drive, power supply, disc drive, motherboard, heat sink, and fan. Complete disassembly is carried out to remove each component, and the order of disassembly stations follows the path defined by precedence relationships. This requires a six-station disassembly line where six components are disassembled. Disassembly can be performed in two forms, depending on the condition of components: non-destructive or destructive. Non-destructive disassembly for components with decent shape and functioning, whereas destructive disassembly is for components with zero remaining lifetime or broken components. Since the non-destructive process ensures that the working component is not damaged during disassembly, the time and cost for non-destructive disassembly is higher than destructive disassembly. Each component in an EOLP is checked for its function before deciding on the form of disassembly.

Disassembled functioning components are sent to component inventory bins based on their quality level—high (Bin 1), medium (Bin 2), or low (Bin 3)—to meet the demand for the in-plant remanufacturing line. Broken components are sold to third party recyclers for material recovery. Demand for material recovery is assumed as constant and at a high rate. In case high- and medium-quality component bins are full, the components are placed to lower quality bins and are insufficiently utilized. If a low-quality component bin is full, components are sold for their parts or are transported to third-party recyclers. A generic flowchart of the disassembly process is exhibited in Figure 6.3.

Following dismantling, additional component testing becomes unnecessary since component conditions are quantified through RFID sensors. OEM eradicates the uncertainty in disassembly yield, eliminating inspection and sorting steps, and reducing the total disassembly time.

Recovered components are used to fulfill two types of demand: internal and external. Internal demand is the number of components required for the in-plant remanufacturing line, whereas external demand is for the second-hand components market. The component inventory is primarily used to satisfy the internal demand for the remanufactured product line. Excess component inventory is used to fulfill external demand. To ensure environmental compliance and zero disposal, additional inventory is assumed to be sold to recyclers for material recovery. Backorder cost occurs once inventory falls short in meeting external component demand; the manufacturer can generate revenue through the sales of used components and recyclable materials.

Following their disassembly, EOL components are directed to a six-station remanufacturing line. Products are remanufactured on a quality-dependent basis such as high-, medium-, and low-quality products and stored in their inventory bins accordingly. When reprocessing like-new condition products, only high-quality component inventory is utilized. Medium- and low-quality components are used for medium- and low-quality product remanufacturing, respectively. Recovered components are assembled in reverse order of disassembly stations, depicted in Figure 3.2. In case the inventory is insufficient to fulfill internal demand, components are procured from outside suppliers.

In this study, the DES's behavior is tested under various experimental conditions through a design-of-experiment study, namely, three-level (3^k) factorial design. The

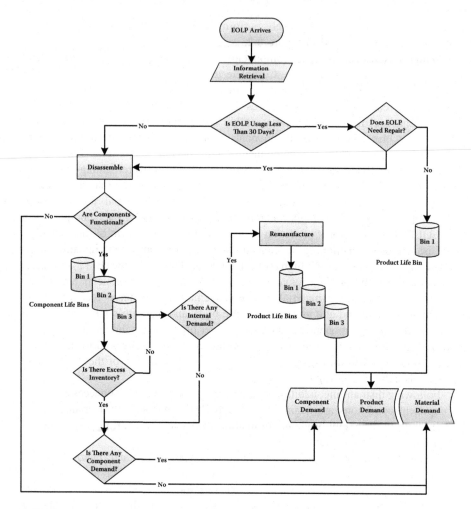

FIGURE 6.2 A General Flow of Product Recovery Processes.

simulation model for product recovery operations is based on 50 factors, each at three levels. Implementing a full factorial design with 50 factors for three levels requires an extensive number of experiments due to the number of possible combinations between the factors. To reduce the number of experiments to a practical level, the Taguchi Orthogonal Arrays (OAs) design is built to obtain the subset of the minimum number of combinations that yield the maximum information about the performance of the response function.

4   TRADE-IN-TO-UPGRADE POLICY MODEL

There are different trade-in modes that OEMs present to their existing and future customers as their marketing strategies. Such strategies can be categorized as trade-ups, gift cards, store credits, or discount programs. This study investigates

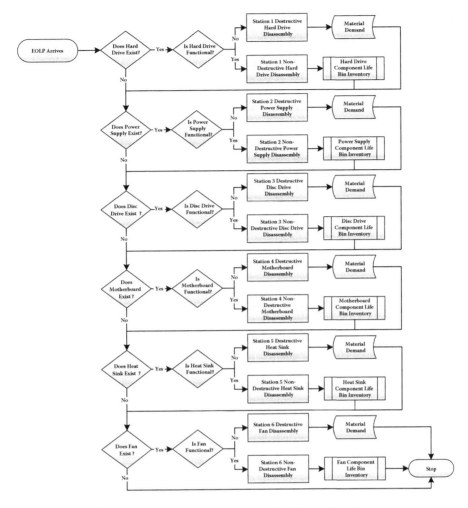

FIGURE 6.3   Disassembly Flow of High-Quality EOLP Game Consoles.

the optimal pricing decision for trade-up incentives, where replacement customers are offered to exchange their discarded products with their successive versions. By embedding the cost parameters obtained from the discrete-event product recovery model, a general pricing scheme for EOL GCs at varying quality conditions is introduced. In some events, fulfilling customers' expectations, however, grows into a complex problem, where theoretical acquisition prices become insufficient in offering valid rebates from the customers' perspective. To achieve an optimal strategy, the proposed pricing model is integrated into a simulation model as part of the digital twin environment. The model allows the manufacturer to mimic an individual customer behavior towards the allocated incentive, and the outcomes, in a dynamic platform. This helps OEM construct a predictive model to propose best-fitted rebates.

The pricing scheme is a profit-maximizing policy model for quality-dependent pricing offers adopted from the model proposed by Ray et al. (2005).

The model notations and formulation are elaborated in the following sections.

4.1 NOTATIONS

i	:	Quality index ($i = \{1, 2, 3\}$)
t	:	Age of a returned product within the useful lifespan ($0 \leq t \leq 3$)
$p_{tr}(t)$:		Age-dependent new product price for any customer returning a product at age t
p_n	:	Market price for a new product
p_p	:	Perceived discounted price of any customer joining trade-in for new product purchases
φ	:	Maximum price that any customer pays for new product purchases
c_{dto}	:	Unit cost of disassembly-to-order system
c_{rem}	:	Unit cost of remanufacturing
c_h	:	Unit cost of product handling including inventory
$R(t)$:		Return revenue function of an EOLP at age t

4.2 MATHEMATICAL FORMULATION

The OEM obtains the true quality of the products distinguished between three quality indexes (i), viz. high quality, medium quality, and low quality, presenting product age profiles within the range of $t \in [0, 3]$. Replacement customers trading their GCs are charged a discounted price $p_{tr}(t)$ ($\leq p_n$), where the discount amount depends on the age of their products. End-users' behavior towards accepting the offered incentive is strongly reliant on the surplus between their perceived discounted price (p_p) and the actual discounted price. The buyer's surplus can be shown as:

$$p_p - p_{tr}(t). \tag{1}$$

To allow analytical tractability, p_p is assumed as heterogeneous following uniform distribution ($p_p \sim U[0, \varphi]$) due to its large degree of variability. As a result, the probability of a randomly chosen product holder with a device at age t accepting price $p_{tr}(t)$ can be expressed in Equation (2). Considering $p_{tr}(t)$ varies between 0 and φ ($0 \leq p_{tr}(t) \leq \varphi$), the probability of a buyer returning the product remains positive.

$$P(trade) = P(p_p - p_{tr}(t) > 0) = \frac{\varphi - p_r(t)}{\varphi}. \tag{2}$$

The selling price of remanufactured products in the secondary market differs based on their quality levels. As the deterioration rates of older components increases, the profitability of remanufacturing those items and reselling in the marketing decreases. Therefore, the manufacturer incurs a return revenue $R(t)$ associated with the reusability of each returned product at age t and quality index i at the remanufacturing line as defined below:

$$R(t) = c_{rem} - c_h - it. \tag{3}$$

Hinging on above expressions, the objective function for trade-in-to-upgrade incentives becomes:

$$E\left[\pi_{trade}(p_{tr}(t))\right] = \frac{1}{\varphi} \int_0^\varphi (p_{tr}(t) - c_{dto} + R(t))(\varphi - p_{tr}(t))f(t)dt. \tag{4}$$

The optimal price $p_{tr}(t)$ for any product at age t that maximizes the expected profit is derived by the first-order condition $\partial E\left[\pi_{trade}(p_{tr}(t))\right]/\partial p_{tr}(t) = 0$. This can be interpreted as:

$$p_{tr}^*(t) = \frac{\varphi + c_{dto} - R(t)}{2}. \tag{5}$$

4.3 DYNAMIC TRADE-IN POLICYMAKING MODEL

A DTO digital twin platform enables the OEM to compute lucrative rebates for product returns by extracting the actual cost of product recovery. However, in some cases, the exact condition of the device can be more depreciated than the product holder anticipates, which leads buyers to reject the incentive program. Predicting the individual customer behavior, the producer can estimate the effects of offering an engaging incentive to each customer and increase its profitability. Focusing on this, the pricing model is embedded in a dynamic DES model.

The DES model evaluates real-time return on investment for product submissions by considering the conceptual depreciation value of customers. Based on this assumption, the simulation mimics the customer behavior against the proposed discounted price. Figure 6.4 depicts the dynamic simulation-based game model.

5 NUMERICAL ANALYSIS

Relevant data for factors were collected from the maintenance and remanufacturing facility of one of the leading Japan-based consumer electronics companies. The simulation models are constructed through ARENA v.15.1 to compute the expected cost of product recovery operations and predict customers' behavior against offered rebates and its resulting payoffs for the OEM.

Based on historical data, the product handling times follow a triangular distribution between [0, 3] with a mode of 1.35. Moreover, the DES model is tested under varying experimental conditions through a three-level full factorial design, where factors are constructed and a level value is assigned to each factor based on the collected data. Motivated by this, 50 factors are employed, each at three levels of low, intermediate, and high. Tables 6.1 and 6.2 demonstrate factors with level values and parameters used in the simulation, respectively.

Implementing a full factorial design for 50 factors, however, yields an extensive number of combinations (3^{50} = 7.179E + 23) between factors. This hinders the practicality of this study. To reduce the number of experiments to a practical level, a subset of the minimum number of combinations that yield the maximum information possible is selected. This design-of-experiments study is defined as the Taguchi's Orthogonal Arrays (OAs) design (Taguchi, 1987). In selecting the minimum set of the combinations, the number of OAs must be greater than or equal to the system's degree-of-freedom, also noted as $L_{101}(3^{50})$ ([(Number of levels − 1) × Number of Factors)] + 1 = 101).

The period to run each experiment obtained through the OAs covers eight hours per shift, one shift per day, five days per week for six months, over 100 replications. ARENA model calculates the total cost during the simulation runtime using the following equation:

$$Expected\ Cost = DC + RMC + HC + BC + TC - (CS + MS), \qquad (6)$$

where DC is the total disassembly cost of components; RMC is the total re-manufacturing cost of products; HC is the total holding cost of components; BC is the total backorder cost of components in the secondary and recycling market; TC is the total transportation cost of products to the main facility; CS is the total revenue generated through component sales for component demand; MS is the total revenue generated through component sales for material demand. The price for components depends on the quality level of the parts, whereas the revenue obtained from components sold to the recyclers is regardless of the condition of the components. The scrap value of components is calculated by multiplying the weight of the component by the unit scrap revenue. It is assumed that all surpass inventory is sold to recyclers for material recovery. The transportation cost is assumed to be $50 for each trip. The model investigates the expected overall cost of DTO en bloc. This helps derive the expected unit cost of processing an EOLP at a certain quality level in the entire product recovery system. This outcome is utilized in the dynamic pricing model to attain the expected margin of the entry of an individual product at a specific condition into the system.

5.1 ANALYSIS FOR DYNAMIC TRADE-IN POLICYMAKING

Employing the findings of the product recovery DES model, this analysis deals with trade-in policymaking to examine an engaging quotation for varying quality of returned products from the perspectives of all parties engaged in the transaction. The ARENA simulation program is used to determine the actual quality of components, revealing the exact condition of return products. Depending on the characteristics of EOL GCs, take-back incentives are assessed according to various criteria from the manufacturer's perspective — the age and the quality index of the EOL product, return revenue, and cost of the disassembly-to-order system. Based on these findings, the model provides the variables for customers' decision as

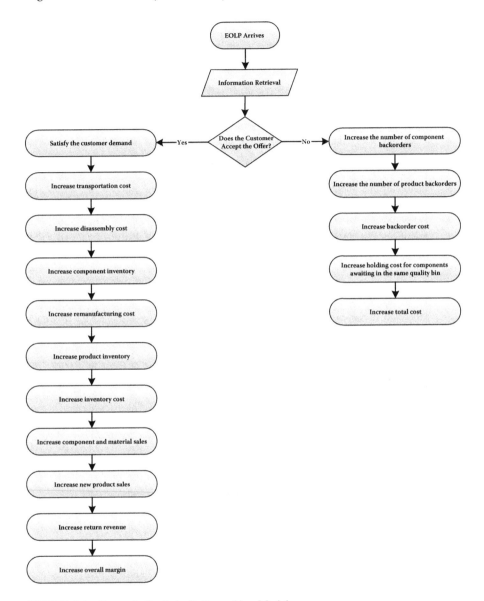

FIGURE 6.4   Dynamic Trade-in Policymaking Model.

"accept" or "reject" based on the expected residual value of a particular customer, the price offered, the conceived age, and the actual age of the device. Additionally, the simulation model extracts the cost and revenue parameters for each submission.

Specifically, each customer has a perceived value for their products and a relative expected discounted price for a new product purchase. Assume the new product price p_n = \$400 and the expected discount value for an individual product holder paying for a new device $p_p \sim U[0, 550]$. The achievable margin for a returned item is reliant on the customer's decision as "accept" or "reject" depending

TABLE 6.1

Factors with Three-Level Values Used in the Design-of-Experiments Study

No.	Factor	Unit	Levels		
			1	2	3
1	Mean EOLP Game Consoles Arrival Rate	Prod./hr	10	20	30
2	Probability of EOLP Game Consoles for Repair	%	5	10	15
3	Probability of a Missing Hard Drive	%	5	10	15
4	Probability of a Missing Power Supply	%	5	10	15
5	Probability of a Missing Disc Drive	%	5	10	15
6	Probability of a Missing Motherboard	%	5	10	15
7	Probability of a Missing Heat Sink	%	5	10	15
8	Probability of a Missing Fan	%	5	10	15
9	Probability of Non-functional High-Quality Hard Drive	%	5	10	15
10	Probability of Non-functional High-Quality Power Supply	%	5	10	15
11	Probability of Non-functional High-Quality Disc Drive	%	5	10	15
12	Probability of Non-functional High-Quality Motherboard	%	5	10	15
13	Probability of Non-functional High-Quality Heat Sink	%	5	10	15
14	Probability of Non-functional High-Quality Fan	%	5	10	15
15	Probability of Non-functional Medium Quality Hard Drive	%	10	15	20
16	Probability of Non-functional Medium Quality Power Supply	%	10	15	20
17	Probability of Non-functional Medium Quality Disc Drive	%	10	15	20
18	Probability of Non-functional Medium Quality Motherboard	%	10	15	20
19	Probability of Non-functional Medium Quality Heat Sink	%	10	15	20
20	Probability of Non-functional Medium Quality Fan	%	10	15	20
21	Probability of Non-functional Low-Quality Hard Drive	%	15	20	25
22	Probability of Non-functional Low-Quality Power Supply	%	15	20	25
23	Probability of Non-functional Low-Quality Disc Drive	%	15	20	25
24	Probability of Non-functional Low-Quality Motherboard	%	15	20	25
25	Probability of Non-functional Low-Quality Heat Sink	%	15	20	25
26	Probability of Non-functional Low-Quality Fan	%	15	20	25
27	Mean Non-Destructive Disassembly Time for Station 1	Min.	0.6	0.8	1
28	Mean Non-Destructive Disassembly Time for Station 2	Min.	0.8	1	1.3
29	Mean Non-Destructive Disassembly Time for Station 3	Min.	0.8	1	1.3
30	Mean Non-Destructive Disassembly Time for Station 4	Min.	1	1.5	2
31	Mean Non-Destructive Disassembly Time for Station 5	Min.	0.8	1	1.3
32	Mean Non-Destructive Disassembly Time for Station 6	Min.	0.6	0.8	1
33	Mean Destructive Disassembly Time for Station 1	Min.	0.3	0.5	0.6
34	Mean Destructive Disassembly Time for Station 2	Min.	0.4	0.5	0.7
35	Mean Destructive Disassembly Time for Station 3	Min.	0.4	0.5	0.7
36	Mean Destructive Disassembly Time for Station 4	Min.	0.6	0.8	1
37	Mean Destructive Disassembly Time for Station 5	Min.	0.4	0.5	0.7
38	Mean Destructive Disassembly Time for Station 6	Min.	0.3	0.5	0.6
39	Mean Assembly Time for Station 1	Min.	0.6	0.8	1

TABLE 6.1 (*continued*)

No.	Factor	Unit	Levels		
			1	2	3
40	Mean Assembly Time for Station 2	Min.	0.8	1	1.3
41	Mean Assembly Time for Station 3	Min.	1.3	1.5	1.8
42	Mean Assembly Time for Station 4	Min.	1	1.3	1.5
43	Mean Assembly Time for Station 5	Min.	1	1.3	1.5
44	Mean Assembly Time for Station 6	Min.	0.8	1	1.3
45	Mean Demand Rate for Hard Drive	Parts/hr.	10	15	20
46	Mean Demand Rate for Power Supply	Parts/hr.	10	15	20
47	Mean Demand Rate for Disc Drive	Parts/hr.	10	15	20
48	Mean Demand Rate for Motherboard	Parts/hr.	10	15	20
49	Mean Demand Rate for Heat Sink	Parts/hr.	10	15	20
50	Mean Demand Rate for Fan	Parts/hr.	10	15	20

on the perceived surplus. Therefore, the OEM is expected to improve the predictive model towards maximizing the likelihood of a customer to accept the offer.

The time frame for the simulation spans a period of eight hours per shift, one shift per day, five days a week for six months with 100 replications. A total of 28,851 and 28,861 data points are assembled. A sample output is exhibited in Table 6.3, where the end-users' decisions to accept the incentive is presented as 1 ($P(Y = 1)$), and as 0 for reject ($P(Y = 0)$).

Hinging on these findings, an empirical analysis is implemented to estimate individual consumer behavior. Since the customer decision for accepting the proposed offer is a strict "yes" or "no", the simulation results are fed into a logistic regression model to obtain the probability of a customer accepting the offer based on the expected rebate and the actual offer. This technique gives manufacturers the ability to grant unique offers to each customer, leading to a faster and more efficient decision mechanism.

In this regard, the customer's decision to trade-in-to-upgrade price is dependent on the surplus between the perceived price and the actual price offered:

$$\alpha = p_n - p_{tr}(t), \tag{7}$$

The probability function for the customer's acceptance decision can be formulated as:

$$P(\alpha) = P(Y = 1) = \frac{e^{\beta_0 + \beta_1 \alpha}}{1 + e^{\beta_0 + \beta_1 \alpha}}. \tag{8}$$

Based on the extracted probability function, a binary logistic regression model is implemented for each take-back strategy. IBM SPSS Statistics v.20 is used to run the analysis. To derive the best-fitting covariates β_0 and β_1 for the given dataset, the

TABLE 6.2

Parameters Used in the DTO Simulation System

No.	Parameter	Unit	Value
1	Backordering Cost Rate	%	40
2	Holding Cost Rate	%	10
3	Disassembly Cost	$/min.	1
4	Assembly Cost	$/min.	1
5	Price for High-Quality Hard Drive	$	75
6	Price for High-Quality Power Supply	$	70
7	Price for High-Quality Disc Drive	$	90
8	Price for High-Quality Motherboard	$	180
9	Price for High-Quality Heat Sink	$	45
10	Price for High-Quality Fan	$	45
11	Price for Medium Quality Hard Drive	$	55
12	Price for Medium Quality Power Supply	$	50
13	Price for Medium Quality Disc Drive	$	70
14	Price for Medium Quality Motherboard	$	150
15	Price for Medium Quality Heat Sink	$	25
16	Price for Medium Quality Fan	$	20
17	Price for Low-Quality Hard Drive	$	30
18	Price for Low-Quality Power Supply	$	25
19	Price for Low-Quality Disc Drive	$	35
20	Price for Low-Quality Motherboard	$	50
21	Price for Low-Quality Heat Sink	$	15
22	Price for Low-Quality Fan	$	15
23	Price for Unit Material Recycling	$/lbs.	0.6
24	Weight for Hard Drive	lbs.	0.36
25	Weight for Power Supply	lbs.	1
26	Weight for Disc Drive	lbs.	1
27	Weight for Motherboard	lbs.	0.5
28	Weight for Heat Sink	lbs.	0.4
29	Weight for Fan	lbs.	0.36

maximum likelihood function is performed. Tables 6.4 and 6.5 presents the classification table and the estimations for regression coefficients β_0 and β_1, respectively.

With the help of the approximated parameters, Figure 6.5 depicts the range of expected profit for each submission based on the price difference α, where the expected profit ranges between [−88.74, 191.16] for trade-in-to-upgrade program.

The empirical analysis demonstrates several interesting findings. The approximation of a customer's decision is statistically significant (p-value < 0.001), where the correction prediction is calculated as 78%. Considering the cutoff value of 0.5

TABLE 6.3

Sample Simulation Output for Customer Decisions in Trade-in-to-Upgrade Offers

Sample	Perceived Age (yr)	Actual Age (yr)	Quality Index	Return Revenue ($)	Expected Price ($)	Offered Price ($)	Decision (Y)
1	0.86	1.28	2	49.23	202.59	287.88	1
2	1.51	1.51	2	48.77	318.06	288.1141	0
3	0.81	2.03	3	28.90	230.67	298.05	1
4	2.46	3.00	3	23.93	311.49	300.53	0
5	2.15	2.58	3	27.27	342.70	298.86	0
				...			
28847	1.02	1.33	2	49.14	269.70	287.93	1
28848	1.40	1.96	2	47.88	338.92	288.56	0
28849	0.90	0.90	1	57.90	252.93	283.55	1
28850	1.49	2.24	3	28.29	345.19	298.36	0
28851	0.38	0.38	1	58.42	252.23	283.29	1

for the regression analysis, the proportion of customers accepting and rejecting the offer is almost equally distributed. This implies that higher discounts deliver higher product submissions. Therefore, the manufacturer carries low risks of offering lower prices due to the increasing volume of product returns, achieving sustainable profit generation by attaining long-term customer loyalty.

6 CONCLUSION

This chapter introduced a novel approach in obtaining optimal trade-in incentives for individual product submissions in a digital-twin DTO system. Focusing on this,

TABLE 6.4

Classification Table for Trade-in-to-Upgrade Program

			Classification Table[a]		
	Observed		Predicted		
			Decision		Percentage Correct
			0	1	
Step 1	Decision	0	14395	2882	83.3
		1	3648	7926	68.5
	Overall Percentage				77.4

[a]The cut value is .500.

TABLE 6.5

Coefficient Estimation for Trade-in-to-Upgrade Program

		β	SE	Wald	DF	$p >$ Chi-Sq	Exp (β)	95% CI for Exp (β)	
								Lower	Upper
Step 1[a]	β_1	.048	.001	7088.54	1	.000	1.049	1.048	1.050
	β_0	−.900	.017	2698.81	1	.000	.407		

Variables in the Equation

[a]Variable(s) entered on step 1:α.

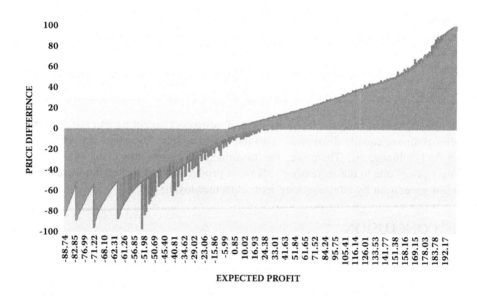

FIGURE 6.5   The Expected Profit for Each Product Submission in Trade-in Program.

a discrete event simulation model was constructed to mirror the concept of digital twin technology in an IoT-enabled blockchain platform. Such smart infrastructure helps companies eradicate the unpredictability in the disassembly yield and boost autonomy along the CLSC. This also enables manufacturers to compute the cost of product recovery on a real-time basis. In this regard, the simulation model was tested under varying experimental conditions using the Taguchi's Orthogonal Arrays design as a design-of-experiments study. By determining the total cost of the system, a dynamic trade-in pricing policy model was implemented and analyzed through logistic regression to approximate the specific product holder's decision against the offered trade-in-to-upgrade discount.

REFERENCES

Agrawal, V. V., Ferguson, M., & Souza, G. C. 2016. "Trade-in rebates for price discrimination and product recovery." *IEEE Transactions on Engineering Management*, 63(3), 326–339. doi:10.1109/TEM.2016.2574244.

Aizawa, H., Yoshida, H., & Sakai, S.-i. 2008. "Current results and future perspectives for Japanese recycling of home electrical appliances." *Resources, Conservation and Recycling*, 52(12), 1399–1410. doi:10.1016/j.resconrec.2008.07.013.

Alqahtani, A. Y., & Gupta, S. M. 2017a. "Evaluating two-dimensional warranty policies for remanufactured products." *Journal of Remanufacturing*, 7(1), 19–47. doi:10.1007/s13243-017-0032-8.

Alqahtani, A. Y., & Gupta, S. M. 2017b. "Warranty as a marketing strategy for remanufactured products." *Journal of Cleaner Production*, 161, 1294–1307. doi:10.1016/j.jclepro.2017.06.193.

Alqahtani, A. Y., & Gupta, S. M. 2018. *Warranty and Preventive Maintenance for Remanufactured Products: Modeling and Analysis*. Taylor & Francis.

Alqahtani, A. Y., Kongar, E., Pochampally, K. K., & Gupta, S. M. 2019. *Responsible Manufacturing: Issues Pertaining to Sustainability*. Taylor & Francis.

Alshibli, M., El Sayed, A., Kongar, E., Sobh, T. M., & Gupta, S. M. 2016. "Disassembly sequencing using tabu search." *Journal of Intelligent & Robotic Systems*, 82(1), 69–79.

Alshibli, M., El Sayed, A., Tozanli, O., Kongar, E., Sobh, T. M., & Gupta, S. M. 2018. "A decision maker-centered end-of-life product recovery system for robot task sequencing." *Journal of Intelligent & Robotic Systems*, 91(3), 603–616. doi:10.1007/s10846-017-0749-5.

Bartodziej, C. J. 2016. *The Concept Industry 4.0: An Empirical Analysis of Technologies and Applications in Production Logistics*. Springer.

Battaïa, O., Dolgui, A., Heragu, S. S., Meerkov, S. M., & Tiwari, M. K. 2018. "Design for manufacturing and assembly/disassembly: joint design of products and production systems." *International Journal of Production Research*, 56(24), 7181–7189. doi:10.1080/00207543.2018.1549795.

Bottani, E., Cammardella, A., Murino, T., & Vespoli, S. 2017. "From the cyber-physical system to the digital twin: the process development for behaviour modelling of a Cyber Guided Vehicle in M2M logic." In: Proceedings of the XXII Summer School "Francesco Turco" Industrial Systems Engineering, Palermo, Italy, September13–15, pp. 96–102, ISBN:978-88-908649-0-2.

Brettel, M., Friederichsen, N., Keller, M., & Rosenberg, M. 2014. "How virtualization, decentralization and network building change the manufacturing landscape: an industry 4.0 perspective." *International Journal of Mechanical, Industrial Science and Engineering*, 8(1), 37–44.

Bumblauskas, D., Gemmill, D., Igou, A., & Anzengruber, J. 2017. "Smart maintenance decision support systems (Smdss) based on corporate big data analytics." *Expert Systems with Applications*, 90, 303–317. doi:10.1016/j.eswa.2017.08.025.

Cao, K., Wang, J., Dou, G., & Zhang, Q. 2018. "Optimal trade-in strategy of retailers with online and offline sales channels." *Computers & Industrial Engineering*, 123, 148–156. doi:10.1016/j.cie.2018.05.028.

Cao, K., Xu, X., Bian, Y., & Sun, Y. 2019. "Optimal trade-in strategy of business-to-consumer platform with dual-format retailing model." *Omega*, 82, 181–192. doi:10.1016/j.omega.2018.01.004.

Chen, J.-M., & Hsu, Y.-T. 2015. "Trade-in strategy for a durable goods firm with recovery cost." *Journal of Industrial and Production Engineering*, 32(6), 396–407. doi:10.1080/21681015.2015.1071288.

Christidis, K., & Devetsikiotis, M. 2016. "Blockchains and smart contracts for the Internet of Things." *IEEE Access, 4,* 2292–2303. doi:10.1109/ACCESS.2016.2566339.

Cole, D., Mahapatra, S., & Webster, S. 2017. "A comparison of buyback and trade-in policies to acquire used products for remanufacturing." *Journal of Business Logistics, 38*(3), 217–232. doi:10.1111/jbl.12159.

Dulman, M. T., & Gupta, S. M. 2018a. "Evaluation of maintenance and EOL operation performance of sensor-embedded laptops." *Logistics, 2*(1), 3. doi:https://doi.org/10. 3390/logistics2010003.

Dulman, M. T., & Gupta, S. M. 2018b. "Maintenance and remanufacturing strategy: using sensors to predict the status of wind turbines." *Journal of Remanufacturing, 8*(3), 131–152. doi:10.1007/s13243-018-0050-1.

Duman, G. M., Kongar, E., & Gupta, S. M. 2019. "Estimation of electronic waste using optimized multivariate grey models." *Waste Management, 95,* 241–249. doi:10.1016/j. wasman.2019.06.023.

Ehm, F. 2018. "A data-driven modeling approach for integrated disassembly planning and scheduling." *Journal of Remanufacturing.* doi:10.1007/s13243-018-0058-6.

ElSayed, A., Kongar, E., & Gupta, S. M. 2012. "An evolutionary algorithm for selective disassembly of end-of-life products." *International Journal of Swarm Intelligence and Evolutionary Computation, 1,* 1–7. doi:10.4303/ijsiec/Z110601.

Fiksel, J. 2009. *Design for Environment: A Guide to Sustainable Product Development.* McGraw Hill Professional.

Fudenberg, D., & Tirole, J. 1998. "Upgrades, tradeins, and buybacks." *The RAND Journal of Economics, 29*(2), 235–258. doi:10.2307/2555887.

Guide, J. V. D. R., Teunter, R. H., & van Wassenhove, L. N. 2003. "Matching demand and supply to maximize profits from remanufacturing." *Manufacturing & Service Operations Management, 5*(4), 303–316. doi:10.1287/msom.5.4.303.24883.

Gungor, A., & Gupta, S. M. 1998. "Disassembly sequence planning for products with defective parts in product recovery." *Computers & Industrial Engineering, 35*(1), 161–164. doi:10.1016/S0360-8352(98)00047-3.

Gupta, S. M., & Taleb, K. N. 1994. "Scheduling disassembly." *International Journal of Production Research, 32*(8), 1857–1866. doi:10.1080/00207549408957046.

Gupta, S. M., Imtanavanich, P., & Nakashima, K. 2010. "Using neural networks to solve a disassembly-to-order problem." *International Journal of Biomedical Soft Computing and Human Sciences: The Official Journal of the Biomedical Fuzzy Systems Association, 15*(1), 65–69.

Hahler, S., & Fleischmann, M. 2017. "Strategic grading in the product acquisition process of a reverse supply chain." *Production and Operations Management, 26*(8), 1498–1511. doi:10.1111/poms.12699.

Han, X., Yang, Q., Shang, J., & Pu, X. 2017. "Optimal strategies for trade-old-for-remanufactured programs: Receptivity, durability, and subsidy." *International Journal of Production Economics, 193,* 602–616. doi:10.1016/j.ijpe.2017.07.025.

Heese, H. S., Cattani, K., Ferrer, G., Gilland, W., & Roth, A. V. 2005. "Competitive advantage through take-back of used products." *European Journal of Operational Research, 164*(1), 143–157. doi:10.1016/j.ejor.2003.11.008.

Hehenberger, P., Vogel-Heuser, B., Bradley, D., Eynard, B., Tomiyama, T., & Achiche, S. 2016. "Design, modelling, simulation and integration of cyber physical systems: methods and applications." *Computers in Industry, 82,* 273–289. doi:10.1016/j. compind.2016.05.006.

Hofmann, E., & Rüsch, M. 2017. "Industry 4.0 and the current status as well as future prospects on logistics." *Computers in Industry, 89,* 23–34. doi:10.1016/j.compind. 2017.04.002.

Ilgin, M. A., & Gupta, S. M. 2010. "Environmentally conscious manufacturing and product recovery (ECMPRO): a review of the state of the art." *Journal of Environmental Management, 91*(3), 563–591. doi:10.1016/j.jenvman.2009.09.037.

Johnson, M. R., & McCarthy, I. P. 2014. "Product recovery decisions within the context of extended producer responsibility." *Journal of Engineering and Technology Management, 34*, 9–28. doi:10.1016/j.jengtecman.2013.11.002.

Joshi, A. D., & Gupta, S. M. 2019. "Evaluation of design alternatives of end-of-life products using Internet of Things." *International Journal of Production Economics, 208*, 281–293. doi:10.1016/j.ijpe.2018.12.010.

Kalayci, C. B., Hancilar, A., Gungor, A., & Gupta, S. M. 2015. "Multi-objective fuzzy disassembly line balancing using a hybrid discrete artificial bee colony algorithm." *Journal of Manufacturing Systems, 37*, 672–682. doi:10.1016/j.jmsy.2014.11.015.

Karanjkar, N., Joglekar, A., Mohanty, S., Prabhu, V., Raghunath, D., & Sundaresan, R. 2018, 1-3 Nov. 2018. *Digital Twin for Energy Optimization in an SMT-PCB Assembly Line.* Paper presented at the 2018 IEEE International Conference on Internet of Things and Intelligence System (IOTAIS).

Kim, J., Rao, R. S., Kim, K., & Rao, A. R. 2011. "More or less: a model and empirical evidence on preferences for under- and overpayment in trade-in transactions." *Journal of Marketing Research, 48*(1), 157–171. doi:10.1509/jmkr.48.1.157.

Kongar, E., & Gupta, S. M. 2002. "A multi-criteria decision making approach for disassembly-to-order systems." *Journal of Electronics Manufacturing, 11*(02), 171–183. doi:10.1142/S0960313102000345.

Lambert, A. J. D., & Gupta, S. M. 2002. "Demand-driven disassembly optimization for electronic products package reliability." *Journal of Electronics Manufacturing, 11*(02), 121–135. doi:10.1142/s0960313102000436.

Lambert, A. J. D., & Gupta, S. M. 2005. *Disassembly Modeling for Assembly, Maintenance, Reuse and Recycling.* CRC press, Boca Raton, FL.

Liu, K., & Zhang, Z.-H. 2018. "Capacitated disassembly scheduling under stochastic yield and demand." *European Journal of Operational Research, 269*(1), 244–257. doi:10.1016/j.ejor.2017.08.032.

Madni, A. M., Madni, C. C., & Lucero, S. D. 2019. "Leveraging digital twin technology in model-based systems engineering." *Systems, 7*(1), 7.

Mandolla, C., Petruzzelli, A. M., Percoco, G., & Urbinati, A. 2019. "Building a digital twin for additive manufacturing through the exploitation of blockchain: a case analysis of the aircraft industry." *Computers in Industry, 109*, 134–152. doi:10.1016/j.compind.2019.04.011.

McGovern, S. M., & Gupta, S. M. 2007. "A balancing method and genetic algorithm for disassembly line balancing." *European Journal of Operational Research, 179*(3), 692–708. doi:10.1016/j.ejor.2005.03.055.

Negri, E., Fumagalli, L., & Macchi, M. 2017. "A review of the roles of digital twin in CPS-based production systems." *Procedia Manufacturing, 11*, 939–948. doi:10.1016/j.promfg.2017.07.198.

Ondemir, O., & Gupta, S. M. 2014a. "A multi-criteria decision making model for advanced repair-to-order and disassembly-to-order system." *European Journal of Operational Research, 233*(2), 408–419. doi:10.1016/j.ejor.2013.09.003.

Ondemir, O., & Gupta, S. M. 2014b. "Quality management in product recovery using the Internet of Things: an optimization approach." *Computers in Industry, 65*(3), 491–504. doi:10.1016/j.compind.2013.11.006.

Oraiopoulos, N., Ferguson, M. E., & Toktay, L. B. 2012. "Relicensing as a secondary market strategy." *Management Science, 58*(5), 1022–1037. doi:10.1287/mnsc.1110.1456.

Özceylan, E., Kalayci, C. B., Güngör, A., & Gupta, S. M. 2019. "Disassembly line balancing problem: a review of the state of the art and future directions." *International Journal of Production Research, 57*(15-16), 4805–4827. doi:10.1080/00207543.2018.1428775.

Phadke, M. S. 1989. *Quality Engineering Using Robust Design*. Prentice Hall, New Jersey.

Ray, S., Boyaci, T., & Aras, N. 2005. "Optimal prices and trade-in rebates for durable, remanufacturable products." *Manufacturing & Service Operations Management*, 7(3), 208–228. doi:10.1287/msom.1050.0080.

Rios, F. C., Chong, W. K., & Grau, D. 2015. Design for disassembly and deconstruction – challenges and opportunities. *Procedia Engineering*, *118*, 1296–1304. doi:10.1016/j.proeng.2015.08.485.

Rodič, B. 2017. "Industry 4.0 and the new simulation modelling paradigm." *Organizacija*, *50*(3), 193–207.

Smetana, S., Seebold, C., & Heinz, V. 2018. "Neural network, blockchain, and modular complex system: the evolution of cyber-physical systems for material flow analysis and life cycle assessment." *Resources, Conservation and Recycling*, *133*, 229–230. doi:10.1016/j.resconrec.2018.02.020.

Soh, S. L., Ong, S. K., & Nee, A. Y. C. 2015. "Application of design for disassembly from remanufacturing perspective." *Procedia CIRP*, *26*, 577–582. doi:10.1016/j.procir.2014.07.028.

Stark, R., Kind, S., & Neumeyer, S. 2017. "Innovations in digital modelling for next generation manufacturing system design." *CIRP Annals*, *66*(1), 169–172. doi:10.1016/j.cirp.2017.04.045.

Subic, A., Xiang, Y., Pai, S., & de La Serve, E. 2017. *Blockchain and Industry 4.0: Why Blockchain is at the heart of the Fourth Industrial Revolution and Digital Economy?* Retrieved from Capgemini: https://www.capgemini.com/au-en/wp-content/uploads/sites/9/2018/10/Blockchain-and-Industry-4.0.pdf

Taguchi, G. 1987. *System of Experimental Design: Engineering Methods to Optimize Quality and Minimize Costs*. Unipub Kraus International, New York, NY, USA.

Teslya, N., & Ryabchikov, I. 2017, 6–10 Nov. 2017. *Blockchain-based Platform Architecture for Industrial IoT*. Paper presented at the 2017 21st Conference of Open Innovations Association (FRUCT).

Tian, X., & Zhang, Z.-H. 2019. "Capacitated disassembly scheduling and pricing of returned products with price-dependent yield." *Omega*, *84*, 160–174. doi:10.1016/j.omega.2018.04.010.

Torrente-Velásquez, J. M., Ripa, M., Chifari, R., Bukkens, S., & Giampietro, M. 2020. "A waste lexicon to negotiate extended producer responsibility in free trade agreements." *Resources, Conservation and Recycling*, *156*, 104711. doi:10.1016/j.resconrec.2020.104711.

Tozanli, O., Duman, G. M., Kongar, E., & Gupta, S. M. 2017. "Environmentally concerned logistics operations in fuzzy environment: a literature survey." *Logistics*, *1*(1), 4. doi:10.3390/logistics1010004.

Tozanlı, Ö., & Kongar, E. 2020. "Integration of industry 4.0 principles into reverse logistics operations for improved value creation: a case study of a mattress recycling company." In A. Erkollar (Ed.), *Enterprise & Business Management: A Handbook for Educators, Consultants, and Practitioners* (pp. 1–17). Enterprise & Business Management: Tectum Wissenschaftsverlag.

Tozanlı, Ö., Kongar, E., & Gupta, S. M. 2020. "Trade-in-to-upgrade as a marketing strategy in disassembly-to-order systems at the edge of blockchain technology." *International Journal of Production Research*, *58*(23), 7183–7200. doi:10.1080/00207543.2020.1712489.

Wang, Y., Singgih, M., Wang, J., & Rit, M. 2019. "Making sense of blockchain technology: How will it transform supply chains?" *International Journal of Production Economics*, *211*, 221–236. doi:10.1016/j.ijpe.2019.02.002.

Wu, C.-H. 2015. "Strategic and operational decisions under sales competition and collection competition for end-of-use products in remanufacturing." *International Journal of Production Economics*, *169*, 11–20. doi:10.1016/j.ijpe.2015.07.020.

Wu, C.-H., & Wu, H.-H. 2016. "Competitive remanufacturing strategy and take-back decision with OEM remanufacturing." *Computers & Industrial Engineering*, *98*, 149–163. doi:10.1016/j.cie.2016.05.033.

Yadav, S., & Singh, S. P. 2020. "Blockchain critical success factors for sustainable supply chain." *Resources, Conservation and Recycling*, *152*, 104505. doi:10.1016/j.resconrec. 2019.104505.

Yin, R., Li, H., & Tang, C. S. 2015. "Optimal pricing of two successive-generation products with trade-in options under uncertainty." *Decision Sciences*, *46*(3), 565–595. doi:10. 1111/deci.12139.

Zhang, A., Zhong, R. Y., Farooque, M., Kang, K., & Venkatesh, V. G. 2020. "Blockchain-based life cycle assessment: An implementation framework and system architecture." *Resources, Conservation and Recycling*, *152*, 104512. doi:10.1016/j.resconrec.2019. 104512.

Zhang, F., & Zhang, R. 2015. "Trade-in remanufacturing, strategic customer behavior, and government subsidies." (October 7, 2017). Available at SSRN: https://ssrn.com/ abstract=2571560 or http://dx.doi.org/10.2139/ssrn.2571560.

Zheng, Z., Xu, W., Zhou, Z., Pham, D. T., Qu, Y., & Zhou, J. 2017. "Dynamic modeling of manufacturing capability for robotic disassembly in remanufacturing." *Procedia Manufacturing*, *10*, 15–25. doi:10.1016/j.promfg.2017.07.005.

Zhou, L., & Gupta, S. M. 2019. "A pricing and acquisition strategy for new and remanufactured high-technology products." *Logistics*, *3*(1), 8. doi:10.3390/logistics3010008.

7 Use of Metaheuristics in Reverse Logistics and De-manufacturing

Seamus M. McGovern[1] and Surendra M. Gupta[2]
[1]U.S. DOT National Transportation Systems Center, 55 Broadway, Cambridge, MA 02142, USA
[2]Department of Mechanical and Industrial Engineering, Northeastern University, 360 Huntington Avenue, Boston, MA 02115, USA

1 INTRODUCTION

1.1 BACKGROUND

Research by Álvarez-Miranda and Pereira (2019) on the complexity of assembly-line balancing (ALB) problems found over 50 publications since 1995 having the error of declaring ALB as being NP-complete or NP-hard and attributed to Gutjahr and Nemhauser (1964) which cannot be correct as Gutjahr and Nemhauser's paper predates Cook (1971) and Karp's (1972) defining work on complexity theory by seven years.

Similarly, the reverse logistics and de-manufacturing literature is rich with applications of the sub-optimal solution techniques of heuristics and metaheuristics. Our research in this area based on the surveys compiled by Ilgin and Gupta (2010), Battaïa and Dolgui (2013), and Ozceylan et al. (2019) has revealed analogous observations as those by Alvarez-Miranda and Pereira, where metaheuristics are selected for application without first confirming that conventional, optimal-solution generating techniques are not applicable, along with other items related to the use and application of metaheuristics such as their modification and benchmark data set selection. In this chapter, we review our findings related to complexity theory and the use of metaheuristics in reverse logistics and de-manufacturing using the disassembly-line balancing problem (DLBP) as a case study.

1.2 DLBP CASE STUDY

Assembly and disassembly have many obvious similarities and, in this chapter, the DLBP is used as the particular reverse logistics and de-manufacturing case study. Just as the assembly line is considered to be the most efficient way to manufacture large numbers of products, the disassembly line has been successfully used in the reverse manufacturing of products at their end of life. A significant amount of research has addressed the challenges associated with sequencing disassembly on a paced line to minimize the number of workstations needed and to ensure that idle times at each workstation are similar.

DLBP seeks to fulfill five objectives:

1. Minimize the number of disassembly workstations, hence, minimize the total idle time.
2. Ensure the sum of the idle times at each workstation is similar.
3. Remove hazardous components early in the disassembly sequence.
4. Remove high-demand components before low-demand components.
5. Minimize the number of direction changes required for disassembly.

A major constraint is the requirement to provide a feasible disassembly sequence for the product being investigated. The result is an integer, deterministic, n-dimensional (where n represents the number of parts for removal), multiple-criteria decision-making problem with an exponentially growing search space. Solutions consist of an ordered sequence (i.e. n-tuple) of elements; if a DLBP solution consisted of the eight-tuple $\langle 5, 2, 8, 1, 4, 7, 6, 3 \rangle$, then component 5 would be removed first, followed by component 2, then component 8, and so on.

While different authors use a variety of definitions for the term "balanced" in reference to assembly (Elsayed and Boucher, 1994) and disassembly lines, we apply the following definition (McGovern and Gupta, 2011) that considers the total number of workstations NWS and the station times STj (i.e. the total processing time required in workstation j); the following definition will be used consistently throughout this chapter:

Definition: A line is optimally balanced when the fewest possible number of workstations are needed and the variation in idle times between all workstations is minimized while observing all constraints. This is mathematically described by

$$\text{Minimize } NWS$$

then

$$\text{Minimize } [\max (ST_x) - \min (ST_y)] \; \forall \; x, \, y \in \{1, \, 2, ..., NWS\}$$

Both DLBP and ALB are similar to the familiar bin-packing problem. In that problem, given a collection of differently sized items (the sizes could be physical dimensions, weight, time, etc.), the objective is to place all of them into equally sized bins, while using the fewest number of bins. The bin-packing problem is a well-known NP-complete problem. DLBP has many similarities to ALB, but several defining differences as detailed by McGovern and Gupta (2019).

1.3 Complexity Theory

Complexity theory provides the basis for understanding the level of computational effort required to obtain a problem solution and, ultimately, for needed alternative solution-generating techniques such as metaheuristics. A great deal of the explanations in this section can be attributed to Garey and Johnson (1979), Papadimitriou and Steiglitz (1998), and Tovey (2002), and is further detailed in relation to DLBP by McGovern and Gupta (2011).

Computational complexity is the measure of how much work—typically measured in time or computer memory—is required to solve a given problem. If the problem is easy, it can probably be solved as a linear program, network model, or some other similar method. If a problem is hard, finding an exact solution can be time-intensive or even impractical, requiring the use of enumerative methods or accepting an approximate solution obtained with heuristics or metaheuristics.

In its most basic form, the goal of DLBP is to minimize and equalize the idle times found at each flow-shop machine (i.e. workstation). Since finding the optimal solution could require investigation of all permutations of the sequence of n part-removal times, there are $n!$ possible solutions. Using the big O notation of complexity theory, the time complexity of searching this space is $O(n!)$. This is known as factorial complexity and typically referred to as exponential time. Though a possible solution to an instance (i.e. a specific set of data) of this problem can be verified (i.e. checked against a given threshold) in polynomial time (i.e. quickly), it cannot always be optimally solved. This problem is classified as nondeterministic polynomial (NP). If the problem could be solved in polynomial time, it would be classified as polynomial (P). P denotes the class of easy problems; NP-hard is a larger class than NP-complete and includes NP-complete. Being NP-hard or NP-complete effectively means that it takes a long time to exactly solve all cases of sufficiently large sizes (Tovey, 2002). The formality of NP-completeness is of special interest since, as put by Garey and Johnson (1979), there is a certain theoretical satisfaction in precisely locating the complexity of a problem via a "completeness" result that cannot be obtained with a "hardness" result.

The theory of NP-completeness is applied only to decision problems. Adding a threshold value to an optimization problem converts it to a "yes-no" (i.e. decision) form that is the standard format used in complexity analysis and NP-completeness proofs. The yes-no version of a problem is analyzed to classify its optimization version. Therefore, hard problems should stay hard, and easy problems should stay easy when converted to the yes-no form. That is, the yes-no version should be hard if and only if the optimization version is hard. If the problem has been correctly converted to the yes-no form, it will never be more difficult to solve than the original version. In the class NP, if lucky guesswork is allowed, then the problems can be solved in polynomial time. Most reasonable and practical problems are in NP.

Cook provided the foundations for NP-completeness in 1971 by showing that a polynomial-time reduction from one problem to another ensures any polynomial-time algorithm used on one can be used to solve the other. In addition, Cook emphasized the class of NP-decision problems that can be solved on a nondeterministic computer (a fictitious computer that operates in a manner that is not predictable; that is, a computer that guesses solutions). The final significant contribution of that paper was the proof that every problem in NP can be reduced to one particular problem in NP (the satisfiability problem or SAT) and the suggestion that other problems in NP may share the SAT's property of being the hardest problem in NP. This group of the hardest problems in NP is the class of NP-complete problems.

Though it is suspected that NP-complete problems are intractable (where a problem cannot be solved by any known polynomial-time algorithm), this has never been proven or disproved. Once determined and proven, the answer to the question "is P = NP?" will affect every problem that is NP-complete. If and when answered, it will be known that no NP-complete problem will ever be solved in polynomial time or that all NP-complete problems can be optimally solved.

The process of devising an NP-completeness proof for a decision problem Π consists of the following four steps:

1. Show that $\Pi \in$ NP.
2. Select a known NP-complete problem Π'.
3. Construct a transform g from Π' to Π.
4. Prove that g is a polynomial transformation.

The main idea in an NP-completeness proof is to model a known NP-complete problem as the problem under consideration (not the reverse, which is what might be done to solve a problem).

To summarize the classes informally, a problem is in the class NP if a guessed answer can be checked against the instance's threshold(s) in polynomial time. A problem is in the class NP-hard if a known NP-complete problem can be transformed to it in polynomial time. A problem is in the class NP-complete if both these requirements are met.

McGovern and Gupta (2011) have proven that DLBP is NP-complete, NP-hard, and unary NP-complete.

1.4 COMPLEXITY THEORY AND JUSTIFICATION OF METAHEURISTICS

There are normally considered to be two general categories of techniques in addressing NP-complete problems (Garey and Johnson, 1979; note that Papadimitriou and Steiglitz, 1998 further expand on these to allow for six alternatives).

The first category includes approaches that attempt to improve upon exhaustive search as much as possible. If realistic cases that are modeled using integer programming are not very large, commercial math-programming software may be effective. Among the most widely used approaches to reduce the search effort of problems with exponential time complexity are those based on branch-and-bound or implicit enumeration techniques (e.g., Garfinkel and Nemhauser, 1972). These generate partial solutions within a tree-structured search format and utilize powerful bounding methods to recognize partial solutions that cannot possibly be extended to actual solutions, thereby eliminating entire branches of the search in a single step.

Other approaches include dynamic programming, cutting-plane methods (see Hu, 1969; Garfinkel and Nemhauser, 1972), and Lagrangian techniques (Geoffrion, 1974; Held and Karp, 1971). Dynamic programming can solve knapsack, partition, and similar problems as long as the instance is not too large. However, other NP-hard problems including 3-partition and bin packing are not susceptible to quick solution by dynamic programming.

The second category allows attainment of suboptimal (though, ideally near-optimal or optimal) solutions in a reasonable amount of time using heuristics and metaheuristics. These are considered to be especially effective when the instances are exceptionally large, the problem is exceedingly difficult in practice (e.g., the job-shop scheduling problem or the quadratic assignment problem), or when solution speed is critical.

This second category has grown dramatically in popularity in many areas, including end-of-life product processing as substantiated in multiple published surveys (Ilgin and Gupta, 2010, Battaïa and Dolgui, 2013, and Ozceylan et al., 2019). This review of published reverse logistics and de-manufacturing research has revealed findings like those of Álvarez-Miranda and Pereira (2019). Specifically, some research appears to use metaheuristics when other operations-research techniques would be better suited when applied to the problem and would result in an optimal solution—which is not assured with metaheuristics. In some cases, it can be seen that the problem is not shown to be NP-complete or NP-hard (which would readily justify the application of metaheuristics) while in others, additional restrictions placed on the problem end up rendering it less complex in terms of computational complexity. The term complexity theory is often construed to have a meaning similar to the conventional understanding of complexity, where adding specificity or restrictions to a problem makes it more complex. In fact, in terms of complexity theory, the opposite is typically the case. While adding restrictions makes a problem more complex in the general sense that an additional constraint is something else to consider, it does not necessarily make it more complex in the computational sense if it does not increase the size of the search space. For example, adding precedence constraints to the bin-packing problem not only makes it similar in structure to an assembly- or disassembly-sequencing problem, it also would seem to add complexity (e.g., something else to consider in any potential solution). However, adding precedence constraints cannot increase the computational complexity because it does not increase the number of solutions to consider; in fact, it can be expected to reduce them (McGovern and Gupta, 2019). There cannot be any more than $n!$ orderings (with this being the case of there being no precedence constraints) and adding any precedence constraints reduces this, the extreme being a solution that has precedence constraints entirely in series, giving only one possible solution (e.g., only one feasible solution such as $\langle 1, 2, 3, 4, 5 \rangle$ where 2's predecessor is 1, 3's predecessor is 2, etc.) even though there are n parts having a maximum search space of $n!$ (here, 5! or 120).

George Dantzig, the father of the simplex method used in solving and analyzing linear programming models, is claimed to have indicated in 1979 (Tovey, 2002) that "complexity theory is usually an excuse for laziness. Give me a problem: I'll solve it." In some sense, this seems prophetic when problems that can be optimally solved are addressed by researchers using typically sub-optimal metaheuristics. Just as complexity theory can justify the use of metaheuristics (since solution alternatives could be limited), if a problem is not shown to be NP-complete or at least NP-hard, solving it using metaheuristics may be simply an exercise in using a metaheuristic, as there may exist some other, conventional technique from operations research that will work (e.g., linear programming, dynamic programming, etc.) while having the added benefit of providing an optimal solution.

2 METAHEURISTIC DESIGN AND APPLICATION

2.1 Overview

The review of reverse logistics and de-manufacturing research (Ilgin and Gupta, 2010, Battaïa and Dolgui, 2013, and Ozceylan et al., 2019) revealed not just a proliferation of metaheuristics in solving these types of problems, but also a significant percentage that exhibited modifications from a metaheuristic's original design. These modifications are sometimes necessary to accommodate the problem under study or due to improvements found through rigorous testing, but it often appeared that neither of these were the case. Published modifications were typically subtle and appeared to be made to improve metaheuristic performance. However, upon a more detailed study, these changes could often be seen doing just the opposite, with changes made that excessively reinforced good solutions to the detriment of exploring potentially better solutions that do not look like the best found to that point in the search. In this section, modifications are reviewed using the DLBP as an example.

2.2 Core Metaheuristic Concepts

Understanding an individual metaheuristic's function and theory can ensure modifications are made correctly and only when necessary, enhancing rather than restricting performance.

A heuristic, or "best guess," is a single, normally deterministic process—typically tailored to one problem—to find a near-optimal solution. An example would be first-fit decreasing for the bin-packing problem where items to be put into bins are sorted by size; the largest is put into the first bin, then the second-largest is put into the first bin it fits into (e.g. the second bin if there is no room for it in the first), and so on, until all items are assigned to bins (see McGovern and Gupta, 2011 for more details). While not ensuring the minimum number of bins, the process should typically come quite close, is intuitive to understand, and is extremely fast.

Metaheuristics are an extension of the processes presented by heuristics. Typically, a metaheuristic will consist of multiple, duplicate heuristic processes, each having its own stochastic component. This adding of uncertainty allows each of these multiple processes to explore lower-performing areas of the search space with the idea that it could ultimately lead to a better solution. During the entire process (which is normally repeated until a time limit is reached or until the best solution remains unchanged for some predetermined length of time), the best solution, or pool of best solutions, found at that point is stored so there is no risk of finding worse solutions. In addition, some metaheuristics (e.g., genetic algorithm) use these best solutions as a starting point for subsequent searches.

It is also important to recognize that many metaheuristics at their core are simply multiple greedy algorithms with randomness added, so it is helpful to understand the benefits and limitations of those algorithms. Ant colony optimization (ACO) is a good example of this: in a simplified sense, ACO makes sequential decisions as it adds to each individual solution, with each decision being greedy (i.e. the best decision that can be made at that time) but having a stochastic element (i.e. it does not always pick the best). The genetic algorithm (GA) is an example of a metaheuristic that does not use a greedy structure: GAs randomly pick two solutions from the pool of best solutions, breaks those into two parts, and recombine them while occasionally making minor modification—known as a mutation—to a solution. Also, since many metaheuristics are variations of this multiple stochastic greedy algorithm structure, it is not practical to expect significant differences in solution performance between them either in runtime speed or solution quality. This idea is captured by the no-free-lunch theorem, where the computational cost of finding a solution, when averaged over all similar problems, is the same for any solution technique; therefore, none offers a consistent, significant advantage.

Finally, it is also important to understand that naming conventions are not necessarily a precise description of an algorithm's process. Terms like genetic algorithm, simulated annealing (annealing is a heating and cooling sequence intended to reduce a metal's hardness), and artificial bee colony algorithm allude to how the search process mimics some system in nature or industry, but a deeper understanding of the underlying mathematics and assumptions are necessary to properly apply, and certainly to correctly modify, any metaheuristic. Hillier and Lieberman (2015) provided an overall introduction to the types, characteristics, and design considerations of metaheuristics.

2.3   METAHEURISTIC MODIFICATIONS TO ADDRESS PROBLEM STRUCTURE

Occasionally, metaheuristics require minor modifications to accommodate the problem structure and how its solution would be appropriately searched for. This type of modification requires understanding—not just in the metaheuristic's search process, but also in the specific problem as well. This is especially important since the unmodified algorithm will find an answer due to the random nature of metaheuristic searches, but it might be able to find a better or find an answer more quickly if adjustments are made. An example of this using the DLBP case study would be the application of ACO where several slight modifications were in order (McGovern and Gupta, 2011), three of which are highlighted here.

Due to the nature of the DLBP (i.e. whether or not a task should be added to a workstation depends on that workstation's current idle time and precedence constraints), the ACO probability calculation—that determines the probability of an individual ant selecting one path given several options—was changed from being calculated once at the beginning of each tour (typical with ACO) to being calculated dynamically, generating new probabilities at each time increment during a search cycle.

In addition, the ACO algorithm step "Place the m ants on the n nodes" was moved from step 1 to 2 in the process (and "nodes" was changed to "vertex" to recognize the DLBP's formulation as a digraph). This was done so that the resetting of each ant to each vertex would be repeated in each cycle, prohibiting the potential accumulation of ants on a single ending vertex resulting in the subsequent restarting of all of the ants from that one vertex in successive cycles—resetting the ants would ensure solution diversity. This was also necessary due to the nature of the DLBP where the sequence is a critical and unique element of any solution. A part with numerous precedence constraints will typically be unable to be positioned in the front of the sequence (i.e. removed early) and will often be found toward the end. In DLBP, not resetting each ant could potentially result in a situation where all ants choose the same final part due to precedence constraints—when each ant attempts to initiate a subsequent search (i.e. the next cycle) from that last vertex, all fail due to the infeasibility of attempting to remove a final part first, effectively terminating the search in just one cycle.

Another problem-structure-related item had to do with the representation of edges in the search algorithm network and their effect on the process. It can be inferred from the original ACO paper (Dorigo et al., 1996) that edge p, q is equivalent to edge q, p. Although this is acceptable and desirable in a traveling salesperson-type problem, in DLBP, sequence is an essential element of an n-tuple and DLBP solution (due to precedence limitations and size constraints at each workstation). For this reason, edges p, q and q, p in DLBP ACO are directed (i.e., arcs) and therefore distinct and unique and as such, when the equivalent of a notional ant's pheromone is added to one, it is not added to the other as would normally be done in ACO.

Modifications can be dictated by the limitations posed by the problem a meta-heuristic is being applied.

2.4 METAHEURISTIC MODIFICATIONS DUE UNDERSTANDING OF THE METAHEURISTIC'S PROCESS

Metaheuristics may also require minor modifications to accommodate the structure of the metaheuristic's search process, the selection of coefficients, or other constants. For example, the genetic algorithm used for DLBP was in several ways modified from a general GA (McGovern and Gupta, 2011). Instead of the worst portion of the population being selected for crossover, in DLBP GA, all the population was randomly considered. This better enabled the selection of nearby solutions (i.e. solutions similar to the best ones to date) common in many scheduling problems. Also, mutation was performed only on children, not the worst parents. This was done to address the small population chosen in DLBP GA and to counter the crossover algorithm's (in this case, the crossover algorithm known as PPX) tendency to duplicate parents. Finally, duplicate children were sorted to make their deletion from the population likely since there is a tendency to create duplicate solutions (due to PPX) and due to the small size of the population that was saved from generation to generation.

Items for modification could be discovered based on an understanding of the metaheuristic and its limitations or, alternatively, by observing a small solution population while the metaheuristic is running as part of its development prior to a final application to the problem.

2.5 METAHEURISTIC COEFFICIENT SELECTION

A metaheuristic's coefficients and other constants typically are directly related to the metaheuristic's willingness to explore solutions that are not as good as current solutions in the hope of finding improvement. The review of papers that used metaheuristics for reverse logistics and de-manufacturing problems revealed several authors that changed the metaheuristic's coefficients, usually adjusting probability related items. In these cases, they were typically increased to exhibit a search preference towards any better solutions found. Coefficients that deviate from those designed and tested by the algorithm developer run the risk of reinforcing good answers too much—effectively eliminating the algorithm's ability to find better answers by going through a series of worse answers—and defeating the purpose of using the stochastic search capabilities of a metaheuristic.

However, modifications of coefficients can be of value in research if they are adjusted because of testing rather than artificial increases that overly reinforce good answers. For example, DLBP GA saw several coefficient or variable changes that provide examples of post-solution problem analysis- and testing-based performance improvements. As part of this type of development, (McGovern and Gupta, 2011) a small population was used (20 versus the more typical 10,000 to 100,000) in DLBP GA to minimize data storage requirements and simplify analysis while a large number of generations waswere used (10,000 versus the more typical 10 to 1000) to compensate for this small population while not being so large as to take an excessive amount of processing time. This was also done to avoid solving all cases to optimality since it can be desirable to determine the point at which an algorithm's performance begins to break down and how it manifests itself. Lower than the recommended 90% (Koza, 1992), a 60% crossover was selected based on test and analysis, since developmental testing indicated that a 60% crossover provided better solutions and did so with one-third less processing time. Previous assembly-line balancing literature that indicated best results have typically been found with crossover rates of from 0.5–0.7 also substantiated the selection of this lower crossover rate. A mutation was performed about 1% of the time; although some texts recommend 0.01% mutation, while applications in journal papers have used as much as 100% mutation, 1.0% gave excellent algorithm performance for DLBP.

2.6 METAHEURISTIC STOPPING CRITERIA

Another design consideration for metaheuristics is the stopping criteria. As a metaheuristic is a sub-optimal solution-generating algorithm, unlike optimization techniques such as simplex for linear programming problems, it can neither identify the optimal solution nor can it know that it has achieved it—without some stopping criteria, it would run indefinitely. There are two conventions for stopping criteria (see Hillier and Lieberman, 2015 for a summary): solution convergence and runtime. Convergence describes a situation where a better solution is not found for a pre-selected number of iterations, at which point the metaheuristic terminates; runtime refers to the process by which the metaheuristic terminates after a given amount of time or, similarly, a given number of searches or iterations.

The review of research in metaheuristics applications to end-of-life product processing problems overwhelmingly found convergence to be the preferred stopping criteria. However, this preference can potentially restrict the performance of a given metaheuristic. As metaheuristic is intended to be used in the situation where no optimal-solution generating technique would be practical due to the exorbitant amount of time even a relatively small NP-hard problem could be expected to take when solved by exhaustive search (at the expense of typically generating a sub- or near-optimal solution), the selection of a metaheuristic is for speed at the possible expense of solution quality. Since runtime is the overriding consideration, determining an acceptable search time to allow the algorithm to run is a more logical alternative than convergence. In addition, it is not unusual for the best solution found at each iteration to change rapidly early in the search then more and more slowly as the search continues. Due to the stochastic nature of metaheuristics, allowing additional time—regardless of how long it has been since a better solution—could enable improved results, assuming the solution found to that point is not, in fact, the optimal solution. A purported NP-hard problem that converges quickly may simply be an indication that the problem itself is not that difficult—possibly due to restrictions that make a problem more complex in terms of solution requirements but not in terms of computation complexity since the feasible search space is reduced. It may also be indicative of a data set that is overly simplistic. In either case, the result is to artificially reward metaheuristic structures or coefficients that restrict the exploration of diverse search spaces. A difficult problem—and an associated, appropriately difficult data set—will see a metaheuristic's solution performance drop rapidly with instance size. All metaheuristics and heuristics tested on DLBP using a challenging data set were unable to provide the optimal solution for the number of workstations or the measure of balance no later than an instance size of just $n = 16$ when averaged (due to the stochastic nature of metaheuristic-generated solutions) over multiple runs (McGovern and Gupta, 2011).

Using the DLBP case study, the maximum number of cycles was set at 300 for DLBP ACO (i.e. $NC_{max} = 300$), since larger problems than those demonstrated in the DLBP research had been shown to reach their best solution by that count (Dorigo et al., 1996) and as a result of additional experimentation using DLBP (McGovern and Gupta, 2011). The process was not run until no new improvements were seen (i.e. convergence or, using the ACO terminology, stagnation behavior) but was run continuously on subsequent solutions until NC_{max} was reached. This enabled the probabilistic component of ACO to leave potential local minima. Repeating the DLBP ACO method in this way improved line balance over time.

3 METAHEURISTIC PERFORMANCE EVALUATION

New methodologies must first have their developed software thoroughly tested by undergoing a verification and validation process. Verification consists of providing a wide range of inputs to a module of the software to ensure proper operation of an individual software component, while validation determines whether the program provides a correct output for a given input. In addition, any heuristic solution methodology needs to be applied to a collection of test cases to demonstrate its performance and identify its limitations.

Benchmark data sets are common for many NP-complete problems, such as Oliver30 and RY48P for application to the traveling salesperson problem and Nugent15/20/30, Elshafei19, and Krarup30 as the quadratic assignment problem. Unfortunately, because of their size and their design, most of these existing data sets have no known optimal answer and new solutions are not compared to the optimal solution, but the best-known solution to date.

Another alternative for the evaluation of metaheuristics is the measurement of their worst-case and average-case performance (Tovey, 2002). Average-case analysis often involves simply devising a benchmark (a set of supposedly typical instances), running the heuristic and its competitors on them, and comparing the results. Some experiments that have been done tend to confirm the suspicion that average behavior is generally much better than worst-case behavior (Garey and Johnson, 1979). It should be noted that designing an adequate benchmark is considered to be difficult. However, to compute average-case performance, some probability distribution on the instances must first be assumed and it is often not clear which one to choose (one way to avoid this is to use algorithms that do their own randomizing; Rabin, 1976). The distributions under which heuristics can be analyzed with currently available techniques can be very different from the actual distributions that occur in practice, where instances tend to be highly structured (including biases that can be difficult to capture mathematically) and distributions can change in unpredictable ways as time passes. Moreover, average-case results do not reveal anything about how heuristics will perform for particular instances, whereas worst-case results guarantee at least a bound on this performance. Therefore, it is generally preferred to analyze them in as many ways as possible, including both worst-case and average-case perspectives.

As metaheuristics may only generate a sub-optimal solution, it can be of great value to have a computationally intensive data set where the optimal solution is always known regardless of instance size. This would avoid the need to a probabilistic analysis and allow to both measure a metaheuristic's answer quality and its time to attain that answer. Generating a data set for a problem that has a known optimal solution for any instance size and is not so simplistic that it becomes trivial for the metaheuristic to solve may not always be possible, but an example of this can be seen using the DLBP case study (McGovern and Gupta, 2011).

3.1 FORMULATION

A size-independent a priori benchmark data set was generated based on the following. Since, in general, solutions to larger and larger DLBP instances cannot be verified as optimal (due to the time complexity of exhaustive search), it was proposed that instances be generated in such a way that always provides a known solution. This was done by using part removal times *PRT* consisting exclusively of prime numbers further selected to ensure that no combinations of these are allowed for any equal summations (to reduce the number of possible optimal solutions). For example, part removal times 1, 3, 5, and 7 and a cycle time (the maximum amount of time available to complete all tasks assigned to each workstation) of $CT = 16$ would have minimum idle-time solutions of not only one 1, one 3, one 5, and one 7 at each workstation, but various additional combinations of these as well since $1 + 7 = 3 + 5 = ½ CT$. Subsequently, the chosen instances were made up of parts with removal times of 3, 5, 7, and 11 and $CT = 26$. As a result, the optimal balance for all subsequent instances would result in a perfect balance and consist of combinations of 3, 5, 7, and 11 at each workstation and idle times of zero. In general, the size of this a priori data set is then given as

$$n = x \,|PRT|: \; x \in \mathbf{Z}^+ \tag{7.1}$$

where $\mathbf{Z}^+$ represents the set of positive integers (i.e. $\{1, 2, ...\}$).

To further complicate the data (i.e. provide a large, feasible search space), only one part was listed as hazardous and this was one with the largest part removal time (the last one listed in the original data). In addition, one part (the last listed, second largest part removal time component) was listed as being demanded. This was done so that only the hazardous sequencing and the demand sequencing would be demonstrated while providing a slight solution sequence disadvantage to any purely greedy methodology (since two parts with part removal times of 3 and 5 are needed with the larger part removal time parts to reach the optimal line balance F^*, assigning hazardous and demanded parts to those smaller-part-removal-time parts may allow some methodologies to artificially obtain the initial F^* single workstation sequence). From each part-removal-time size, the first listed part was selected to have a removal direction differing from the other parts with the same part removal time. This was to demonstrate direction selection while requiring any solution-generating methodology to move these first parts of each part-removal-time size encountered to the end of the sequence (i.e. into the last workstation) to obtain the optimal direction value of $R^* = 1$ (i.e. if the solution technique being evaluated can successfully place the hazardous and demanded parts towards the front of the sequence). Also, there were no precedence constraints placed on the sequence, a deletion that further challenges any method's ability to attain an optimal solution.

Known optimal results include optimal balance, hazardous part placement, demanded part placement, and direction changes for part removal as $F^* = 0$, $H^* = 1$, $D^* = 2$, $R^* = 1$. While this chapter considers the use of data with $|PRT| = 4$ unique part removal times, in general for any n parts consisting of this type of data, the following can be calculated

$$NWS^* = \frac{n}{|PRT|} \tag{7.2}$$

$$NWS_{nom} = n \qquad (7.3)$$

$$I^* = 0 \qquad (7.4)$$

$$I_{nom} = \frac{n \cdot CT \cdot (|PRT| - 1)}{|PRT|} \qquad (7.5)$$

$$F^* = 0 \qquad (7.6)$$

where I represents the total idle time, while balance F is calculated using

$$F = \sum_{j=1}^{NWS} (CT - ST_j)^2 \qquad (7.7)$$

and with the worst case balance F_{nom} as

$$F_{nom} = \sum_{k=1}^{n} (CT - PRT_k)^2. \qquad (7.8)$$

3.2   GENERATING THE BENCHMARK

Formulae have been developed to generate all data parameters, as well as calculate optimal and nominal measures, for any size instance (as constrained by Formula (7.1)).

Hazard values h_k and measure H are given by (where k represents the kth of a total of n parts)

$$h_k = \begin{cases} 1, & k = n \\ 0, & otherwise \end{cases} \qquad (7.9)$$

$$H^* = 1 \qquad (7.10)$$

$$H_{nom} = n \qquad (7.11)$$

with demand values d_k and measure D given by

$$d_k = \begin{cases} 1, & k = \frac{n \cdot (|PRT| - 1)}{|PRT|} \\ 0, & otherwise \end{cases} \qquad (7.12)$$

$$D^* = \begin{cases} 2, & H = 1 \\ 1, & otherwise \end{cases} \tag{7.13}$$

$$D_{nom} = \begin{cases} n - 1, & H = n \\ n, & otherwise \end{cases} \tag{7.14}$$

and part removal direction values r_k and measure R given by

$$r_k = \begin{cases} 1, & k = 1, \frac{n}{|PRT|} + 1, \frac{2n}{|PRT|} + 1, \ldots, \frac{(|PRT|-1)\cdot n}{|PRT|} + 1 \\ 0, & otherwise \end{cases} \tag{7.15}$$

$$R^* = 1 \tag{7.16}$$

$$R_{nom} = \begin{cases} 0, & n = |PRT| \\ 2\cdot|PRT| - 1, & n = 2\cdot|PRT| \\ 2\cdot|PRT|, & otherwise \end{cases} \tag{7.17}$$

Since $|PRT| = 4$ in this chapter, each part removal time is generated by

$$PRT_k = \begin{cases} 3, & 0 < k \le \frac{n}{4} \\ 5, & \frac{n}{4} < k \le \frac{n}{2} \\ 7, & \frac{n}{2} < k \le \frac{3n}{4} \\ 11, & \frac{3n}{4} < k \le n \end{cases} \tag{7.18}$$

While the demand values as generated by Formula (7.12) are the preferred representation (due to the small numerical values make it easy to interpret demand efficacy since $D = k$), algorithms that allow incomplete disassembly may terminate after placing the single demanded part in the solution sequence. In this case, Formulae (7.12)–(7.14) may be modified to give

$$d_k = \begin{cases} 2, & k = \frac{n\cdot(|PRT|-1)}{|PRT|} \\ 1, & otherwise \end{cases} \tag{7.19}$$

$$D^* = \begin{cases} 2 + \sum_{p=1}^{n} p, & H = 1 \\ 1 + \sum_{p=1}^{n} p, & otherwise \end{cases} \tag{7.20}$$

$$D_{nom} = \begin{cases} n - 1 + \sum_{p=1}^{n} p, & H = n \\ n + \sum_{p=1}^{n} p, & otherwise \end{cases} \tag{7.21}$$

3.3 ANALYSIS

A data set such as this one, containing parts with equal *PRT*s and no precedence constraints, will unfortunately have more than one optimal solution. To properly gauge the performance of any metaheuristic on the DLBP a priori data, the number of optimal solutions needs to be quantified. From probability theory it is known that, for example, with $n = 12$ and $|PRT| = 4$, the size of the set of optimally balanced solutions $|F^*|$ for the optimal value of three workstations (i.e., $NWS^* = 3$) when using the DLBP a priori data could be calculated as $(12 \cdot 9 \cdot 6 \cdot 3) \cdot (8 \cdot 6 \cdot 4 \cdot 2) \cdot (4 \cdot 3 \cdot 2 \cdot 1) =$ 17,915,904 as seen in Table 7.1 (see McGovern and Gupta, 2011 for all calculations, including formulas for performing these calculations using any size n).

Although the sizes of both DLBP a priori optimal solution sets (i.e. optimal only in balance F, and optimal in F, H, D, and R) are quite large in the $n = 12$ instance Table 7.1 examples, they are also significantly smaller than the search space of $n! = 479,001,600$. As seen in Table 7.1, the number of solutions that are optimal in balance alone goes from 100% of n at $n = 4$ to 22.9% at $n = 8$, and to less than 1% at $n = 16$; as n grows, this percentage gets closer and closer to 0%. The number of solutions optimal in all objectives goes from less than 8.3% of n at $n = 4$, to 0.12% at $n = 8$, dropping to effectively 0% at $n = 16$; again, as n grows, the percentage of optimal solutions gets closer and closer to zero.

Not only does this engineered data set allow knowing what the optimal solution is for any instance size, but it is also a difficult data set for metaheuristics to optimally solve. Previous studies (McGovern and Gupta, 2011) have shown that all six metaheuristics and heuristics evaluated were, on average, not able to provide the optimal solution either for the number of workstations or for the measure of balance, some as early as an instance size of $n = 8$ and none later than $n = 16$. These are extremely small data set instances (note that $n = 8$ would consist of just two workstations optimally, while $n = 16$ would have four) to not be able to consistently find an optimal solution, and this provides a good example of the challenges that can be posed by

---

TABLE 7.1

Comparison of Possible Solutions to Optimal Solutions for a Given n Using the DLBP a priori Data

N	n!	Number Optimal in Balance	Number Optimal in All	Percentage Optimal in Balance	Percentage Optimal in All
4	24	24	2	100.00%	8.33%
8	40,320	9,216	48	22.86%	0.12%
12	479,001,600	17,915,904	10,368	3.74%	0.00%
16	2.09228E+13	1.10075E+11	7,077,888	0.53%	0.00%

using this data set design. This also demonstrates how some larger data sets that seem to be optimally solved in literature may, in fact, have a significantly reduced search space due to restrictions presented by precedence constraints or may simply be getting stuck at a local optima due to metaheuristic modifications that are overly restrictive in allowing the algorithm to seek out diverse solutions.

4   SUMMARY

Research by Álvarez-Miranda and Pereira (2019) found over 50 publications containing the error of declaring assembly-line balancing problems as being NP-complete or NP-hard per attribution to Gutjahr and Nemhauser's 1964 work—which cannot be correct as that paper predates the defining work on complexity theory by seven years. The findings from our research based on the survey efforts by Ilgin and Gupta (2010), Battaïa and Dolgui (2013), and Ozceylan et al. (2019) have yielded similar observations while helping provide guidelines for properly and successfully selecting and using metaheuristics in addressing reverse logistics and de-manufacturing problems. These guidelines, based on a review of the current literature, include

- An NP-complete proof can provide an indication of the need for metaheuristics in the case of a particular problem. Cook provided the foundations for NP-completeness in 1971 by showing that a polynomial-time reduction from one problem to another ensures any polynomial-time algorithm used on one can be used to solve the other. In addition, Cook provided proof that every problem in NP can be reduced to one particular problem in NP (the satisfiability problem or SAT) and the suggestion that other problems in NP may share SAT's property of being the hardest problem in NP. This group of the hardest problems in NP is the class of NP-complete problems. Proof of a problem as being NP-complete, NP-hard, unary NP-complete, etc. suggests consideration of the use of metaheuristics rather than conventional optimal operations-research solution techniques. If the problem cannot be proven to be in one of these complexity groups, an alternative (i.e. conventional) operations-research technique should be investigated for use, as these can ensure an optimal solution where a metaheuristic cannot.
- Adding complexity to a problem (in the conventional sense; i.e. in terms of restrictions, constraints, or other considerations) can make the problem less computationally complex (in the mathematical sense; i.e. in terms of complexity theory). A more complex problem description can actually simplify the solution search space for what was an NP-complete or other computationally complex problem by moving it to the realm of problems solvable by conventional optimization techniques. Adding complexity in the sense of additional restrictions or requirements does not necessarily make a problem more complex from a complexity theory perspective and, in fact, often has the opposite effect, often rendering the problem to one that can be optimally solved using traditional operations-research techniques.

- Some metaheuristics require slight tuning or modifications to properly address a given problem, necessitating a full understanding of both the problem and the theory of the chosen metaheuristic.
- Alternatively, ad hoc modifications of metaheuristics done without any substantiated justification can decrease the search algorithm's performance, pushing many metaheuristics to mimic multiple, greedy search algorithms, thereby negating the benefit of the stochastic search capabilities of the original metaheuristic.
- By their nature (i.e., as often being effectively multiple, greedy search algorithms having a stochastic component to the greedy operation) and as captured in the notion of the no-free-lunch theorem (i.e. the computational cost of finding a solution, when averaged over all similar problems, is the same for any solution technique; therefore, none offers a consistent, significant advantage, on average), there is no consensus on any metaheuristic as being superior to another.
- As metaheuristics do not assure an optimal solution but enable some near-optimal solution in a reasonable amount of time, dealing with a problem that has an exponential search space—any data set that the metaheuristic can solve to what is considered to be the best-known solution or what is known to be the optimal solution—may be indicative of a data set that does not realistically challenge the metaheuristic or, more likely, a problem that is not as computationally complex as originally thought.
- To measure a metaheuristic's performance, benchmark data sets are required. Ideally, these data sets would have a known optimality for all set sizes as well as the ability to not be so simple as to be solvable for even moderately sized instances.

These guidelines, grounded in the insights of Álvarez-Miranda and Pereira (2019) and based on the state-of-the-art research compiled by Ilgin and Gupta (2010), Battaïa and Dolgui (2013), and Ozceylan et al. (2019), should allow efficient and appropriate application of metaheuristics, where justified, in addressing computationally complex reverse logistics and de-manufacturing problems.

REFERENCES

Álvarez-Miranda, E., and Pereira, J. (2019). "On the complexity of assembly line balancing problems." *Computers & Operations Research*, 108, 182–186.

Battaïa, O. and Dolgui. A. (2013). "A taxonomy of line balancing problems and their solution approaches." *International Journal of Production Economics*, 142, 259–277.

Cook, S. A. (1971). "The complexity of theorem-proving procedures." *Proceedings 3rd Annual Association for Computing Machinery Symposium on Theory of Computing*, New York, N.Y., 151–158.

Dorigo, M., Maniezzo, V., and Colorni, A. (1996). "The ant system: optimization by a colony of cooperating agents." *IEEE Transactions on Systems, Man, and Cybernetics–Part B*, 26(1), 1–13.

Elsayed, E. A. and Boucher, T. O. (1994). *Analysis and Control of Production Systems*, Prentice Hall, Upper Saddle River, N.J.

Garey, M. and Johnson, D. (1979). *Computers and Intractability: A Guide to the Theory of NP Completeness*, W. H. Freeman and Company, San Francisco, Calif.

Garfinkel, R. S. and Nemhauser, G. L. (1972). *Integer Programming*, John Wiley & Sons, New York, N.Y.

Geoffrion, A. M. (1974). "Lagrangian relaxation and its uses in integer programming." *Mathematical Programming Study*, 2, 82–114.

Gutjahr, A. L. and Nemhauser, G. L. (1964). "An algorithm for the line balancing problem." *Management Science*, 11(2), 308–315.

Held, M. and Karp, R. M. (1971). "The traveling salesman problem and minimum spanning trees: Part II." *Mathematical Programming*, 6, 62–88.

Hillier, F. S. and Lieberman, G. J. (2015). *Introduction to Operations Research*, 10th ed., McGraw-Hill, New York, N.Y.

Hu, T. C. (1969). *Integer Programming and Network Flows*, Addison-Wesley, Reading, Mass.

Ilgin, M. A. and Gupta, S. M. (2010). "Environmentally conscious manufacturing and product recovery (ECMPRO): A review of the state of the art." *Journal of Environmental Management*, 91(3), 563–591.

Karp, R. M. (1972). "Reducibility among combinatorial problems." *Complexity of Computer Computations*, R. E. Miller and J. W. Thatcher, eds., Plenum Press, New York, N.Y., 85–103.

Koza, J. R. (1992). *Genetic Programming: On the Programming of Computers by the Means of Natural Selection*, The MIT Press, Cambridge, Mass.

McGovern, S. M. and Gupta, S. M. (2011). *The Disassembly Line: Balancing and Modeling*. McGraw-Hill, New York, N.Y.

McGovern, S. M. and Gupta, S. M. (2019). "Categorical and mathematical comparisons of assembly and disassembly lines." *Responsible Manufacturing: Issues Pertaining to Sustainability*, A. Y. Alqahtani, E. Kongar, K. Pochampally, and S. M. Gupta, eds., CRC Press, Boca Raton, Florida, 119–152.

Ozceylan, E., Kalayci, C. B., Gungor, A. and Gupta, S. M. (2019). "Disassembly line balancing problem: a review of the state of the art and future directions." *International Journal of Production Research*, 57(15-16), 4805–4827.

Papadimitriou, C. H. and Steiglitz, K. (1998). *Combinatorial Optimization: Algorithms and Complexity*, Dover Publications, Mineola, N.Y.

Rabin, M. O. (1976). "Probabilistic algorithms." *Algorithms and Complexity: New Directions and Recent Results*, J. F. Traub, ed., Academic Press, New York, N.Y., 21–39.

Tovey, C. A. (2002). "Tutorial on computational complexity." *Interfaces*, 32(3), 30–61.

8 Reverse Logistics Network Design for Recycling of Packaging Waste
A Case Study

Neslihan Demirel[1], Eray Demirel[2], and H. Gökçen[3]
[1]Faculty of Applied Sciences, Department of International Trade and Logistics Kayseri University, Kayseri, Turkey
[2]12th Regional Directorate of State Hydraulic Works, Kayseri, Turkey
[3]Faculty of Engineering, Department of Industrial Engineering Gazi University, Ankara, Turkey

1 INTRODUCTION

Due to the rising population and changing consumption habits, solid waste from consumption is also increasing rapidly. Because of decreasing solid waste landfills in urban areas and increasing health and environmental concerns, the management of solid waste has become important. To overcome challenging issues in waste management, the policy to be implemented should be as in Figure 8.1. As a priority, waste generation should be prevented; when unavoidable, it should be reduced. The waste generated should be reused and when not possible, should be recycled. If recycling is not possible (from a technical or economic point of view), it should be recovered for energy production by incineration. If none of these options can be realized, waste should be disposed of by applying certain procedures that do not harm the environment and human health in landfills.

In Turkey, the collection of solid waste and disposal operations are implemented according to the Regulation on Waste Management and other related regulations—Regulation on Control of Medical Waste, Hazardous Waste, Packaging Waste, Waste Batteries and Accumulators, and Construction and Demolition Waste. Municipalities are held responsible for the collection, transportation, storage, recycling, and disposal of waste. According to statistics of the Turkish Statistical Institute, more than 32.21 million tons of municipal solid waste (MSW) was collected

147

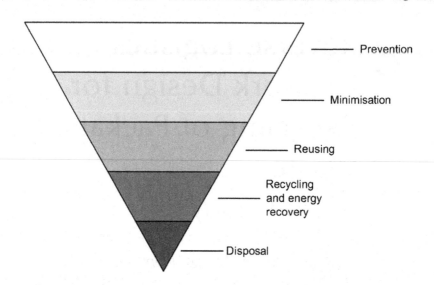

FIGURE 8.1   Waste Management Hierarchy (Source: EU Directive 2008/98/EC).

in the year 2018 in Turkey (TurkStat, 2018). Although most of the waste is recycled and transferred to an economic asset in developed countries, in 2018, only a small portion of the waste stream was recovered or composted in Turkey. About 20.2% of the waste was disposed into municipality dumping sites, 67.2% in controlled landfill sites, 0.38% of waste was composted, 0.019% burned, 0.002% discharged into rivers, 0.006% was buried, 0.2% was disposed of using other methods, and 11.9% was delivered to recovery facilities (TurkStat, 2018).

Packaging waste is recyclable and has an economic value—it constitutes an important percentage of weight and volume in solid waste. Decisions about the design of packaging and selection of materials should be addressed first as they affect the amount of packaging waste generated. After that, separate collection of packaging waste and recycling are important processes, as they ensure the reduction of the amount of waste sent to the landfills, besides the economic benefits. Thus, the separate collection of recyclable packaging waste will not only eliminate environmental problems related to waste management, but also provide raw materials for the industry and create potential employment in the recycling sector by increased investments in this field (Exposito and Velasco, 2018). Until the year 2018 in Turkey, packaging waste was handled according to the Regulation on Control of Packaging Waste published in the Official Gazette dated 24.08.2011. This regulation was abrogated by the Regulation on the Control of the Packaging Waste published in the Official Gazette dated 27.12.2017; the packaging waste management was amended by a new regulation on the 1st of January 2018 considering the European Parliament and Council Directive 94/62/EC of 20 December 1994 on packaging and packaging waste. Finally, a few changes were made on the targets with Regulation on Amendment of the Packaging Waste Control Regulation published in the Official Gazette dated 13.03.2020. The main purpose of the regulation is the production of environmental-friendly packaging, prevention of packaging waste generation, and reduction of packaging waste to be disposed of through reuse, recycle, and recovery methods. The regulation also sets the standards for separate collection at the source, transportation, separation, and recycling of packaging waste. The regulation covers all the packages put on the market and their waste. Although municipalities are responsible for the collection, transportation, storage, recycling, and disposal of packaging waste, those who produce them and put it on the market must cover the costs. The authorized institution and/or marketers are obliged to achieve the recycling targets for packaging waste given in Table 8.1. On the other hand, the responsibility of the municipalities in Turkey has increased with the published Regulation in the Official Gazette dated 12.07.2019 on zero waste. The zero-waste management system aims to protect the environment, human health, and all resources in waste management processes with the effective management of raw materials and natural resources. The purpose of the Zero-Waste Regulation is to determine the establishment, dissemination, development, monitoring, financing, record, and certification of the zero-waste management system. According to this regulation, metropolitan municipalities in Turkey with a population of 250,000 above must establish a zero-waste management system before 31 December 2020. The regulation holds the municipalities responsible for the separate collection of non-hazardous recyclable paper, glass, metal, and plastic wastes from the houses; and the placing of a sufficient number and capacity of waste bins in easily accessible places on the streets (Regulation on the Zero Waste Management, 2019).

To cope with the problems created by the increased generation of packaging waste, the European Union (EU) began works much earlier and adopted the Packaging and Packaging Waste Directive (1994/62/EC) on 20 December 1994. The first recycling target rates commanded in EU Directive 94/62/EC were raised in the recent Directive EU 2018/852 amending Directive 94/62/EC on packaging and packaging waste (Directive EU 2018/852). Table 8.2 shows the obligatory targets for the recycling of packaging waste in Turkey and member states of the EU. Targets are also set for each material type of recyclable waste,

TABLE 8.1

Recycling Targets of Packaging Waste by Years (Data from Regulation on Amendment of the Packaging Waste Control Regulation, 2020)

Date	Glass (%)	Plastics (%)	Metal (%)	Paper/Cardboard (%)	Wood (%)
By 2026	70	55	60	75	25
By 2031	75	55	70	85	30
2031 and later	75	55	70	85	30

TABLE 8.2

Recycling Targets for Packaging Waste in Turkey and EU

Turkey		EU	
Date	Total Recycling Target	Date	Total Recycling Target
2018	54%	2008	55%
2019	54%	2025	65%
2020–2025	55%	2030	70%
2026–2030	65%	–	–
2031 and later	70%	–	–

including glass, paper, plastics, wood, and metals in both EU and Turkish directives.

Packaging waste must be collected separately from other waste where they are created to reduce environmental pollution, utilize landfills, contribute to the economy, and oblige the related regulations in many developed and developing countries. In this study, a generic network is designed, and a mathematical model is formulated for recycling packaging and textile waste in Kayseri in the Central Anatolian Region, and one of the biggest cities of Turkey. The proposed model aims to gain economic value from the packaging and textile waste and minimize the harmful effect of this recyclable waste on the environment by reducing landfilled quantities. In addition, this study is expected to constitute an example for practitioners regarding the separate collection of recyclable waste from the source and their processing environmentally-friendly within the framework of zero-waste management.

2 LITERATURE SURVEY

Enacted obligations by governments on waste minimization and end-of-life products make reverse logistics processes more important in industries such as automotive, electrical, and electronic equipment and battery (Demirel et al., 2016). Packaging is part of these industries since packaging is necessary and important for all industries in advertising and informing. Packaging waste increases as the production and use of packaging materials rise due to the growing population, urbanization, and changing consumption habits. Therefore, the selection of packaging material is a crucial decision for all sectors in terms of the environmental impacts of packaging as well as costs (Jarupan et al., 2004). When the function of the package ends and it becomes waste, proper management takes an important place in the responsibilities of municipalities. The economic value that packaging waste includes, as well as its ability to cope with environmental and capacity problems by efficient waste management, can be shown in the interest of the government on this issue and its popularity in literature. This section gives a brief overview of relevant recent studies and advancements in the recycling of solid waste, especially packaging waste.

Grodzinska-Jurczak et al. investigated the management of solid waste and what to do to improve the current system in Poland. They expected a constant rise in the amount of municipal solid waste due to increasing urbanization, increasing standards of living, changing patterns of consumption, and changes in the waste composition patterns (Grodzinska-Jurczak et al., 2001). Metin et al. evaluated MSW statistics and management practices, including waste recovery and recycling initiatives, in Turkey (Metin et al., 2003). Pokhrel and Viraraghavan analyzed solid waste generation, compositions, and management practices as well as impacts on the environment and public health in Nepal (Pokhrel and Viraraghavan, 2005). Tınmaz and Demir addressed the current waste management practices and problems in Çorlu (a town in Turkey) and evaluated various management methods for suitable solid waste management practices (Tınmaz and Demir, 2006). Kofoworola researched recovery and recycling practices in MSW management in Lagos, Nigeria (Kofoworola, 2007). Banar et al. investigated waste management alternatives from an environmental point of view for Eskisehir (a town in Turkey) (Banar, Cokaygil, and Ozkan 2009). Kaushal et al. analyzed the changing trend in MSW quantities and characteristics in major urban agglomerations of India over the last four decades (Kaushal et al., 2012). Sharholy et al. and Hazra and Goel also investigated MSW management practices and challenges in India (Sharholy et al., 2008, Hazra and Geol, 2009). Pires et al. presented a literature review of waste management practices and the pros and cons of these practices in each member state of the EU (Pires et al., 2011). Bari et al. (2012) aimed to investigate the waste reuse patterns in Khulna (a city in Bangladesh) to further improve the established waste management system (Bari et al., 2012). Ezeah and Roberts (2012) investigated the barriers as well as success factors that affect solid waste management in Abuja, Nigeria by a questionnaire-interview methodology. Groot et al. proposed a model for comparing collection costs, incorporating fixed and variable vehicle cost, labor cost, container cost, and emission cost for different collection strategies such as source separation and post-separation of post-consumer plastics packaging waste in the Netherlands (Groot et al., 2014). Marques et al. compared the economic viability of waste management systems of Belgium and Portugal (Marques et al., 2014). Bing et al. performed a comparison of current practices in various EU countries and identified the characteristics and key issues of waste recycling from waste management and reverse logistics' point of view. The authors concluded that waste recycling is a multi-disciplinary problem that needs to be considered at different decision levels simultaneously (Bing et al., 2016). Marhinto et al. compared two different systems (the drop-off and curbside plus drop-off) of collection of packaging waste in three neighborhoods in Portugal to provide insights into their performance and operation (Marhinto et al., 2017) and, in another study, the authors compared the mixed system (curbside plus drop-off) with drop-off and curbside collection systems with respect to environmental and economic aspects (Pires et al., 2017). Eren and Erdoğan examined the situation of packaging waste in Turkey and the widespread effects of packaging waste management plan implemented in Erzurum (a city in Turkey) (Eren and Erdoğan, 2018).

Although many researchers have addressed the recycling of solid waste in different cities and countries, there are only a few studies that consider the efficient design of a recycling network for solid waste and formulate a mathematical model for optimized flows. The number of studies is further reduced when the subject is limited by network design and modeling for packaging waste recovery. Diamadopoulos et al. aimed to develop a decision support tool for the optimal design of an MSW recycling system in Chania city of Greece. They proposed an integer linear programming approach considering all the costs arising from recycling of products, disposal of solid waste, closure and monitoring of the old landfill, and opening of a new one (Diamadopoulos et al., 1995). Bautista and Pereira focused on modeling the problem of locating collection areas for urban waste management and presented an application to the metropolitan area of Barcelona. The authors presented a heuristic approach to solve the proposed mathematical model (Bautista and Pereira, 2006). Demirel and Gökçen investigated the MSW management practices in Ankara and proposed a mathematical model to minimize the total costs of the system via proper management of solid waste in the city (Demirel and Gökçen, 2013). Nakatani et al. aimed to develop a framework for the variability-based optimal design of plastics recycling to construct a robust recycling system over the external changes in the market such as fluctuations of material prices (Nakatani et al., 2017). Bing et al. proposed a mixed-integer linear programming (MILP) model for recycling household plastics waste to minimize the transportation cost and environmental impact simultaneously and choose the most suitable combination of separation methods in the Netherlands (Bing et al., 2014). Olapiriyakul et al. proposed a multi-objective optimization model for sustainable waste management network design. The proposed model was applied to a case in Pathum Thani (Thailand) (Olapiriyakul et al., 2019). Zhao and Huang proposed a multi-period, multi-objective network design problem for hazardous waste considering the location of waste facilities, allocation, and transportation of waste and residue for minimization of the total cost and risk (Zhao and Huang, 2019).

In this study, to comply with related regulations and manage the recovery of packaging and textile waste efficiently, we presented a multi-period MILP model for a generic recycling network, including the collection of recyclable waste separately from the houses, their transportation to the authorized collection and separation facilities, transportation of non-recyclable residue to the landfills, and selling obtained paper, plastics, glass, metal, textile, and wood to authorized recycling facilities. Costs arising from holding the collection and separation facility open, collection of recyclable waste separately from household organic waste, transportation, CO_2 emissions, separation of packaging waste by type in the collection and separation facilities, and disposal are considered in the objective function aside from the revenue obtained from selling the paper, plastics, glass, metal, textile, and wood to recycling facilities. The proposed model was applied to a case in Kayseri, Turkey.

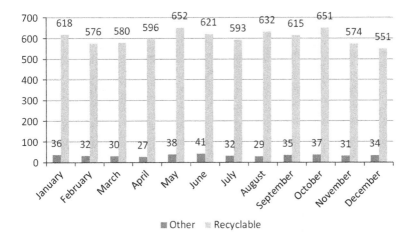

FIGURE 8.2 Recyclable Waste Collected in Blue Bags and Other Household Waste Thrown into These Bags in 2018 (ton).

3 MANAGEMENT OF NON-HAZARDOUS RECYCLABLE WASTE IN KAYSERI

We considered Melikgazi, one of the biggest districts in Kayseri and composed of 32 neighborhoods, where the process of collecting recyclable household waste separately at source is carried out. In the considered district, municipal officials distribute blue plastic bags to approximately 3,400 apartments and ask occupants to throw away their recyclable waste in these bags, separate from their organic and hazardous waste. In this system, a special recyclable waste collection vehicle collects the recyclable waste once or twice a week, separately from the daily collection process of solid waste. Figure 8.2 shows the recyclable waste quantities collected by the municipality in 2018 and non-recyclable quantities thrown into distributed blue bags by occupants.

FIGURE 8.3 Collection and Separation Facility.

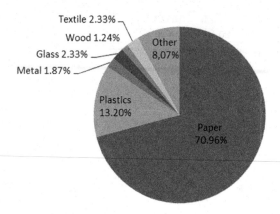

FIGURE 8.4   Composition of Recyclable Waste in Kayseri.

Figure 8.2 shows that the amount of other household waste getting out of the distributed bags for recyclable waste is below the 10% limit every month. This means that the municipality successfully carries out recyclable waste management in terms of Turkey standards. Figure 8.3 shows the municipality's collection and separation facility.

Figure 8.4 shows the collected composition of recyclable waste in Kayseri that has economic value and is a key element of decreasing the landfilled quantities of MSW.

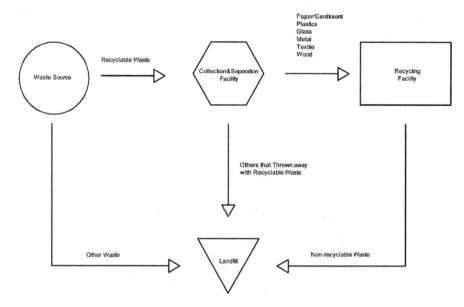

FIGURE 8.5   A Generic Recycling Network for Packaging Waste.

4   RECYCLING NETWORK DESIGN AND MODELING

In this section, a generic recycling network design regarding the separate collection of recyclable waste from other household waste, their processing, and proper disposal is proposed, and a mathematical model is formulated. The proposed network is shown in Figure 8.5.

Waste that includes economic value is collected separately in the source and they are transported to collection and separation facilities. In these facilities, recyclable waste is separated by type and sent to the authorized recycling facilities to be used as raw material. The non-recyclable residues are sent to landfills for proper disposal. The proposed network provides not only economic but also environmental benefits via a decrease in solid waste quantities sent to the landfills.

4.1   Modeling

In this section, a multi-period, multi-echelon MILP model is formulated for the proposed recycling network. In the model, we assume that all collected quantities are transported to the collection and separation facility and processed there. We considered environmental costs caused by transportation on top of transportation costs; there are different environmental costs for different types of vehicles, depending on the model and capacity. For calculation, we used the formula given below:

$$ec = \frac{price\ of\ Eu\ carbon\ allowances\ (\text{€} / ton) \times carbon\ equivalent\ conversion\ factor\ (kgCO_2e / ton.km)}{1000\ (kg / ton)}$$

For the initial input value for calculating, we used 25.15 €/ton as the price of EU carbon allowances (CO2 European Emission Allowances, 2019), 0.25773 kgCO$_2$e/ ton·km as the carbon equivalent conversion factor for freighting goods with heavy goods vehicle (HGV) 100% laden (Greenhouse gas reporting: conversion factors, 2018). Total transportation and environmental costs regarded in the mathematical model for each vehicle and each period are calculated as:

$$tec_{vt} = tc_{vt} + ec_{vt}$$

Costs arising from collection, transportation, facility operation, separation, and disposal as well as revenues gained from selling paper, plastics, glass, metal, textile, and wood are considered in the objective.

4.1.1   Notations

Indices
    i: set of neighborhoods
    j: set of collection and separation facilities
    k: set of paper and cardboard recycling facilities
    l: set of plastics recycling facilities
    m: set of glass recycling facilities

n: set of metal recycling facilities
r: set of textile recycling facilities
s: set of wood recycling facilities
p: set of landfills
v: set of vehicles
t: set of periods

Parameters:

R_{it}: waste collected from neighborhood i in period t
oc_{jt}: cost of holding the collection and separation facility j opened in period t
cc_t: unit collection cost of waste from neighborhoods in period t
sc_t: unit separation cost at collection and separation facilities in period t
lc_t: unit disposal cost at landfills in period t
tc_{vt}: unit transportation cost of vehicle v per km in period t
ec_{vt}: unit environment cost of vehicle v per km in period t
tec_{vt}: unit transportation and environment cost of vehicle v per km in period t
cap_{jt}: capacity of collection and separation facility j in period t
D_{kt}: demand for paper recycling facility k in period t
D_{lt}: demand for plastics recycling facility l in period t
D_{mt}: demand for glass recycling facility m in period t
D_{nt}: demand for metal recycling facility n in period t
D_{rt}: demand of textile recycling facility r in period t
D_{st}: demand for wood recycling facility s in period t
cap_{pt}: capacity of landfill p in period t
d_{ij}: distance between the neighborhood i and collection and separation facility j
d_{jk}: distance between the collection and separation facility j and paper recycling facility k
d_{jl}: distance between the collection and separation facility j and plastics recycling facility l
d_{jm}: distance between the collection and separation facility j and glass recycling facility m
d_{jn}: distance between the collection and separation facility j and metal recycling facility n
d_{jr}: distance between the collection and separation facility j and textile recycling facility r
d_{js}: distance between the collection and separation facility j and wood recycling facility s
d_{jp}: distance between the collection and separation facility j and landfill p
pp_t: selling price of paper per kg in period t
plp_t: selling price of plastics per kg in period t
gp_t: selling price of glass per kg in period t
mp_t: selling price of metal per kg in period t
tp_t: selling price of textile per kg in period t
wp_t: selling price of wood per kg in period t
α: percentage of paper in collected recyclable waste
β: percentage of plastics in collected recyclable waste

γ: percentage of glass in collected recyclable waste

η: percentage of metal in collected recyclable waste

δ: percentage of textile in collected recyclable waste

μ: percentage of wood in recyclable waste

σ: percentage of non-recyclable waste transported to landfills by recycling facilities

φ: percentage of organic waste intervened in collected recyclable waste

Decision Variables:

x_{ijvt}: amount shipped from neighborhood i to collection and separation facility j with vehicle v in period t

y_{jkvt}: amount shipped from collection and separation facility j to paper recycling facility k with vehicle v in period t

w_{jlvt}: amount shipped from collection and separation facility j to plastics recycling facility l with vehicle v in period t

z_{jmvt}: amount shipped from collection and separation facility j to glass recycling facility m with vehicle v in period t

u_{jnvt}: amount shipped from collection and separation facility j to metal recycling facility n with vehicle v in period t

g_{jrvt}: amount shipped from collection and separation facility j to textile recycling facility r with vehicle v in period t

h_{jsvt}: amount shipped from collection and separation facility j to wood recycling facility s with vehicle v in period t

f_{jpvt}: amount shipped from collection and separation facility j to landfill p with vehicle v in period t

re_{pt}: total amount of waste sent to landfill p from recycling facilities in period t

A_{jt}: if collection and separation facility j is processing in period t, 1; otherwise, 0

4.1.2 Formulation

Objective Function:

Max

$$
\sum_j \sum_k \sum_v \sum_t pp_t \cdot y_{jkvt} + \sum_j \sum_l \sum_v \sum_t plp_t \cdot w_{jlvt} + \sum_j \sum_m \sum_v \sum_t gp_t \cdot z_{jmvt} +
$$
$$
\sum_j \sum_n \sum_v \sum_t mp_t \cdot u_{jnvt} + \sum_j \sum_r \sum_v \sum_t tp_t \cdot g_{jrvt} + \sum_j \sum_s \sum_v \sum_t wp_t \cdot h_{jsvt} - \sum_j \sum_t oc_{jt} \cdot A_{jt} -
$$
$$
\sum_i \sum_t cc_t \cdot R_{it} - \sum_i \sum_j \sum_v \sum_t sc_t \cdot x_{ijvt} - \sum_i \sum_j \sum_v \sum_t tec_{vt} \cdot d_{ij} \cdot x_{ijvt} -
$$
$$
\sum_j \sum_k \sum_v \sum_t tec_{vt} \cdot d_{jk} \cdot y_{jkvt} - \sum_j \sum_l \sum_v \sum_t tec_{vt} \cdot d_{jl} \cdot w_{jlvt} - \sum_j \sum_m \sum_v \sum_t tec_{vt} \cdot d_{jm} \cdot z_{jmvt} -
$$
$$
\sum_j \sum_n \sum_v \sum_t tec_{vt} \cdot d_{jn} \cdot u_{jnvt} - \sum_j \sum_r \sum_v \sum_t tec_{vt} \cdot d_{jr} \cdot g_{jrvt} - \sum_j \sum_s \sum_v \sum_t tec_{vt} \cdot d_{js} \cdot h_{jsvt} -
$$
$$
\sum_j \sum_p \sum_v \sum_t tec_{vt} \cdot d_{jp} \cdot f_{jpvt} - \sum_j \sum_p \sum_v \sum_t lc_t \cdot (f_{jpvt} + re_{pt})
$$

$$(8.1)$$

Constraints:

$$
R_{it} = \sum_j \sum_v x_{ijvt} \quad \forall \ i, \ t \tag{8.2}
$$

$$\alpha. \sum_i \sum_v x_{ijvt} = \sum_v \sum_k y_{jkvt} \quad \forall j, t \tag{8.3}$$

$$\beta. \sum_i \sum_v x_{ijvt} = \sum_v \sum_l w_{jlvt} \quad \forall j, t \tag{8.4}$$

$$\gamma. \sum_i \sum_v x_{ijvt} = \sum_v \sum_m z_{jmvt} \quad \forall j, t \tag{8.5}$$

$$\eta. \sum_i \sum_v x_{ijvt} = \sum_v \sum_n u_{jnvt} \quad \forall j, t \tag{8.6}$$

$$\delta. \sum_i \sum_v x_{ijvt} = \sum_v \sum_r g_{jrvt} \quad \forall j, t \tag{8.7}$$

$$\mu. \sum_i \sum_v x_{ijvt} = \sum_v \sum_s h_{jsvt} \quad \forall j, t \tag{8.8}$$

$$\varphi. \sum_i \sum_v x_{ijvt} = \sum_v \sum_p f_{jpvt} \quad \forall j, t \tag{8.9}$$

$$\sum_p re_{pt} = \sigma. \left(\sum_j \sum_k \sum_v y_{jkvt} + \sum_j \sum_l \sum_v w_{jlvt} + \sum_j \sum_m \sum_v z_{jmvt} + \sum_j \sum_n \sum_v u_{jnvt} + \sum_j \sum_r \sum_v g_{jrvt} + \sum_j \sum_s \sum_v h_{jsvt} \right) \quad \forall t \tag{8.10}$$

$$\sum_i \sum_v x_{ijvt} \le A_{jt} \cdot cap_{jt} \quad \forall j, t \tag{8.11}$$

$$\sum_j \sum_v y_{jkvt} \le D_{kt} \quad \forall k, t \tag{8.12}$$

$$\sum_j \sum_v w_{jlvt} \le D_{lt} \quad \forall l, t \tag{8.13}$$

$$\sum_j \sum_v z_{jmvt} \le D_{mt} \quad \forall m, t \tag{8.14}$$

$$\sum_j \sum_v u_{jnvt} \le D_{nt} \quad \forall n, t \tag{8.15}$$

$$\sum_j \sum_v g_{jrvt} \le D_{rt} \quad \forall r, t \tag{8.16}$$

$$\sum_j \sum_v h_{jsvt} \le D_{st} \quad \forall s, t \tag{8.17}$$

$$\sum_j \sum_v f_{jpvt} + re_{pt} \le cap_{pt} \quad \forall p, t \tag{8.18}$$

$$\sum_i \sum_j x_{ijvt} + \sum_j \sum_k y_{jkvt} + \sum_j \sum_l w_{jlvt} + \sum_j \sum_m z_{jmvt} + \sum_j \sum_n u_{jnvt} + \sum_j \sum_r g_{jrvt} + \sum_j \sum_s h_{jsvt} + \sum_j \sum_p f_{jpvt} \le cap_{vt} \quad \forall v, t \tag{8.19}$$

TABLE 8.3
Monthly Collected Waste Quantities from Neighborhoods (kg)

Neighborhood						Period						
	1	2	3	4	5	6	7	8	9	10	11	12
1	4620	0	27660	50100	36000	12950	38650	29000	35400	39850	25050	1500
2	38300	62090	65770	51950	53600	23450	43700	38950	43400	41700	29500	4550
3	11260	13700	14000	17450	9550	5950	9050	11000	15250	10850	6650	7150
4	9800	10100	9050	13200	7450	8600	8150	7700	7800	10800	5200	3100
5	24850	20100	21450	21400	21150	17800	20550	19550	23650	26000	15100	2400
6	16050	14300	13900	18350	16150	10250	12350	15450	14450	17150	6400	0
7	13300	10800	18350	15050	15150	6650	12950	14050	13350	12900	15350	3200
8	2350	0	2560	0	3300	2800	0	0	5900	5700	0	0
9	38610	37260	37650	40290	41610	12950	35600	27750	38500	39250	22100	3750
10	16200	18000	17400	22300	19900	9450	15200	9800	17150	14600	10600	6750
11	12100	9000	13350	13450	17950	6350	7050	15250	15300	13100	7900	0
12	13100	12200	16350	14100	20550	6150	7900	10100	12950	12550	12350	0
13	6900	7650	12650	9850	10350	9000	6600	2600	8300	10200	9300	0
14	2350	20010	10900	40500	31650	19150	25200	13300	5400	7300	27600	1000
15	10550	10800	10400	13150	9900	4600	7500	7150	9950	7050	5150	4650
16	7100	9550	8650	10750	7400	4900	5600	4800	6450	5650	6300	2000
17	9300	9800	9300	9400	10000	6800	4800	7250	9500	8000	8350	2200
18	7150	7300	7500	7500	9200	5550	3200	8150	7750	9500	5400	0
19	6250	6550	10600	8050	10850	4200	4850	6150	8750	8000	7500	2000
20	0	240	2290	0	1350	0	0	0	1850	0	1850	0
21	25550	34150	36850	38450	29700	18150	34700	26150	37250	36500	21800	11500

(Continued)

TABLE 8.3 (Continued)

Neighborhood	1	2	3	4	5	6	7	8	9	10	11	12
						Period						
22	14800	11800	13430	13900	17550	9350	12450	14400	9850	10800	7100	0
23	0	0	0	0	4450	4700	0	2100	0	1800	2300	0
24	10850	11050	10600	14650	12800	4750	10600	7300	11400	8600	5750	6150
25	5650	7050	7500	7150	8000	5850	9150	5900	8200	8600	6100	1850
26	0	0	0	3650	3300	4250	6400	4100	5850	8000	8100	1450
27	10100	10200	10800	10350	14300	7600	9650	13750	11300	14250	8250	0
28	9750	10700	14000	11400	17300	6600	7090	8250	7700	8550	7600	2250
29	6400	8450	7250	8650	9350	5000	6050	2200	4700	4700	7100	0
30	22490	9960	10500	0	3700	3350	5350	3200	0	0	1250	0
31	9570	14400	20430	0	29800	3500	11700	4200	10600	17200	10800	0
32	3870	3850	0	2040	0	0	0	0	9100	0	0	0

TABLE 8.4

Distances between the Neighborhoods and Collection and Separation Center (km)

Neighborhood	1	2	3	4	5	6	7	8	9	10
	7.9	24.3	16.1	8.5	11.6	25.0	20.0	7.2	8.9	12.2
Neighborhood	11	12	13	14	15	16	17	18	19	20
	12.4	4.5	12.7	14.4	15.2	13.2	12.3	10.9	26.1	7.2
Neighborhood	21	22	23	24	25	26	27	28	29	30
	7.9	12.6	23.4	12.6	13.9	11.2	12.1	24.1	23.5	9.3
Neighborhood	31	32	33							
	16.9	24.0	8.5							

$$A_{jt} \in \{0, 1\} \ \forall j, t \qquad (8.20)$$

$$x_{ijvt}, y_{jkvt}, w_{jlvt}, z_{jmvt}, u_{jnvt}, g_{jrvt}, h_{jsvt}, f_{jpvt} \geq 0 \ \forall \ i, j, k, l, m, n, r, s, p, v, t$$

$$(8.21)$$

The objective of the model given in (8.1) is to maximize the profit of the system. It consists of the revenue obtained by selling the separated materials to recycling facilities, on top of the costs of collection and separation facilities maintenance, collection, transportation, separation, disposal, and environmental impact. Constraint (8.2) ensures that all collected waste is transported to collection and separation facilities. Constraints (8.3)–(8.8) determine the quantities transported from the collection and separation facilities to paper recycling, plastics recycling, glass recycling, metal recycling, textile recycling, wood recycling facilities, respectively. Constraints (8.9) and (8.10) establish the transported quantities to landfills from collection and separation and recycling facilities. Constraints (8.11) and (8.18) stipulate that the transportation amounts must not exceed the capacity of collection and separation facilities and landfills, respectively. Constraints (8.12)–(8.17) prevent sending more materials to recycling facilities than demanded. Constraint (8.19) restricts the capacity of vehicles. Constraint (8.20) represents the binary variable. Lastly, Constraint (8.21) enforces the non-negativity restriction on the decision variables.

4.2 CASE STUDY

Data collection to build the model has been conducted in cooperation with the municipality through interviews, industrial reports, and the literature. There are 32 neighborhoods within the municipal boundaries where recyclable household materials are collected separately. One authorized collection and separation facility and landfill exists in Kayseri. In total, there are 13 licensed recycling facilities where the municipality sells materials—including three for papers, four for plastics, one for glass, one for metal, three for textiles, and one for wood. Table 8.3 shows

TABLE 8.5
Distances between the Collection and Separation Facility and Recycling Facilities/Landfill (km)

Collection and Separation Center	Recycling Facility													
	Paper			Plastics				Textile			Glass	Wood	Metal	Landfill
	1	2	3	1	2	3	4	1	2	3	1	1	1	1
1	9.6	12.7	14	13	13	15.3	13.9	6.2	9.3	15.7	16.1	13.9	13.7	10

TABLE 8.6

Selling Prices of Recyclable Materials (TL/kg)

Material	Period											
	1	2	3	4	5	6	7	8	9	10	11	12
Paper	0.55	0.58	0.66	0.73	0.73	0.88	0.6	0.58	0.56	0.63	0.56	0.59
Plastics	1.25	1.21	1.23	1.27	1.21	1.45	1.30	1.19	1.15	1.17	1.16	1.14
Glass	0.12	0.12	0.12	0.12	0.12	0.12	0.12	0.12	0.12	0.12	0.12	0.12
Metal	2.00	2.00	2.00	2.00	2.00	3.38	3.38	3.38	3.28	2.68	3.00	3.85
Textile	0.80	0.80	0.80	0.80	0.65	0.65	0.65	0.65	0.80	0.93	0.83	0.78
Wood	0.25	0.25	0.25	0.25	0.25	0.25	0.25	0.25	0.25	0.25	0.25	0.25

the collected amounts of recyclable waste consisting of packaging and textile waste from the 32 neighborhoods per month in 2019.

Two types of special trucks with different capacities are available to collect recyclable waste from neighborhoods and transfer them from the collection and separation facility to recycling facilities and landfill. The older trucks have higher transportation costs as well as higher CO_2 emissions due to lower engine efficiency. Unit transportation and environmental costs per km are calculated by the formula given in Section 4.1. Demands are determined as 150,000 kg, 22,000 kg, 4,750 kg, and 10,000 kg for paper recycling, plastics recycling, textile recycling, and wood recycling facilities, respectively and accepted as fixed by periods. Demands are set as 15,000 kg for glass and metal recycling facilities for all periods. The geographical position of every locality (neighborhoods, collection and separation facility, recycling facilities, and landfill) is defined by cartographic coordinates (longitude and latitude) and the real distances between the facilities in the network are calculated via Google Maps. Distances between the neighborhoods and collection and separation facility are given in Table 8.4. Distances between the collection and separation facility and recycling facilities/landfill are given in Table 8.5.

Table 8.6 shows the monthly average selling price of collected and separated materials in the market. Prices of glass and wood have remained the same for all months in 2019, while others have changed per month.

4.3 RESULTS AND DISCUSSIONS

The model is solved by using a GAMS-CPLEX solver to obtain optimal values. The objective is found out to be 930478.78 TL for 2019. The total cost of the system is 2.0994×10^6 and the total income of the system is 3.0299×10^6 in the optimal solution. Monthly transported quantities of recyclable waste from the first eight of the 32 to the collection and separation facility with different vehicles are given in Table 8.7. Data and results of the model that are not provided in the study are available upon request. Transported quantities from collection and separation facility to recycling facilities and to landfill are given in Table 8.8.

TABLE 8.7

Transported Quantities from Neighborhoods to Collection and Separation Facility (kg)

Neig.	Vehicle Type	1	2	3	4	5	6	7	8	9	10	11	12
											Period		
1	1	4620	–	–	–	–	12950	38650	29000	–	–	25050	1500
	2	–	–	27660	50100	36000	–	–	–	35400	39850	–	–
2	1	38300	62090	65770	51950	53600	23450	43700	38950	43400	41700	29500	4550
	2	–	–	–	–	–	–	–	–	–	–	–	–
3	1	11260	13700	14000	17450	9550	5950	9050	11000	15250	10850	6650	7150
	2	–	–	–	–	–	–	–	–	–	–	–	–
4	1	9800	–	–	–	–	8600	8150	7700	–	–	5200	3100
	2	–	10100	9050	13200	7450	–	–	–	7800	10800	–	–
5	1	24850	20100	21450	21400	21150	17800	20550	19550	23650	26000	15100	2400
	2	–	–	–	–	–	–	–	–	–	–	–	–
6	1	16050	14300	13900	18350	16150	10250	12350	15450	14450	17150	6400	–
	2	–	–	–	–	–	–	–	–	–	–	–	–
7	1	13000	10800	18350	15050	15150	6650	12950	14050	13350	12900	15350	3200
	2	–	–	–	–	–	–	–	–	–	–	–	–
8	1	2350	–	–	–	–	2800	–	–	–	–	–	–
	2	–	–	2560	–	3300	–	–	–	5900	2700	–	–

TABLE 8.8

Transported Quantities to Recycling Facilities and to Landfill (kg)

Variable	Value	Variable	Value	Variable	Value	Variable	Value	Variable	Value	Variable	Value
$y_{1,1,1,1}$	150000	$y_{1,3,1,4}$	45632	$w_{1,4,1,1}$	4730	$u_{1,1,1,6}$	4687	$g_{1,2,1,10}$	4750	$f_{1,1,1,5}$	40617
$y_{1,1,1,2}$	150000	$y_{1,3,1,5}$	57149	$w_{1,4,1,2}$	8940	$u_{1,1,1,7}$	7144	$g_{1,2,1,11}$	2562	$f_{1,1,1,6}$	20227
$y_{1,1,1,3}$	106630	$w_{1,1,1,1}$	22000	$w_{1,4,1,3}$	16870	$u_{1,1,1,8}$	6350	$g_{1,2,2,3}$	4750	$f_{1,1,1,7}$	30831
$y_{1,1,1,4}$	77770	$w_{1,1,1,2}$	22000	$w_{1,4,1,4}$	20295	$u_{1,1,1,9}$	7798	$g_{1,2,2,4}$	4750	$f_{1,1,1,8}$	27402
$y_{1,1,1,5}$	39340	$w_{1,1,1,3}$	22000	$w_{1,4,1,5}$	22000	$u_{1,1,1,10}$	7838	$g_{1,2,2,5}$	4750	$f_{1,1,1,9}$	33652
$y_{1,1,1,6}$	150000	$w_{1,1,1,4}$	22000	$w_{1,4,1,7}$	6429	$u_{1,1,1,11}$	5868	$g_{1,3,1,3}$	1245	$f_{1,1,1,10}$	33825
$y_{1,1,1,7}$	150000	$w_{1,1,1,5}$	22000	$w_{1,4,1,8}$	821	$u_{1,1,1,12}$	1261	$g_{1,3,1,4}$	1849	$f_{1,1,1,11}$	25324
$y_{1,1,1,8}$	150000	$w_{1,1,1,6}$	22000	$w_{1,4,1,9}$	11044	$g_{1,1,1,1}$	4750	$g_{1,3,1,5}$	2227	$f_{1,1,1,12}$	5443
$y_{1,1,1,9}$	150000	$w_{1,1,1,7}$	22000	$w_{1,4,1,10}$	11328	$g_{1,1,1,6}$	4570	$g_{1,3,1,9}$	216	$re_{1,1}$	50907
$y_{1,1,1,10}$	150000	$w_{1,1,1,8}$	22000	$z_{1,1,1,1}$	8602	$g_{1,1,1,7}$	4750	$g_{1,3,1,10}$	266	$re_{1,2}$	55304
$y_{1,1,1,11}$	150000	$w_{1,1,1,9}$	22000	$z_{1,1,1,2}$	9345	$g_{1,1,1,8}$	4750	$h_{1,1,1,1}$	4578	$re_{1,3}$	63589
$y_{1,1,1,12}$	47863	$w_{1,1,1,10}$	22000	$z_{1,1,1,3}$	10745	$g_{1,1,1,11}$	1572	$h_{1,1,1,2}$	4973	$re_{1,4}$	67166
$y_{1,1,2,3}$	43370	$w_{1,1,1,11}$	22000	$z_{1,1,1,4}$	11349	$g_{1,1,1,12}$	4750	$h_{1,1,1,3}$	5718	$re_{1,5}$	69404
$y_{1,1,2,4}$	72230	$w_{1,2,1,1}$	22000	$z_{1,1,1,5}$	11727	$g_{1,1,2,2}$	4750	$h_{1,1,1,4}$	6040	$re_{1,6}$	34563
$y_{1,1,2,5}$	110660	$w_{1,2,1,2}$	22000	$z_{1,1,1,6}$	5840	$g_{1,1,2,3}$	4750	$h_{1,1,1,5}$	6241	$re_{1,7}$	52681
$y_{1,2,1,1}$	111960	$w_{1,2,1,3}$	22000	$z_{1,1,1,7}$	8902	$g_{1,1,2,4}$	4750	$h_{1,1,1,6}$	3108	$re_{1,8}$	46822
$y_{1,2,1,2}$	134590	$w_{1,2,1,4}$	22000	$z_{1,1,1,8}$	7912	$g_{1,1,2,5}$	4750	$h_{1,1,1,7}$	4737	$re_{1,9}$	57502
$y_{1,2,1,3}$	150000	$w_{1,2,1,5}$	22000	$z_{1,1,1,9}$	9716	$g_{1,1,2,7}$	180	$h_{1,1,1,8}$	4210	$re_{1,10}$	57799
$y_{1,2,1,4}$	150000	$w_{1,2,1,6}$	11086	$z_{1,1,1,10}$	9766	$g_{1,1,2,9}$	4750	$h_{1,1,1,9}$	5171	$re_{1,11}$	43271
$y_{1,2,1,5}$	150000	$w_{1,2,1,7}$	22000	$z_{1,1,1,11}$	7312	$g_{1,1,2,10}$	4750	$h_{1,1,1,10}$	5197	$re_{1,12}$	9301
$y_{1,2,1,6}$	27861	$w_{1,2,1,8}$	22000	$z_{1,1,1,12}$	1572	$g_{1,2,1,1}$	3852	$h_{1,1,1,11}$	3891		
$y_{1,2,1,7}$	90945	$w_{1,2,1,9}$	22000	$u_{1,1,1,1}$	6903	$g_{1,2,1,2}$	4595	$h_{1,1,1,12}$	836		

(Continued)

TABLE 8.8 (Continued)

Variable	Value	Variable	Value	Variable	Value	Variable	Value	Variable	Value	Variable	Value
$y_{1,2,1,8}$	145900	$w_{1,2,1,10}$	22000	$u_{1,1,1,2}$	7500	$g_{1,2,1,6}$	1090	$f_{1,1,1,1}$	29792		
$y_{1,2,1,9}$	147430	$w_{1,2,1,11}$	19421	$u_{1,1,1,3}$	8623	$g_{1,2,1,7}$	4152	$f_{1,1,1,2}$	32366		
$y_{1,2,1,10}$	72672	$w_{1,2,1,12}$	8903	$u_{1,1,1,4}$	9108	$g_{1,2,1,8}$	3162	$f_{1,1,1,3}$	37214		
$y_{1,3,1,3}$	27225	$w_{1,3,1,5}$	437	$u_{1,1,1,5}$	9412	$g_{1,2,1,9}$	4750	$f_{1,1,1,4}$	39307		

TABLE 8.9
Income of the System from Different Material Sales

Definition	Value (TL) ($\times 10^6$)	Percentage of Total Income
Total income of the system	3.02990	100
Income from paper sales	1.99700	65.9
Income from plastics sales	0.71499	23.6
Income from glass sales	0.01233	0.41
Income from textile sales	0.07844	2.60
Income from wood sales	0.01368	0.45
Income from metal sales	0.21342	7.04

The total collection cost is determined as 8.8228×10^5, which constitutes 42% of the total cost of the system. The collection cost (which is approximately 13 times greater than disposal cost—the smallest cost item) is followed by the separation cost with 21%, transportation and environment cost with 19,4%, facility processing cost with 14,3%, respectively. In the income frame, while the maximum contribution is provided by paper sales with 65.9%, other contributions are provided by plastics, metal, and textile sales, respectively. Glass and wood sales make up less than 1% of the total income (Table 8.9).

5 CONCLUSION

With the rapid growth in world population and consumption, production is also increasing, causing the depletion of natural resources. The increase in waste parallel to the increase in consumption and depletion of natural resources has begun to threaten the environment and human health. To cope with these negative results, humanity has sought to consume renewable energy resources instead of using non-renewable energy resources, to reduce the amount of waste generated and reuse the waste to gain both economic and environmental benefits.

In this study, a generic recycling network is designed, and a mathematical model is formulated for recycling packaging waste to gain economic value from the packaging and textile waste and minimize the harmful effects of recyclable waste on the environment. The economic value of the materials obtained from the collection and separation of recyclable waste and costs from facility operation, collection, transportation, CO_2 emissions, separation, and disposal are considered in the model. The proposed model is applied to a case in Kayseri, Turkey. Separate collection of recyclable waste from other household waste at source has caused the highest cost, while the highest income has been obtained from the sales of paper/ cardboard in the optimum solution.

ACKNOWLEDGMENTS

The authors are thankful to the Melikgazi Municipality for sharing and allowing data to be used in this study. Special thanks to Mr. Ramazan AYDIN working in the municipality as an environmental engineer for his kind assistance.

REFERENCES

Banar M., Cokaygil Z., Ozkan A. (2009), "Life cycle assessment of solid waste management options for Eskisehir, Turkey", *Waste Management*, 29, 54-52.

Bari Q. H., Hassan K. M., Haque R. (2012), "Scenario of solid waste reuse in Khulna city of Bangladesh", *Waste Management*, 32, 2526–2534.

Bautista J., Pereira J. (2006), "Modeling the problem of locating collection areas for urbanwaste management. An application to the metropolitan area of Barcelona", *Omega*, 34, 617–629.

Bing X., Bloemhof-Ruwaard J. M., Vorst J. G. A. J. (2014), "Sustainable reverse logistics network design for household plastic waste", *Flexible Services and Manufacturing Journal*, 26, 119–142.

Bing X., Bloemhof J. M., Ramos T. R. P., Barbosa-Povoa A. P., Wong C. Y., Vorst J. G. A. J. (2016), "Research challenges in municipal solid waste logistics management", *Waste Management*, 48, 584–592.

CO_2 European Emission Allowances, (2019), Available at: https://markets.businessinsider.com/commodities/co2-european-emission-allowances.

Demirel N., Gökçen H. (2013), "Recycling network design for municipal waste management in Ankara", *XI. International Logistics and Supply Chain Congress*, November 07-09, Kayseri, Turkey.

Demirel E., Demirel N., Gökçen H. (2016), "A mixed integer linear programming model to optimize reverse logistics activities of end-of-life vehicles in Turkey", *Journal of Cleaner Production*, 112, 2101–2113.

Diamadopoulos E., Koutsantonakis Y., Zaglara V. (1995), "Optimal design of municipal solid waste recycling systems", *Resources, Conservation and Recycling*, 14, 21–34.

Directive (EU) 2018/852 of the European Parliament and of the Council of 30 May. (2018) Amending Directive 94/62/EC on Packaging and Packaging Waste, Available at: http://v4.pimailer.com/mailer/UserFiles/%7B80227E75-0A1F-447F-A338-D35F1D4D7849%7D/files/eu-2018-852.pdf.

Disposal/recovery methods and amount of municipal waste, 1994-(2018). TurkStat, Available at: http://www.tuik.gov.tr/PreIstatistikTablo.do?istab_id=1577.

Eren Z., Erdoğan E. (2018), "Separate collection of packaging wastes within the scope of integrated solid waste management: The case study of Yakutiye district in Erzurum city", *Artvin Çoruh University Natural Hazards Application and Research Center Journal of Natural Hazards and Environment*, 4(2), 82–88.

European Parliament and of the Council of the European Union, Directive 2008/98/EC on waste and repealing certain Directives. (2008), Available at: https://eur-lex.europa.eu/legal-content/EN/TXT/?uri=celex%3A32008L0098.

Exposito A., Velasco F. (2018), "Municipal solid-waste recycling market and the European 2020 Horizon Strategy: A regional efficiency analysis in Spain", *Journal of Cleaner Production*, 172, 938–948.

Ezeah C., Roberts C. L. (2012), "Analysis of barriers and success factors affecting the adoption of sustainable management of municipal solid waste in Nigeria", *Journal of Environmental Management*, 103, 9–14.

Greenhouse Gas Reporting: Conversion Factors. (2018), Available at: https://www.gov.uk/government/publications/greenhouse-gas-reporting-conversion-factors-2018.

Grodzinska-Jurczak M. (2001), "Management of industrial and municipal solid wastes in Poland", *Resources, Conservation and Recycling*, 32, 85–103.

Groot J., Bing X., Bos-Brouwers H., Bloemhof-Ruwaard J. (2014), "A comprehensive waste collection cost model applied to post-consumer plastic packaging waste", *Resources, Conservation and Recycling*, 85, 79–87.

Hazra T., Geol S. (2009), "Solid waste management in Kolkata, India: Practices and challenges", *Waste Management*, 29, 470–478.

Jarupan, L., Kamarthi, S. V. and Gupta, S. M. (2004), "Application of combinatorial approach in packaging material selection", Proceedings of the SPIE International Conference on Environmentally Conscious Manufacturing IV, October 26–27, Philadelphia, Pennsylvania.

Kaushal R. K., Varghese G. K., Chabukdhara M. (2012), "Municipal solid waste management in India – current state and future challenges: A review", *International Journal of Engineering Science and Technology*, 4, 1473–1489.

Kofoworola O. F. (2007), "Recovery and recycling practices in municipal solid waste management in Lagos, Nigeria", *Waste Management*, 27, 1139–1143.

Marhinto G., Gomes A., Santos P., Ramos M., Cardoso J., Silveira A., Pires A. (2017), "A case study of packaging waste collection systems in Portugal – Part I: Performance and operation analysis", *Waste Management*, 61, 96–107.

Marques R. C., Cruz N. F., Simoes P., Ferreira S. F., Pereira M. C., Jaeger S. D. (2014), "Economic viability of packaging waste recycling systems: A comparison between Belgium and Portugal", *Resources, Conservation and Recycling*, 85, 22–33.

Metin E., Eröztürk A., Neyim C. (2003), "Solid waste management practices and review of recovery and recycling operations in Turkey", *Waste Management*, 23, 425–432.

Nakatani J., Konno K., Moriguchi Y. (2017), "Variability-based optimal design for robust plastic recycling systems", *Resources, Conservation and Recycling*, 116, 53–60.

Olapiriyakul S., Pannakkong W., Kachapanya W., Starita S. (2019), "Multi-objective optimization model for sustainable waste management network design", *Journal of Advanced Transportation*, 2019(2), 1–15.

Pires A., Martinho G., Chang N. (2011), "Solid waste management in European countries: A review of systems analysis Techniques", *Journal of Environmental Management*, 92, 1033–1050.

Pires A., Sargedas J., Miguel M., Pina J., Martihno G. (2017), "A case study of packaging waste collection systems in Portugal – Part II: Environmental and economic analysis", *Waste Management*, 61, 108–116.

Pokhrel D., Viraraghavan T. (2005), "Municipal solid waste management in Nepal: Practices and challenges", *Waste Management*, 25, 555–562.

Regulation on Zero Waste Management, Official Gazette. (2019). July 12. Available at: https://www.resmigazete.gov.tr/eskiler/2019/07/20190712-9.htm.

Regulation on Amendment of the Packaging Waste Control Regulation, Official Gazette. (2020). March 13. Available at: https://www.resmigazete.gov.tr/eskiler/2020/03/20200313-4.htm.

Sharholy M., Ahmad K., Mahmood G., Trivedi R. C. (2008), "Municipal solid waste management in Indian cities – a review", *Waste Management*, 28, 459–467.

Tınmaz E., Demir İ. (2006), "Research on solid waste management system: To improve existing situation in Corlu Town of Turkey", *Waste Management*, 26, 307–314.

Zhao J., Huang L. (2019), "Multi-period network design problem in regional hazardous waste management systems", *International Journal of Environmental Research and Public Health*, 16(11), 2042.

9 Comparison of Sensor-Embedded Closed-Loop Supply Chain Systems with Regular Systems

Mehmet Talha Dulman[1], Surendra M. Gupta[1], and Tetsuo Yamada[2]
[1]Department of Mechanical and Industrial Engineering, College of Engineering, Northeastern University, Boston, MA 02115, USA
[2]Department of Informatics, The University of Electro-Communications, Chofu, Tokyo

1 INTRODUCTION

The uncertainties that are associated with the quality and quantity of returned household appliances makes managing product recovery very problematic. This study proposes the use of embedded sensors to reduce this. The sensors are embedded in the products and subsequently monitor these products throughout their life cycles. Through collating data related to their performance, the use patterns of the products can be detected and can be used to determine the conditions of the products, and their components. In addition, this can help producers make reprocessing decisions prior to disassembly—if the components are not suitable for reuse, the disassembly can be delayed, resulting in cost savings. Disassembled components will not need further inspection because the sensors provide information about their condition. Moreover, planning for reuse and remanufacturing can be executed earlier, thereby reducing inventory and stock-out costs.

Embedded sensors can also benefit maintenance operations. By monitoring products in use with sensors, failures can be detected before they occur. This enables producers to proactively provide required maintenance such as repairing or replacing deteriorated components before the failure occurs. As such, downtime due to failure can be reduced. Furthermore, an additional inspection for maintenance purposes will not be required since the sensors provide condition information about the products.

In this chapter, the use of sensors in a closed-loop supply chain system—including maintenance of the products and their end-of-life (EOL) processing—is studied. To determine the impact the sensors have on the system, regular (RDWD) and sensor-embedded (SEDWD) dishwasher and dryer systems were developed. These systems were modeled using discrete event simulation. The models were subsequently tested by the design of experiments. Several performance measures, such as disassembly cost, inspection cost, maintenance cost, sales revenues, and total profits of both systems, were compared. Pairwise t-tests were performed to determine the statistical significance of the differences between the performance measures.

2 LITERATURE REVIEW

This section provides a brief overview of the existing literature on remanufacturing, disassembly-to-order systems, maintenance, and sensor-embedded products. Important survey papers about closed-loop supply chains and product recovery were used to generate meaningful insights into the research that has already been conducted in this area.

A systematic review of the studies relevant to this area that were published before 1998 was developed by Gungor and Gupta (1999). This was later supplemented with a further survey that covered the period of 1998-2010 (Ilgin and Gupta, 2010a) and classified the papers published in the area of environmentally conscious manufacturing and product recovery (ECMPRO) into several categories — product design, reverse and closed-loop supply chains, remanufacturing, and disassembly. Gupta (2013) presented an overview of reverse supply chains and issues involved in their implementation. More recently, Ilgin et al. (2015) presented a state-of-the-art survey of existing research that introduced the use of multi-criteria decision-making (MCDM) techniques in ECMPRO. This paper was a more specialized version of Ilgin and Gupta's previous paper. A recent book on MCDM applications in the area of ECMPRO (Gupta and Ilgin, 2018). Govindan and Bouzon (2018) also presented a literature review with a multi-perspective framework for reverse logistics based on stakeholders' theory. Finally, a state-of-the-art review on the disassembly line balancing problem is provided by Ozceylan et al. (2019).

2.1 DISASSEMBLY-TO-ORDER SYSTEMS

Disassembly-to-order (DTO) systems aim to fulfill component demand by disassembling EOL products. Lambert and Gupta (2002) presented a tree network model to solve a DTO problem. To develop this method, the advantages of the two methods—disassembly graph approach and component-disassembly optimization—were combined and their disadvantages were excluded. Kongar and Gupta (2002) introduced a multi-criteria decision-making approach to solve the DTO problem. Their main objective was to find the optimum number of EOL products to be collected and disassembled to fulfill component demands. They also aimed to maximize the total profit and material sales revenue and minimize the disposal and inventory costs. Kongar and Gupta (2006) studied DTO systems under uncertainty and solved the DTO problem by adopting a fuzzy goal programming technique. Kongar and Gupta (2009a) used linear physical programming to solve the DTO problem. Kongar and Gupta (2009b) adopted a multiple objective tabu search approach to solve the DTO problem, while Imtanavanich and Gupta (2006a) attempted to solve it using an evolutionary computation technique with linear physical programming.

Inderfurth and Langella (2006) adapted two heuristics—one-to-one and one-to-many—to obtain a solution to the DTO problem with stochastic yields. Imtanavanich and Gupta (2006b) adopted goal programming to determine the number of returned products to be disassembled to fulfill several objectives and, thereby, solve the DTO problem. Imtanavanich and Gupta (2006c) used a genetic algorithm and fuzzy goal programming, and Imtanavanich et al. (2006) developed an artificial neural network model to solve the DTO problem.

2.2 REMANUFACTURING

Uncertainty about the qualities and quantities of returned products introduces challenges to product recovery operations. To solve this problem, researchers have conducted studies to identify the optimal production plans and schedules for remanufacturing operations. For additional information, see the book by Ilgin and Gupta (2012).

2.2.1 Production Planning

Jayaraman (2006) developed a mathematical model to plan the production and determine the number of cores to be acquired for a remanufacturing system. The model minimized the overall cost of the system—collection, disassembly, remanufacturing, and disposal costs. The model was applied to solve a problem described within a case study on mobile phones. Kim et al. (2006) proposed a mathematical model to solve a supply planning problem for remanufacturing systems. The model aimed to determine the optimal supply for remanufactured goods. According to this model, the components can be purchased from external suppliers or can be acquired by disassembling EOL products and refurbishing the reusable elements of the disassembled components. The optimal quantity helps maximize the cost difference between purchasing new components and refurbishing existing parts. Lu et al. (2006) presented a mathematical model to plan short-term bulk recycling. The model determined which products should be accepted for recycling, what level of processing was needed for the parts, and their cleanup levels. DePuy et al. (2007) developed a model to plan production for remanufacturing products. Supply quality is uncertain in remanufacturing; this paper highlighted this issue by dividing the acquired components into three levels. To be used directly within manufacturing, the components need to be in good condition—they might require some level of repair or may be non-operable. Remanufacturing can only be planned by addressing the issues pertaining to those levels because the conditions are uncertain prior to disassembly. DePuy et al. estimated the expected number of remanufactured products according to variations in the conditions of the components. Xanthopoulos and Iakovou (2009) applied aggregate planning to a remanufacturing problem and developed a model that aimed to optimize the number of EOL products that were collected, disassembled, stored, recycled, and remanufactured.

2.2.2 Scheduling

Guide (1997) proposed the use of priority dispatching rules (PDR), such as first come first served (FCFS), shortest processing time (SPT), longest processing time (LPT), and earliest due date (EDD) to release items to be sent to the remanufacturing area after product disassembly. Guide and Srivastava (1997) analyzed a batching strategy when releasing items after they were disassembled according to the problem discussed in Guide (1997). Guide et al. (1998) later extended the study by incorporating product structure complexity. They analyzed the impact of proactive expediting policies on different product structure complexities. These policies changed the priority of some parts when a certain amount of a product was reassembled. This method was referred to as acceleration. Four rules were analyzed in this article: SPT with acceleration, EDD with acceleration, lowest level BOM with acceleration, and highest level BOM with acceleration. These rules were applied to products of different structural complexities. Guide et al. (1999) also analyzed the effect of lead time variation. In a later study, Guide et al. (2000) developed a new product recovery process that involved repairing disassembled components and using them to remanufacture products and compared the performance of the schedules with different PDRs. Guide et al. (2005) also observed scheduling rules in a two-product disassembly and remanufacturing system used in shared facilities.

2.3 MAINTENANCE

De Carlo and Arleo (2013) proposed an optimization model to find the optimal maintenance strategy for an HVAC system in a pharmaceutical laboratory. They considered three maintenance strategies: corrective, time-based, and condition-based. The results indicated that the corrective maintenance strategy was the cheapest approach. The researchers also pointed out that preventive maintenance could reduce costs if the maintenance intervals are selected carefully. Topal and Ramazan (2010) introduced a new MIP model to identify the optimal mine equipment scheduling while minimizing the maintenance cost. Based on the data acquired, they stated that the application of their optimization model could reduce maintenance costs by 10-25%. Sarker and Faiz (2016) presented a study to determine the preventive maintenance schedules of offshore wind turbines. Maintenance was provided by grouping the wind turbines based on their ages. Different age groups received distinct levels of maintenance.

2.4 SENSOR-EMBEDDED PRODUCTS

Jun et al. (2007) introduced the concept of closed-loop product life cycle management. Based on this concept, the information about the life cycle of a product was captured, stored in a database, and used for real-time and EOL operations. The researchers argued that the data could be used to inform the design phase to develop products that last longer. In addition, real-time information can be useful for predictive maintenance and can reduce maintenance costs. Last, but not least, EOL information can help operators make appropriate decisions about the reuse, remanufacture, or recycle of the products and/or their components. Karlsson (1997) discussed the use of sensors to automate the disassembly process of a product and examined the use of the information collected by the sensors within recycling systems. Karlsson (1998) extended his previous study by applying fuzzy logic to address the uncertainty of the sensors. Klausner et al. (1998) proposed the use of a novel circuit that could collect information about electric motors during their use phase and store this information in an electronic data log. This data could then be used to support reuse decisions. Reusing electric motors can save costs because manufacturing costs significantly higher than reusing. Klausner et al. (1998) suggested embedding sensors into products and using them to collect information about those products. They also discussed the application of this information within reuse or remanufacturing strategies. The study concluded that sensors could reduce manufacturing and misclassification costs.

Vadde et al. (2008) introduced the idea of embedding sensors into the products and subsequently monitoring them to retrieve condition information about their components, thereby facilitating an estimation of their remaining life spans. Ilgin and Gupta (2010b) modeled a sensor-embedded product system on a multi-product disassembly line for refrigerators and washing machines using discrete event simulation. The performance of the sensor-embedded system was compared with conventional systems. A design of experiments study was carried out, and the economic benefits of the sensors were calculated. Ilgin and Gupta (2011a) assessed the benefits of sensors in a single-product disassembly line consisting of sensor-embedded air conditioners. Ilgin and Gupta (2011b) changed the direction of sensor application towards electronic products that are smaller in weight and have different disassembly requirements compared to appliances. They explored the impact of sensors when they are embedded into computers by conducting a design of experiments study. Ilgin and Gupta (2011c) experimented further on the multi-product disassembly line case presented in Ilgin and Gupta's (2010b) paper but tested a single-product disassembly line of washing machines. Ilgin et al. (2011) examined a multi-product disassembly line case involving different appliances that were closely related in terms of mechanism, such as dishwashers and dryers. Ilgin et al. (2014) investigated sensor-embedded systems by exploring a case study of a multi-product disassembly line for three products: ranges, ovens, and cooktops.

A series of papers have described or used sensor-embedded systems (Ondemir & Gupta, 2012; Ondemir et al., 2012; Ondemir & Gupta, 2013a,b; Ondemir & Gupta, 2014a,b; Alqahtani and Gupta, 2017; Zhang et al., 2018). These systems were designed to identify the optimal disassembly plan to fulfill remanufactured product demand, refurbished product demand, and material demand. These studies provided a major contribution to the field by presenting methods where embedded sensors could be used to track the remaining life of the components disassembled from EOL products. The information retrieved from the sensors was used to classify the components into different bins according to their quality, and the market needs—in terms of quality and price—were subsequently fulfilled by remanufacturing the products or selling the components directly.

Dulman and Gupta (2015a,b,c) evaluated the benefits of sensors that were embedded into cell phones. Dulman and Gupta (2016) enhanced the idea of using sensors to improve maintenance operations of the products by predicting failures and providing proper maintenance to the products before they fail.

3   SYSTEM DESCRIPTION

In this study, two systems were considered: SEDWD and RDWD systems. RDWD systems are traditional closed-loop supply chain systems that include maintaining dishwashers and dryers during their life cycles and processing the returned dishwashers and dryers at the end. In SEDWD systems, sensors are embedded into dishwashers and dryers, and condition information about the components is used to improve the performance of several processes—inspection of the products when they fail and disassembly and inspection of the products. Maintenance and EOL processes were considered as two subsystems in this study. Figure 9.1 provides an overview of these two subsystems within closed-loop supply chains designed for dishwashers and dryers.

The production and sales aspects of a supply chain are not included in the systems. In the process modeled as part of this research, systems commence after the dishwashers and dryers are produced and sold. In the first part of the systems, the primary goal is to maintain the products at the lowest possible cost until they complete their useful lives. If the products fail during their useful lives, the associated maintenance and service activities are provided. In the second part of the system, the goal is to enhance EOL profit by remanufacturing, refurbishing, or recycling the returned products. The second part of the process begins when the EOL dishwashers and dryers are collected from the customers.

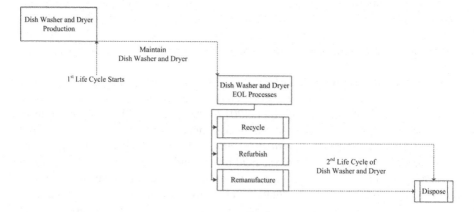

FIGURE 9.1   Maintenance and EOL Life-Cycle Scheme of Dishwasher and Dryers.

3.1 MAINTENANCE OF DISHWASHERS AND DRYERS

In these multi-product systems, dishwashers and dryers are treated with exponential interarrival times when the product use phase begins. The use phase continues until the appliances completes their useful lives. If they fail during that period, maintenance is provided. It is assumed that a corrective maintenance strategy is used in RDWD systems and a predictive maintenance strategy is used in SEDWD systems. When a corrective maintenance strategy is employed, the dishwashers and dryers receive maintenance service only when they fail; they are not serviced periodically. When machines fail, the associated components or subassemblies are replaced instead of repaired. Within the predictive maintenance strategy, the components are replaced with new ones before they fail. The need for replacement is determined using condition-monitoring devices (in this case, sensors). The purpose of replacing failed components with new ones is the performance enhancement of EOL processes by the increase in quality of the returned dishwashers and dryers. At the end of the first part of the system, the products that complete their useful lives are collected and transferred to the EOL facility. Figure 9.2 provides an outlook on the life cycle of dishwashers and dryers.

3.1.1 Maintenance of RDWD Systems

The first process in the maintenance element of the systems involves failure recognition and service activation. These processes cover the inspection of the failed dishwashers and dryers to determine the failed component (or components) and activate the required service operations. Following these processes, failed dishwashers or dryers are transported to the service facility if the failure cannot be taken care of on-site. Otherwise, the service team brings new components to replace the failed ones. The next step involves replacing the failed components, as seen in Figure 9.3. Once the replacement is completed and the products are fixed, they are delivered to the customers (if the service operations are provided in the service facility).

The first part of the system incurs maintenance costs, and these are shown in Figure 9.3. In the first process, recognition of failure, inspection costs occur. Any transportation, such as the transportation of products to the service facility, transportation of the service team, and product deliveries are considered to be logistics costs. Moreover, the replacement of components increases maintenance costs (either new components and subassemblies or their materials are purchased). Finally, a crucial cost item that is included in maintenance costs is the productivity

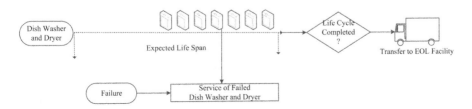

FIGURE 9.2 Life Cycle Outlook of Dishwashers and Dryers.

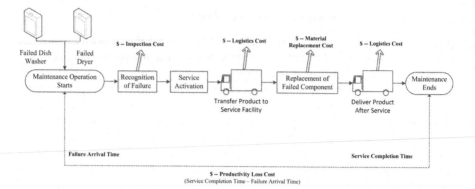

FIGURE 9.3   Maintenance Operations of RDWD Systems.

loss cost. These costs are important because they are directly related to customer satisfaction. When products are awaiting repair or being serviced, customers are not able to use them, and this is expressed as a productivity loss cost in the system. It is calculated as the function of the time between failure arrival and service completion time and product price. By using this function, productivity loss cost due to failure can be calculated.

3.1.2   Maintenance of SEDWD Systems

In SEDWD systems, maintenance operations can be improved by sensors that can track dishwashers and dryers throughout their life cycles and collate condition information about the components and subassemblies. Operators can interpret the information collected by the sensors to identify appropriate actions, such as replacing the components and subassemblies, prior to product failure. This provides the benefit of eliminating inspection processes. As can be seen in Figure 9.4, recognition of failure and service activation processes were removed from the first part of the SEDWD system.

By eliminating the recognition of failure and service activation processes, it is possible to eradicate inspection costs. This would result in lower maintenance costs in SEDWD systems. Moreover, the time products spend during failure would be less in SEDWD systems than in RDWD systems. This can be interpreted as lower productivity loss cost and a higher customer satisfaction rate by providing customers with faster maintenance service.

3.2   Dishwasher and Dryer EOL Processes

EOL dishwashers and dryers are collected either from customers or collection centers and are transported to the EOL facility. They are disassembled via a three-station disassembly line. At each station, the components or subassemblies are disassembled based on the required precedence relationship. For dishwashers, the first component is the metal cover, followed by subassembly of the circuit board. Finally, the motor is disassembled. For the dryer, the first and last

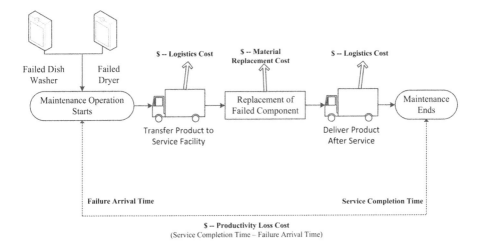

FIGURE 9.4 Maintenance Operations of SEDWD Systems.

components are the same as the dishwashers; however, the drum subassembly is disassembled at Station 2. Figures 9.5 and 9.6 present disassembly and inspection processes for each system.

3.2.1 EOL Processes of RDWD Systems

In RDWD systems, EOL dishwashers and dryers go through several decision gates during their disassembly and inspection processes. These decisions are made by considering the chance of a component to be missing, the reusability of the components, and their quality levels. Products visit stations and components are disassembled if they are available. Otherwise, the products are sent to the next station. The disassembled components are inspected. After the inspection process, reusability and quality levels are determined. The disassembly and inspection processes incur disassembly and inspection costs. The disassembly cost is calculated as a function of disassembly time and labor and operating costs. Inspection costs can be calculated by using parameters such as inspection time, labor cost, and machining cost. Upon completion of the inspection process and classification of the components and subassemblies, it is now possible to make informed reprocessing decisions. The disassembly and inspection processes associated with the RDWD systems are shown in Figure 9.5.

3.2.2 EOL Processes of SEDWD Systems

The disassembly and inspection processes are different between SEDWD and RDWD systems. Disassembly and inspection processes in SEDWD systems can be improved using the information collected by sensors. The disassembly and inspection processes that were assessed for the SEDWD systems can be seen in Figure 9.6.

The decisions that are made in RDWD systems are also made in SEDWD systems; however, they are made by sensors. The missing component and

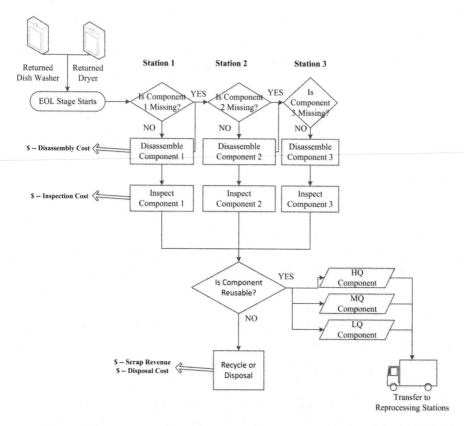

FIGURE 9.5 RDWD Systems Disassembly and Inspection Processes.

subassembly information is known prior to disassembly and proper action—sending the product to the next station without visiting—can be taken. If components are not missing, reusability information is used. If the components are not reusable, they are sent to the next station the same way as the missing parts because there is no need to disassemble them if they are not reusable. If the other component at the next station is not missing and is reusable, they are disassembled together. This reduces the overall disassembly cost. Inventory levels also need to be checked in RDWD systems. After labeling quality levels, inventory levels of each are checked. If there is enough inventory, the components are not disassembled. This can help avoid inventory costs and reduce disassembly costs.

3.2.3   Reprocessing in RDWD and SEDWD Systems

If the components and subassemblies are determined as reusable after disassembly, they are divided into three quality levels: high quality (HQ), medium quality (MQ), and low quality (LQ). Based on these quality levels, the expected life span of the components and subassemblies can be determined. If they are going to be used for resale or remanufacturing, these expected life spans are of importance when determining the price. The expected life span of the HQ-level components is between

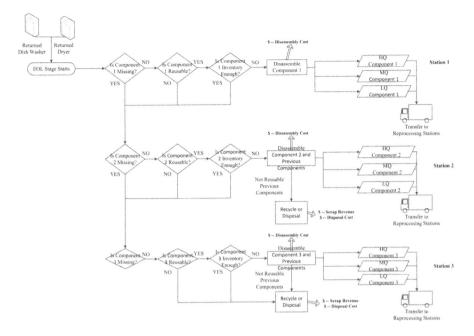

FIGURE 9.6   SEDWD Systems Disassembly and Inspection Processes.

three years and four years. MQ-level components can last around two to three years, and LQ-level components can last around one to two years. Components and subassemblies are reprocessed based on these quality levels. HQ-component flow after disassembly and inspection can be observed in Figures 9.7.

For reprocessing decisions, the inventory levels of each component, subassembly, and remanufactured products are crucial. For the HQ component and subassembly flow, the HQ remanufactured product inventory is checked first. If the maximum HQ remanufactured product inventory level is not reached, the components and subassemblies are transferred to the remanufacturing buffer area. If this level is reached, an alternative option is to resell the components and subassemblies. Within the systems, a further consideration applied within the study was of higher quality level components to replace lower quality parts, as opposed to recycling them if they are not used in their quality levels. For example, if HQ components and subassemblies are not used to remanufacture or resale, they are sent to a lower quality level, MQ. They replace the MQ parts, and the MQ components are sent to recycling. This replacement increases the quality level of the remanufactured products and extends their life span. Similarly, HQ-level components can also be used at LQ levels if they are not required at the MQ level.

Sales revenue, scrap revenue, and disposal cost measures were observed within the component and subassembly flow is assessed in this study according to the following process: sales revenues are generated by selling components and subassemblies; scrap revenue is generated by recycling components and subassemblies

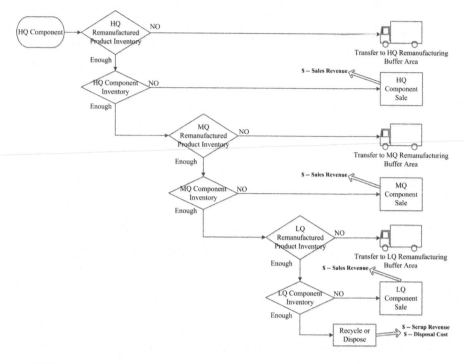

FIGURE 9.7 HQ-Component Flow for Recycle, Resale, and Remanufacturing.

that are no longer functional. Any component that cannot be recycled is disposed of. Figure 9.7 presents the HQ-component flow and relevant performance measures.

The MQ- and LQ-level flows are also similar to the HQ-level flow and are as follows: in the MQ-level flow, components and subassemblies are not sent to higher quality levels because this would reduce the quality level of the HQ products. However, they are sent to LQ-level components as replacement parts. On the other hand, LQ-level parts are only used at LQ-level quality levels.

The renewal-time threshold is used to determine the quality levels of the components and subassemblies. The time that components have been in use—either from the beginning of their life cycles or after their renewal if they were replaced with new ones—is used to compute the renewal-time threshold. If this threshold is exceeded, the expected condition of the components is deemed to be less than that of those that did not exceed the threshold. For the purposes of this study, a renewal-time threshold of four years was applied to both dishwashers and dryers.

The reprocessing options for the HQ, MQ, and LQ component flows have previously been explained; however, there is also a need to elucidate on the remanufacturing process. As previously mentioned, within the process that was examined as part of this study, components and subassemblies are sent to the remanufacturing buffer area. Figure 9.8 shows the remanufacturing element of the systems. The components and subassemblies wait in the remanufacturing buffer area until the full set of components and subassemblies required to assemble a product is complete. If one of the parts is missing, the product cannot be assembled.

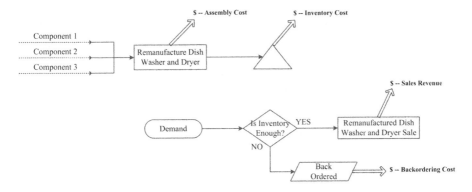

FIGURE 9.8   Remanufacturing and Demand Flow of Dishwashers and Dryers.

The assembly costs of the remanufacturing products can be seen in Figure 9.8. Remanufactured dishwashers and dryers are stored in the inventory if there is no demand. This increases the holding cost.

For this study, demand follows a Poisson distribution. Remanufactured dishwashers and dryers are sold if there is enough inventory to fulfill the demand. Sales revenue is generated by selling these remanufactured dishwashers and dryers. If the inventory is not sufficient, demand is backordered—this increases backordering costs. In addition to remanufacturing, demand flow is also shown in Figure 9.8.

4   DESIGN OF EXPERIMENTS STUDY

A design of experiments study was used to outline the economic differences between the two systems. The primary goal of this study was to understand how the benefits of using sensors can be expressed numerically so that a recommendation can be made for the implementation of these sensor-embedded systems. A total of 63 factors, each with two levels, were determined to have an impact on the economic outcomes of the systems. It was acknowledged that by using these factors, constructing a full factorial design would lead to an excessive number of experiments. An orthogonal arrays method was used for the factorial design that provided a set of experiments that facilitates an unbiased design of experiments study while reducing the required number to a reasonable size. Based on this method, the $L_{64}(2^{63})$ orthogonal array was chosen for the factorial design (Phadke, 1989). In the study, 64 experiments for each system were carried out.

Ilgin et al. (2011) conducted a study that aimed to understand the economic benefits of using sensors in a multi-product disassembly line by examining their use in products such as dishwashers and dryers. Their paper was used as a reference during data collection of the study. The factors that were considered are explained in depth below.

The prices of remanufactured dishwashers, dryers, and their subassembly and component prices are presented in Table 9.1 and the prices of HQ- and MQ-level components were included. LQ-level component prices were excluded because their impact was not deemed to be significant.

TABLE 9.1

Remanufactured Dishwasher, Dryer, Subassembly, and Component Prices

Factor	Level 1	Level 2
HQ Dishwasher ($)	400	360
HQ Dishwasher Metal Cover ($)	75	50
HQ Dishwasher Circuit Board ($)	100	75
HQ Dishwasher Motor ($)	120	80
MQ Dishwasher ($)	320	280
MQ Dishwasher Metal Cover ($)	40	20
MQ Dishwasher Circuit Board ($)	60	40
MQ Dishwasher Motor ($)	70	60
HQ Dryer ($)	550	500
HQ Dryer Metal Cover ($)	75	50
HQ Dryer Drum ($)	30	25
HQ Dryer Motor ($)	180	160
MQ Dryer ($)	450	400
MQ Dryer Metal Cover ($)	40	20
MQ Dryer Drum ($)	20	15
MQ Dryer Motor ($)	150	120

The disassembly times for each station were included in the factorial design, and their values can be seen in Table 9.2.

In addition to the disassembly times, the inspection times at each station were also included in the factorial design. These factors and their levels are presented in Table 9.3.

The probability of missing components, the chance of retrieving the components in a reusable form, and the probability distributions for the quality levels are illustrated in Table 9.4. It was anticipated that performance measures would be directly affected by any change in these probabilities. Thus, HQ- and MQ-level probabilities were included in the factorial design. LQ-level probabilities were excluded from the factorial design because once HQ- and MQ-level probabilities were identified,

TABLE 9.2

Disassembly Time Factors

Factor	Level 1	Level 2
Time at Station 1 (min)	2	1.5
Time at Station 2 (min)	3	2
Time at Station 3 (min)	3	2

TABLE 9.3
Inspection Time Factors

Factor	Level 1	Level 2
Time at Station 1 (min)	2.5	2
Time at Station 2 (min)	6	5
Time at Station 3 (min)	12	10

TABLE 9.4
Probability Factors (%)

Factor	Level 1	Level 2
Missing Dishwasher Circuit Board	12	18
Missing Dishwasher Motor	12	18
Missing Dryer Drum	12	18
Missing Dryer Motor	12	18
Usable Dishwasher Circuit Board	88	76
Usable Dishwasher Motor	88	76
Usable Dryer Drum	88	76
Usable Dryer Motor	88	76
Dishwasher HQ Metal Cover Return	55	50
Dishwasher MQ Metal Cover Return	30	30
Dishwasher HQ Circuit Board Return	55	50
Dishwasher MQ Circuit Board Return	30	30
Dishwasher HQ Motor Return	55	50
Dishwasher MQ Motor Return	30	30
Dryer HQ Metal Cover Return	55	50
Dryer MQ Metal Cover Return	30	30
Dryer HQ Drum Return	55	50
Dryer MQ Drum Return	30	30
Dryer HQ Motor Return	55	50
Dryer MQ Motor Return	30	30

LQ-level probabilities could be calculated through them. LQ-level probabilities are shown in Table 9.5.

The renewal time threshold was explained previously. It was assumed that, if the use period of the components exceeds the renewal-time threshold, the quality levels of these components decreased. The reduced probabilities that were applied to components that exceeded their renewal-time threshold are shown in Table 9.6.

The factorial design also included remanufactured dishwasher and dryer demands and their subassembly and component demand because revenue can be generated by fulfilling them. The LQ-level demands were insignificant factors and

TABLE 9.5

Low-Quality Subassembly and Component Probabilities (%)

Factor	Level 1	Level 2
Dishwasher LQ Metal Cover Return	15	20
Dishwasher LQ Circuit Board Return	15	20
Dishwasher LQ Motor Return	15	20
Dryer LQ Metal Cover Return	15	20
Dryer LQ Drum Return	15	20
Dryer LQ Motor Return	15	20

TABLE 9.6

Quality Level Probabilities When Renewal-Time Threshold is Exceeded (%)

Factor	Level 1	Level 2
Dishwasher HQ Metal Cover Return	25	20
Dishwasher MQ Metal Cover Return	30	30
Dishwasher LQ Metal Cover Return	45	50
Dishwasher HQ Circuit Board Return	25	20
Dishwasher MQ Circuit Board Return	30	30
Dishwasher LQ Circuit Board Return	45	50
Dishwasher HQ Motor Return	25	20
Dishwasher MQ Motor Return	30	30
Dishwasher LQ Motor Return	45	50
Dryer HQ Metal Cover Return	25	20
Dryer MQ Metal Cover Return	30	30
Dryer LQ Metal Cover Return	45	50
Dryer HQ Drum Return	25	20
Dryer MQ Drum Return	30	30
Dryer LQ Drum Return	45	50
Dryer HQ Motor Return	25	20
Dryer MQ Motor Return	30	30
Dryer LQ Motor Return	45	50

were not included in the factorial design. Table 9.7 provides the demand factors and their associated levels.

Additional factors, such as the recognition of failure and service activation, inspection for failure recognition, maintenance labor cost, productivity loss cost, and labor and operation costs of disassembly and inspection were included in the factorial design and can be seen in Table 9.8. The inspection time of the failure was deemed to be normally distributed; the mean and standard deviation values are presented in parentheses in Table 9.8.

TABLE 9.7

Remanufactured Dishwasher, Dryer, and Component Demands

(Follows Poisson Distribution)		
Factor	Level 1	Level 2
HQ Dishwasher (per day)	100	90
MQ Dishwasher (per day)	80	70
HQ Dishwasher Metal Cover (per day)	100	90
MQ Dishwasher Metal Cover (per day)	80	70
HQ Dishwasher Circuit Board (per day)	100	90
MQ Dishwasher Circuit Board (per day)	80	70
HQ Dishwasher Motor (per day)	100	90
MQ Dishwasher Motor (per day)	80	70
HQ Dryer (per day)	100	90
MQ Dryer (per day)	80	70
HQ Dryer Metal Cover (per day)	100	90
MQ Dryer Metal Cover (per day)	80	70
HQ Dryer Drum (per day)	100	90
MQ Dryer Drum (per day)	80	70
HQ Dryer Motor (per day)	100	90
MQ Dryer Motor (per day)	80	70

TABLE 9.8

Additional Factors

Factor	Level 1	Level 2
Recognition and Service Activation	3 days	2 days
Inspection for Recognition (min) (NORM)	(25,2)	(20,2)
Labor Cost for Maintenance	$30/hour	$20/hour
Productivity Loss Cost	$3/day	$2/day
Labor and Operation Cost ($/min)	2.25	1.5

A discrete event simulation was used to model the systems. The discrete event simulation software used in this study was Arena 14.7 (Kelton et al., 2007). Extreme input values were used in testing the outcomes of the models, and the models were verified using this method. The models were validated by plotting some of the performance measures and observing them through simulation runs. The run length of the models was 4,000 days. This was sufficient to cover the expected life span of the products (i.e. seven years for dishwashers and ten years for dryers). In addition to the expected life spans, the run length was sufficient to complete the EOL processes. A total of 64 experiments were carried out for both systems. Additional data was required to run the models, and this is presented in Appendix A.

The purpose of the study was to determine the economic value of embedded sensors by evaluating the contribution they made to closed-loop supply chain systems that included multi-product service operations and a multi-product disassembly line. To determine this economic value, different performance measures, such as maintenance cost, total revenue, total cost, and total profit of the systems, were assessed and observed within the design of experiments study.

$$\textbf{\textit{Maintenance cost}} = \textit{Labor Cost} + \textit{Logistics Cost} + \textit{Productivity Loss Cost}$$
$$+ \textit{Material Replacement Cost} \tag{9.1}$$

$$\textbf{\textit{Total Profit}} = \textit{Total Revenue} - \textit{Total Cost} \tag{9.2}$$

$$\textbf{\textit{Total Revenue}} = \textit{Sales Revenue} + \textit{Scrap Revenue} \tag{9.3}$$

$$\textbf{\textit{Total Cost}} = \textit{Collection Cost} + \textit{Disassembly Cost} + \textit{Inspection Cost}$$
$$+ \textit{Remanufacturing} \quad \textit{Cost} + \textit{Holding Cost} + \textit{Backordering Cost}$$
$$+ \textit{Disposal Cost} \tag{9.4}$$

$$\textbf{\textit{Sensor Value}} = (\textit{Maintenance Cost Savings} + \textit{Total Profit Improvement})/$$
$$\textit{Total Number of Sensors} \tag{9.5}$$

5   RESULTS AND ANALYSIS

The systems assessed in this study were divided into two parts: maintenance and EOL. The maintenance costs were related to the performance of the maintenance aspect of the systems, whereas the EOL profit provided an overview of the economic performance of the EOL processes of the systems. The maintenance costs were $216.65 for the SEDWD systems and $224.08 for the RDWD systems. On the other hand, the mean value of the EOL profit per product was $142.11 for SEDWD systems and $111.34 for RDWD systems. The sensor-embedded systems had lower maintenance costs and generated higher EOL profit.

The disassembly and inspection cost savings associated with the use of sensors can also be observed in Table 9.9. The disassembly costs were calculated to be $421,574.30 for SEDWD systems and $469,912.30 for RDWD systems. As such, disassembly costs were lower in sensor-embedded systems. Moreover, the inspection cost of SEDWD systems was $15,129.14, while it was $1,294,780.00 for RDWD systems—the inspection costs were almost eliminated in SEDWD systems. The total profits of the systems are shown in Table 9.9.

TABLE 9.9
Experiment Results for SEDWD and RDWD Systems

Measure	SEDWD System ($)	RDWD System ($)
EOL Profit/Product	142.11	111.34
Maintenance Cost/Product	216.65	224.08
Total Profit	6122517.00	4796805.00
Disassembly Cost	421574.30	469912.30
Inspection Cost	15129.14	1294780.00

The system performance was assessed by comparing the mean values. However, it was also important to test the significance of the differences between the systems to determine the superiority of the SEDWD systems. Thus, a pairwise *t*-test was used. The pairwise *t*-test was chosen because the factors are unique in each experiment based on the orthogonal array method. The results of the pairwise *t*-test are shown in Table 9.10.

Based on the *t*-test results, the mean difference in EOL profit between the systems was calculated to be $30.77. As such, the SEDWD system was more profitable than the RDWD system. Moreover, the mean difference in maintenance cost between the systems was $-7.43. Based on this calculation, it can be concluded that SEDWD systems have lower maintenance costs than RDWD systems. In addition to the mean differences, the *p*-values are also presented in Table 9.10. The *p*-values were less than 0.0001 and it was concluded that the mean differences between the systems were statistically significant. The disassembly and inspection costs together with the total profit comparisons and their *t*-test results are shown in Table 9.10.

Within this study, the value of a sensor was the combination of the EOL profit improvement and the maintenance cost savings. The increase in EOL profit and savings in maintenance costs associated with the use of the sensor added up to $38.21. If sensors can be obtained at a price lower than $38.21, embedding sensors

TABLE 9.10
Pairwise *t*-Test Results for Mean Difference

Measure	Mean Difference (SEDWD-RDWD) ($)	*p*-Value
Sensor Value	38.21	N/A
EOL Profit/Product	30.77	< 0.0001
Maintenance Cost/Product	−7.43	< 0.0001
Total Profit	1325712.00	< 0.0001
Disassembly Cost	−48338.03	< 0.0001
Inspection Cost	−1279651.00	< 0.0001

TABLE 9.11

95% Confidence Interval of Mean Difference

Measure	95% Confidence Interval of Mean Difference (SEDWD-RDWD)	
	Lower Limit ($)	Upper Limit ($)
Sensor Value	36.47	39.94
EOL Profit/Product	29.15	32.39
Maintenance Cost / Product	−8.06	−6.81
Total Profit	1255845.00	1395579.00
Disassembly Cost	−52456.36	−44219.69
Inspection Cost	−1347476.00	−1211826.00

into dishwashers and dryers will yield a positive return. In addition to the mean differences, the confidence intervals are also presented in Table 9.11.

6 CONCLUSIONS

Embedding sensors into dryers and dishwashers to improve maintenance performance and EOL processing phases of the closed-loop supply chain systems was proposed in this study. Two systems were constructed: RDWD and SEDWD systems. The systems were modeled using discrete event simulations, and the design of experiments was used to carry out the tests for the systems. The results revealed that SEDWD systems perform better than RDWD systems in terms of maintenance costs and EOL profits. This can be attributed to sensors providing valuable information about the condition of the components and subassemblies, thereby providing maintenance prior to failures; disassembly and inspection processes are planned more efficiently. SEDWD systems reduced the maintenance costs associated with the RDWD systems by $7.43 per product. Furthermore, the EOL profit of the SEDWD systems was higher than that of the RDWD systems by as much as $30.77 per product. Combining these values, the total value of the sensor was calculated to be $38.21 per unit.

REFERENCES

Alqahtani, A. Y., and Gupta, S. M. (2017). Warranty as a marketing strategy for re-manufactured products. *Journal of Cleaner Production*, 161, 1294–1307.

De Carlo, F., and Arleo, M. A. (2013). Maintenance cost optimization in condition based maintenance: a case study for critical facilities. *International Journal of Engneering And Technology*, 5(5), 4296–4302.

DePuy, G. W., Usher, J. S., Walker, R. L., and Taylor, G. D. (2007). Production planning for remanufactured products. *Production Planning and Control*, 18(7), 573–583.

Dulman, M. T., and Gupta, S. M. (2015a). Benefits of sensors in cell phone disassembly and remanufacturing. *Proceedings for the Northeast Region Decision Sciences Institute (NEDSI)*, Cambridge, Massachusetts, March 20–22, pp. 1–5.

Dulman, M. T., and Gupta, S. M. (2015b). Use of sensors for cell phones. *Proceedings for the Production and Operations Management Society (POMS)*-WashingtonMeeting, Washington, DC, May 8–May 11, pp. 1–5.

Dulman, M. T., and Gupta, S. M. (2015c). "Disassembling and remanufacturing end-of-life sensor embedded cell phones", *Innovation and Supply Chain Management*, 9(4), 111–117.

Dulman, M. T., and Gupta, S. M. (2016). "Use of sensors for collection of end-of-life products", *Proceedings for the Northeast Region Decision Sciences Institute (NEDSI)*, Alexandria, Virginia, March 31–April 2, pp. 1–6.

Govindan, K., and Bouzon, M. (2018). From a literature review to a multi-perspective framework for reverse logistics barriers and drivers. *Journal of Cleaner Production*, 187(1), 318–337.

Guide Jr., V. D. R. (1997). Scheduling with priority dispatching rules and drum-buffer-rope in a recoverable manufacturing system. *International Journal of Production Economics*, 53(1), 101–116.

Guide Jr., V. D. R., and Srivastava, R. (1997). An evaluation of order release strategies in a remanufacturing environment. *Computers and Operations Research*, 24(1), 37–47.

Guide Jr., V. D. R., Srivastava, R., and Kraus, M. E. (1998). Proactive expediting policies for recoverable manufacturing. *Journal of the Operational Research Society*, 49(5), 479–491.

Guide, V. D. R., Jayaraman, V., and Srivastava, R. (1999). The effect of lead time variation on the performance of disassembly release mechanisms. *Computers and Industrial Engineering*, 36(4), 759–779.

Guide Jr., V. D. R., Srivastava, R., and Kraus, M. E. (2000). Priority scheduling policies for repair shops. *International Journal of Production Research*, 38(4), 929–950.

Guide, V. D. R., Souza, G. C., and van der Laan, E. (2005). Performance of static priority rules for shared facilities in a remanufacturing shop with disassembly and reassembly. *European Journal of Operational Research*, 164(2), 341–353.

Gungor, A., and Gupta, S. M. (1999). Issues in environmentally conscious manufacturing and product recovery: a survey. *Computers and Industrial Engineering*, 36(4), 811–853.

Gupta, S. M. (2013). *Reverse Supply Chains: Issues and Analysis*, CRC Press, Boca Raton, Florida, ISBN: 978-1439899021.

Gupta, S. M., and Ilgin, M. A. (2018). *Multiple Criteria Decision Making Applications in Environmentally Conscious Manufacturing and Product Recovery*, CRC Press, Boca Raton, Florida, ISBN: 978-1498700658.

Ilgin, M. A., and Gupta, S. M. (2010a). Environmentally conscious manufacturing and product recovery (ECMPRO): a review of the state of the art. *Journal of Environmental Management*, 91(3), 563–591.

Ilgin, M. A., and Gupta, S. M. (2010b). Comparison of economic benefits of sensor embedded products and conventional products in a multi-product disassembly line. *Computers and Industrial Engineering*, 59(4), 748–763.

Ilgin, M. A., and Gupta, S. M. (2011a). Evaluating the impact of sensor-embedded products on the performance of an air conditioner disassembly line. *The International Journal of Advanced Manufacturing Technology*, 53(9–12), 1199–1216.

Ilgin, M. A., and Gupta, S. M. (2011b). Performance improvement potential of sensor embedded products in environmental supply chains. *Resources, Conservation and Recycling*, 55(6), 580–592.

Ilgin, M. A., and Gupta, S. M. (2011c). Recovery of sensor embedded washing machines using a multi-kanban controlled disassembly line. *Robotics and Computer-Integrated Manufacturing*, 27(2), 318–334.

Ilgin, M. A., and Gupta, S. M. (2012). *Remanufacturing Modeling and Analysis*, CRC Press, Boca Raton, Florida, ISBN: 9781439863077.

Ilgin, M. A., Gupta, S. M., and Nakashima, K. (2011). Coping with disassembly yield uncertainty in remanufacturing using sensor embedded products. *Journal of Remanufacturing*, 1(1), 1–14.

Ilgin, M. A., Ondemir, O., and Gupta, S. M. (2014). An approach to quantify the financial benefit of embedding sensors into products for end-of-life management: a case study. *Production Planning & Control*, 25(1), 26–43.

Ilgin, M. A., Gupta, S. M., and Battaïa, O. (2015). Use of MCDM techniques in environmentally conscious manufacturing and product recovery: State of the art. *Journal of Manufacturing Systems*, 37, 746–758.

Imtanavanich, P., and Gupta, S. M. (2006a). Evolutionary computation with linear physical programming for solving a disassembly-to-order system. In *the SPIE International Conference on Environmentally Conscious Manufacturing* VI (pp. 30–41). Boston, MA.

Imtanavanich, P., and Gupta, S. M. (2006b). Calculating disassembly yields in a multi-criteria decision making environment for a disassembly-to-order system. Lawrence, K. D., and Klimberg, R. K. eds. *Application of Management Science: In Productivity, Finance, and Operations*. Elseveier Ltd., 109–125.

Imtanavanich, P., and Gupta, S. M. (2006c). Solving a disassembly-to-order system by using genetic algorithm and weighted fuzzy goal programming. In *the SPIE International Conference on Environmentally Conscious Manufacturing* VI (pp. 54–56). Boston, MA.

Imtanavanich, P., Gupta, S. M., and Nakashima, K. (2006). Solving a disassembly-to-order problem using neural networks. In *The Seventh International Conference on EcoBalance*, Tsubuka, Japan.

Inderfurth, K., and Langella, I. M. (2006). Heuristics for solving disassemble-to-order problems with stochastic yields. *OR Spectrum*, 28(1), 73–99.

Jayaraman, V. (2006). Production planning for closed-loop supply chains with product recovery and reuse: An analytical approach. *International Journal of Production Research*, 44(5), 981–998.

Jun, H. B., Kiritsis, D., and Xirouchakis, P. (2007). Research issues on closed-loop PLM. *Computers in Industry*, 58(8), 855–868.

Karlsson, B. (1997, May). A distributed data processing system for industrial recycling. In *Instrumentation and Measurement Technology Conference, 1997. IMTC/97. Proceedings. Sensing, Processing, Networking., IEEE* (Vol. 1, pp. 197–200). IEEE.

Karlsson, B. (1998, May). Fuzzy handling of uncertainty in industrial recycling. In *Instrumentation and Measurement Technology Conference, 1998. IMTC/98. Conference Proceedings. IEEE* (Vol. 2, pp. 832–836). IEEE.

Kelton, D. W., Sadowski R. P., and Sadowski, D. A. (2007). *Simulation with Arena*. (4th ed.). New York: McGraw-Hill.

Kim, K., Song, I., Kim, J., and Jeong, B. (2006). Supply planning model for remanufacturing system in reverse logistics environment. *Computers and Industrial Engineering*, 51(2), 279–287.

Klausner, M., Grimm, W. M., and Hendrickson, C. (1998). Reuse of electric motors in consumer products. *Journal of Industrial Ecology*, 2(2), 89–102.

Klausner, M., Grimm, W. M., Hendrickson, C., and Horvath, A. (1998, May). Sensor-based data recording of use conditions for product takeback. In *Electronics and the Environment, 1998. ISEE-1998. Proceedings of the 1998 IEEE International Symposium on* (pp. 138–143). IEEE.

Kongar, E., and Gupta, S. M. (2002). A multi-criteria decision making approach for disassembly-to-order systems. *Journal of Electronics Manufacturing*, 11(02), 171–183.

Kongar, E., and Gupta, S. M. (2006). Disassembly to order system under uncertainty. *Omega*, 34(6), 550–561.

Kongar, E., and Gupta, S. M. (2009a). Solving the disassembly-to-order problem using linear physical programming. *International Journal of Mathematics in Operational Research*, 1(4), 504–531.

Kongar, E., and Gupta, S. M. (2009b). A multiple objective tabu search approach for end-of-life product disassembly. *International Journal of Advanced Operations Management*, 1(2–3), 177–202.

Lambert, A. J. D., and Gupta, S. M. (2002). Demand-driven disassembly optimization for electronic products package reliability. *Journal of Electronics Manufacturing*, 11(02), 121–135.

Lu, Q., Williams, J. A. S., Posner, M., Bonawi-Tan, W., and Qu, X. (2006). Model-based analysis of capacity and service fees for electronics recyclers. *Journal of manufacturing systems*, 25(1), 45–57.

Ondemir, O., and Gupta, S. M. (2012). Optimal management of reverse supply chains with sensor-embedded end-of-life products. Lawrence K. D., and Kleinman G., eds. *Applications of Management Science*. Emerald Group Publishing Limited, Bingley, UK, 109–129.

Ondemir, O., and Gupta, S. M. (2013a). Advanced remanufacturing-to-order and disassembly-to-order system under demand/decision uncertainty. Gupta, S. M. eds. *Reverse Supply Chains: Issues and Analysis*. CRC Press, BocaRaton, Florida, 203–228.

Ondemir, O., and Gupta, S. M. (2013b). Quality assurance in remanufacturing with sensor embedded products. Nikolaidis Y. eds. *Quality Management in Reverse Logistics*. Springer-Verlag, London, 95–112.

Ondemir, O., and Gupta, S. M. (2014a). A multi-criteria decision making model for advanced repair-to-order and disassembly-to-order system. *European Journal of Operational Research*, 233(2), 408–419.

Ondemir, O., and Gupta, S. M. (2014b). Quality management in product recovery using the Internet of Things: An optimization approach. *Computers in Industry*, 65(3), 491–504.

Ondemir, O., Ilgin, M. A., and Gupta, S. M. (2012). Optimal end-of-life management in closed-loop supply chains using RFID and sensors. *Industrial Informatics, IEEE Transactions on*, 8(3), 719–728.

Ozceylan, E., Kalayci, C. B., Gungor, A., and Gupta, S. M. (2019). Disassembly line balancing problem: A review of the state of the art and future directions, *International Journal of Production Research*, 57(15–16), 4805–4827.

Phadke, M. S. (1989). *Quality Engineering Robust Design*. New Jersey: Prentice Hall.

Sarker, B. R., and Faiz, T. I. (2016). Minimizing maintenance cost for offshore wind turbines following multi-level opportunistic preventive strategy. *Renewable Energy*, 85, 104–113.

Topal, E., and Ramazan, S. (2010). A new MIP model for mine equipment scheduling by minimizing maintenance cost. *European Journal of Operational Research*, 207(2), 1065–1071.

Vadde, S., Kamarthi, S., Gupta, S. M., and Zeid, I. (2008). Product life cycle monitoring via embedded sensors. Gupta S. M., and sLambert A. J. D. Eds.*Environment Conscious Manufacturing*. CRC Press, Boca Raton, FL, 91–103.

Xanthopoulos, A., and Iakovou, E. (2009). On the optimal design of the disassembly and recovery processes. *Waste Management*, 29(5), 1702–1711.

Zhang, Y., Liu, S., Liu, Y., Yang, H., Li, M., Huisingh, D., and Wang, L. (2018). The 'Internet of Things' enabled real-time scheduling for remanufacturing of automobile engines. *Journal of Cleaner Production*, 185, 562–575.

Appendix A

TABLE A1
LQ Remanufactured Dishwasher, Dryer, and Component Prices

LQ Dishwasher ($)	150
LQ Dishwasher Metal Cover ($)	10
LQ Dishwasher Circuit Board ($)	30
LQ Dishwasher Motor ($)	50
LQ Dryer ($)	200
LQ Dryer Metal Cover ($)	10
LQ Dryer Drum ($)	10
LQ Dryer Motor ($)	80

TABLE A2
LQ Remanufactured Dishwasher, Dryer, and Component Demands
(Follows Poisson Distribution)

LQ Dishwasher (per day)	50
LQ Dishwasher Metal Cover (per day)	50
LQ Dishwasher Circuit Board (per day)	50
LQ Dishwasher Motor (per day)	50
LQ Dryer (per day)	50
LQ Dryer Metal Cover (per day)	50
LQ Dryer Drum (per day)	50
LQ Dryer Motor (per day)	50

TABLE A3
Maintenance Data

Dishwasher Failure Interarrival Time (day) (Exponential)	0.0695
Dryer Failure Interarrival Time (day) (Exponential)	0.0973
Dishwasher Expected Lifetime (year)	7
Dryer Expected Lifetime (year)	10
Transportation Before Service (day)	1
Delivery After Service (day) (Triangular Distribution)	(1,2,3)
Transportation Cost ($)	10
Delivery Cost ($)	60

TABLE A4
Subassembly and Component Replacement Times for Maintenance

(Normally Distributed) (Mean, Standard Deviation)

Component 1 (min)	(1,0.1)
Component 2 (min)	(2,0.3)
Component 3 (min)	(2,0.3)

TABLE A5
Subassembly and Component Replacement Costs for Maintenance

Dishwasher Metal Cover ($)	80
Dishwasher Circuit Board ($)	125
Dishwasher Motor ($)	160
Dryer Metal Cover ($)	80
Dryer Drum ($)	40
Dryer Motor ($)	200

TABLE A6
Subassembly and Component Failure Probabilities (%)

Dishwasher Metal Cover Failure	10
Dishwasher Circuit Board Failure	40
Dishwasher Motor Failure	50
Dryer Metal Cover Failure	10
Dryer Drum Failure	40
Dryer Motor Failure	50

TABLE A7
Production and Cost Data

Dishwasher Interarrival Time (min) (Exponentially Distributed)	3
Dryer Interarrival Time (min) (Exponentially Distributed)	3
Disposal Cost ($/lb.)	0.4
Scrap Revenue ($/lb.)	0.6
Holding Cost Rate	0.2
Backordering Cost Rate	0.6

TABLE A8
Remanufacturing Assembly Data

Dishwasher Assembly Time (min)	6
Dryer Assembly Time (min)	5
Assembly Cost($/min)	4

10 DEMATEL-Based Analysis of the Factors Hindering the Effective Use of Industry 4.0 Technologies in Reverse Logistics

Mehmet Ali Ilgin
Department of Industrial Engineering, Manisa Celal Bayar
University, Yunusemre, Manisa, 45140, Turkey

1 INTRODUCTION

Industry 4.0 has emerged as a vital competition factor in recent years. It involves the integrated use of many technologies (e.g., cloud computing, internet of things, and sensor technology) to achieve wireless communication in a network of machines, people, and various mobile devices (Castelo-Branco et al., (2019)). The enhanced communication and collaboration among these entities improve diverse services and lead to better and cost-efficient planning of logistic activities and resources (Kamble et al., (2018)).

Reverse logistics is another concept that receives increasing attention of researchers and industrial practitioners. It involves processes associated with the recovery of products returned by consumers for various reasons (e.g., repair, end-of-life, and warranty). There are two critical issues that increase the importance of reverse logistics in recent years: first, manufacturing companies have started to take back their products to comply with the stricter environmental regulations and have a better environmental image; and second, there is a dramatic increase in consumer returns mainly due to the increased popularity of e-commerce and more liberal policies on consumer returns (Ilgin and Gupta, 2013).

Industry 4.0 technologies have immense potential in reverse logistics systems (Shah et al., (2019)). Internet of things, sensor technology, and radio frequency identification (RFID) are the most used technologies. Condition monitoring of used products and collection of used product information are the main application areas of those in reverse logistics. Although there are studies discussing the use of cloud computing and the internet of services in reverse logistics systems, the use of these in the planning of reverse logistics activities is still in its infancy.

Although there are studies analyzing the barriers associated with the use of industry 4.0 technologies in the manufacturing industry (Türkeş et al., (2019); Singh and Bhanot, 2020), the number of studies analyzing the factors that negatively affect the industry 4.0 initiatives in reverse logistics systems is very limited. In this chapter, the factors that hinder the widespread use of industry 4.0 technologies in reverse logistics systems are analyzed using DEMATEL.

The remainder of the chapter is organized as follows: Section 2 provides brief information on industry 4.0 technologies that have the potential to be used in reverse logistic systems. The factors hindering the widespread use of industry 4.0 technologies in reverse logistic systems are introduced in Section 3. Brief information on DEMATEL methodology is presented in Section 4. Section 5 presents the use of DEMATEL for the analysis of factors that negatively affect the widespread use of industry 4.0 technologies in reverse logistics. Concluding remarks are presented in Section 6.

2   INDUSTRY 4.0 AND REVERSE LOGISTICS

In this section, brief information is provided on industry 4.0 technologies that are used in reverse logistics systems:

- *RFID:* RFID is a technology that uses radio waves to identify objects. Important product information (e.g., model, sale date, disassembly sequence) can be obtained by attaching RFID tags. This information can also be updated when the product is returned for upgrade, repair, or maintenance (Ondemir et al., (2012)). There are many studies on the use of RFID to collect used product information (Ondemir and Gupta, 2014a; Ondemir and Gupta, 2014b; Alqahtani et al., (2019); Shah et al., (2019); Dev et al., (2020)).
- *Sensor Technology:* Sensors are devices that create usable outputs in response to changes in the physical environment (e.g., changes in light, pressure, moisture, etc.) (Gupta and Sikka, 2015). Conditions of the critical components of a product can be monitored by the sensors embedded. The most suitable end of life option (e.g., recycling, remanufacturing) for a used product can be determined based on this information (Ilgin and Gupta, 2010; Ilgin et al., (2011); Ilgin et al., (2014); Alqahtani and Gupta, 2018; Alqahtani and Gupta, 2019).
- *Cloud Computing:* Cloud computing is internet-based technology that enables multiple users to access a variety of computing resources from anywhere at any time (Chaudhry et al., (2019)). The number of publications on the use of cloud computing in reverse logistic systems is limited (Shah et al., (2019); Dev et al., (2020)).
- *The Internet of Things (IoT):* Various smart devices are connected via a network structure in IoT (Bandyopadhyay, 2020). Many IoT-based reverse logistics management systems were developed in recent years (Gu and Liu, 2013; Liu et al., (2018); Garrido-Hidalgo et al., (2019); Thürer et al., (2019); Joshi and Gupta, 2019; Shah et al., 2019; Dev et al., 2020; Garrido-Hidalgo et al., (2020)).
- *Internet of Services (IoS):* The internet of services is based on the idea that everything (e.g., software applications, networks) can be offered as a service

on the internet (Moreno-Vozmediano et al., (2013)). The number of studies on the use of the internet of services in reverse logistics systems is very limited Dev et al., (2020).

- *Cyber-Physical System (CPS):* Cyber-Physical System integrates the physical and virtual world. In CPS, physical processes are monitored by computers while the data and computations carried out are affected by the physical components of the system (Bandyopadhyay, 2020). Dev et al., (2020) provided a good example of the use of CPS in reverse logistic systems. They proposed a framework where real-time information on re-manufactured products and various other operational and policy parameters are provided to the stakeholders of a reverse logistics system by CPS.

3 FACTORS THAT HINDER THE IMPLEMENTATION OF INDUSTRY 4.0 TECHNOLOGIES IN REVERSE LOGISTICS SYSTEMS

Fourteen factors that have a negative impact on industry 4.0 initiatives were identified by reviewing industry 4.0 literature. The factors and the related references are presented in Table 10.1.

The following paragraphs provide brief information on the factors that hinder the implementation of industry 4.0 technologies in reverse logistics systems:

TABLE 10.1

The Factors That Hinder the Implementation of Industry 4.0 in Reverse Logistics Systems

Code	Factor	References
FR1	High-cost	Ding, 2018; Gu et al., (2019)
FR2	Organizational resistance and employee reluctance	Kamble et al., (2018)
FR3	Security-related issues	Ding, 2018
FR4	Low information quality	Ding, 2018; Jabbour et al., (2018)
FR5	Lack of trust between entities	Ding, 2018
FR6	Time-consuming	Ding, 2018
FR7	Lack of skilled workforce	Jabbour et al., (2018)
FR8	Low upper management support	Ding, 2018
FR9	Lack of standards	Horvath and Szabo, 2019
FR10	Integration and compatibility issues	Kamble et al., (2018)
FR11	Lack of understanding of industry 4.0 benefits	Kamble et al., (2018)
FR12	Lack of clear legal procedures	Kamble et al., (2018); Ding, 2018
FR13	Lack of training programs	Ding, 2018); Horvath and Szabo, 2019
FR14	The need for organizational and process changes	Kamble et al., (2018)

- *High Cost:* Implementation of industry 4.0 technologies in reverse logistics systems requires large investment. This can especially be an important barrier for small and medium-sized companies since they have limited financial resources.
- *Organizational Resistance and Employee Reluctance:* Some workers may lose their jobs due to automation capabilities provided by industry 4.0 technologies. This may create resistance towards industry 4.0 technologies throughout the organization and employees may be reluctant to take part in industry 4.0 initiatives.
- *Security-Related Issues:* Enhanced information flow among the partners in an industry 4.0-based reverse logistic system increases the occurrence of cybersecurity threats and data privacy issues.
- *Low Information Quality:* Ensuring data consistency and integrity may be difficult due to frequently changing data and multiple collaborators in industry 4.0-based reverse logistics systems.
- *Lack of Skilled Workforce:* Newly-developed and complex technologies are used in industry 4.0-based reverse logistic systems. Hence, effective implementation of industry 4.0 technologies in a reverse logistic system demands various technical and non-technical skills from the workforce.
- *Low Upper Management Support:* Upper management must provide the necessary motivation and resources for the implementation of industry 4.0 initiatives in reverse logistics systems.
- *Lack of Understanding of Industry 4.0 Benefits:* The benefits of some industry 4.0 technologies may not be clear short term. This may prevent companies from investing in those technologies.
- *Lack of Trust between Entities:* Lack of trust between entities creates problems in various areas including data acquisition and coordination.
- *Time-Consuming:* Implementation of industry 4.0 technologies in reverse logistics systems requires the coordination of different entities. In addition, a substantial amount of time is spent on employee training and technological infrastructure.
- *Lack of Standards:* There is a lack of standards and reference architectures for many industry 4.0 technologies.
- *Integration and Compatibility Issues:* Industry 4.0 requires the interoperability of different technologies and network systems. Hence, various integration and compatibility issues must be solved for the successful operation of an industry 4.0-based reverse logistics system.
- *Lack of Clear Legal Procedures:* The uncertainty on legal procedures regarding data protection and liability for digital operations is an important barrier to the implementation of industry 4.0 initiatives in reverse logistics.
- *Lack of Training Programs:* Companies must provide necessary training programs to ensure that the workforce is ready for the challenges caused by the implementation of industry 4.0 technologies in reverse logistic systems.
- *The Need for Organizational and Process Changes:* Companies must carry out modifications in organizational structure and the working principles of processes to fully utilize the benefits of industry 4.0 in reverse logistics.

TABLE 10.2
The Scale of DEMATEL

Linguistic Variable	Influence Score
No impact	0
Very low impact	1
Low impact	2
High impact	3
Very high impact	4

4 DEMATEL

DEMATEL (Decision Making Trial and Evaluation Laboratory) is a multi-criteria decision-making methodology used to classify multiple criteria into two groups: cause and effect (Gabus and Fontela, 1972; Gabus and Fontela, 1973). DEMATEL steps can be summarized as follows:

Step 1: Formation of the direct relation matrix: DEMATEL scale presented in Table 10.2 is used to construct the direct relation matrix X — an $n \times n$ matrix obtained by pairwise comparisons. An element of this matrix (x_{ij}) represents the degree of factor i's impact on factor j. The values of diagonal elements are equal to zero.

Step 2: Normalize the direct relation matrix: Equations 10.1 and 10.2 are utilized to normalize the direct relation matrix.

$$N = c \times X \tag{10.1}$$

$$c = Min \left(\frac{1}{\max\limits_{1 \le i \le n} \sum\limits_{j=1}^{n} |x_{ij}|}, \frac{1}{\max\limits_{1 \le j \le n} \sum\limits_{i=1}^{n} |x_{ij}|} \right) \tag{10.2}$$

Step 3: Construct the total relation matrix: The total relation matrix S is constructed using the following equation:

$$S = N (I - N)^{-1} \tag{10.3}$$

where N is the normalized direct relation matrix, and I is the identity matrix.

The sum of rows and sum of columns of the total relation matrix S is computed as D and R using the following equations:

$$D = \left[\sum_{i=1}^{n} s_{ij} \right]_{1 \times n} \tag{10.4}$$

$$R = \left[\sum_{j=1}^{n} s_{ij} \right]_{1xn} \qquad (10.5)$$

$(D + R)$ value is a measure of a factor's total relationship with other factors. Factors are categorized based on their $(D - R)$ values. If this value is positive for a factor, it is categorized as "cause". The factors with negative $(D - R)$ values are categorized as "effect".

Step 4: Set the threshold value and obtain the impact-digraph map: An impact-digraph map can be drawn by considering the elements of the total relations matrix. If all elements are considered, the graph will be too complex. That is why a threshold value is determined by decision-makers; the elements with an influence level greater than the threshold value are included in the impact-digraph map.

Step 5: Determine the factor weights: Factor weights are determined using the following equation (Dalalah et al., (2011)):

$$w_i = \sqrt{(D_i + R_i)^2 + (D_i - R_i)^2} \qquad (10.6)$$

Equation 10.7 is used to obtain the normalized weights:

$$W_i = \frac{w_i}{\sum_{i=1}^{n} w_i} \qquad (10.7)$$

There are many applications of the DEMATEL technique in different domains including industry 4.0 (Singh and Bhanot, 2020), green supply chain management (Lin, 2013; Gupta and Ilgin, 2018), and disassembly line balancing (Ilgin, 2019). Readers interested in DEMATEL applications are referred to a recent review paper on DEMATEL by Si et al., (2018).

5   ANALYSIS OF THE FACTORS USING DEMATEL

This section presents the details on the use of DEMATEL to analyze the factors that hinder the use of industry 4.0 technologies in reverse logistics. First, the average direct relation matrix presented in Table 10.3 was constructed based on the opinions of three experts. Then, the normalized direct relation matrix (see Table 10.4) was obtained by using equations 10.1 and 10.2, taking the value of c as 1/22. Finally, equation 10.3 was employed and the total relation matrix given in Table 10.5 was formed.

The row and column sums of the total relation matrix, together with $D_i + R_i$ and $D_i - R_i$ values, are presented in Table 10.6. The last column of this table presents the category of a factor. The factors with a positive $D_i - R_i$ value (FR7, FR9, FR11, FR12, FR13, and FR14) are cause factors. The most important causal factors are FR13 (lack of training programs) and FR9 (lack of standards). The improvements in these two factors will have a significant impact on many other factors. The factors with the highest $D_i + R_i$ values are FR8 (low upper management support), FR6

TABLE 10.3

Average Direct Relation Matrix

	FR1	FR2	FR3	FR4	FR5	FR6	FR7	FR8	FR9	FR10	FR11	FR12	FR13	FR14
FR1	0.00	0.00	0.67	1.00	0.00	1.67	0.33	3.67	0.00	0.67	0.00	0.00	0.33	0.00
FR2	1.67	0.00	0.33	0.67	0.33	2.67	0.67	2.33	0.00	0.67	1.00	0.00	0.33	0.00
FR3	1.67	0.33	0.00	0.33	2.33	0.67	0.00	1.67	0.00	1.67	0.00	0.33	0.00	0.33
FR4	0.00	0.00	0.33	0.00	1.67	0.00	0.00	0.33	0.00	1.33	0.00	0.00	0.00	0.00
FR5	0.00	0.33	0.33	2.00	0.00	1.33	0.00	1.00	0.00	1.00	0.00	0.00	0.00	0.00
FR6	3.33	1.67	0.00	0.00	0.00	0.00	0.00	2.33	0.00	0.00	0.00	0.00	0.00	0.00
FR7	1.67	1.67	2.33	2.33	0.00	2.00	0.00	1.67	0.00	2.33	1.67	0.00	0.67	0.00
FR8	0.00	2.33	0.00	0.67	1.67	1.67	0.33	0.00	0.00	1.33	0.33	0.00	2.00	0.00
FR9	0.67	0.67	1.67	1.67	1.67	1.33	0.00	0.33	0.00	4.00	0.00	0.00	0.00	0.67
FR10	2.33	1.33	1.67	2.33	2.00	3.00	0.00	0.33	0.00	0.00	0.00	0.00	0.00	1.67
FR11	0.00	3.67	0.00	0.00	0.33	0.67	0.33	4.00	0.00	0.00	0.00	0.00	1.33	0.00
FR12	0.33	0.67	1.33	0.33	3.67	1.00	0.00	1.33	0.00	0.67	0.00	0.00	0.00	0.00
FR13	0.00	3.00	1.67	1.33	1.00	1.67	4.00	0.33	0.00	2.33	3.33	0.00	0.00	0.00
FR14	4.00	3.00	0.00	0.33	0.00	3.67	0.00	2.67	0.00	1.33	0.00	0.00	0.00	0.00

(time-consuming), FR10 (integration and compatibility issues), and FR2 (organizational resistance and employee reluctance). These are the factors that have the highest number of interrelationships with other factors.

An impact-digraph map presented in Figure 10.1 was drawn by considering the elements of the total relation matrix that are greater than the threshold value of 0.18. The vertical axis represents $D_i - R_i$ values while the horizontal axis represents $D_i + R_i$ values in Figure 10.1. It is clear from this figure that, the factors placed on the upper part of the graph with positive $D_i - R_i$ values (FR9, FR12, FR13, and FR14) affect the other factors, while the factors placed on the lower part of the graph with negative $D_i - R_i$ values (FR2, FR6, and FR8) are affected by other factors.

Table 10.7 presents the weights and the normalized weights of the factors determined by using equations 10.6 and 10.7, which jointly consider $D_i + R_i$ and $D_i - R_i$ values. According to this table, the most important five factors are FR8 (low upper management support), FR6 (time-consuming), FR2 (organizational resistance and employee reluctance), FR10 (integration and compatibility issues), and FR13 (lack of training programs).

The results of DEMATEL analysis provide several insights that can be useful for managers and policymakers. Upper management should provide necessary support during the implementation of industry 4.0 initiatives. Training programs on the use of industry 4.0 technologies in reverse logistics should be designed for company personnel. Improvements in upper management support and training programs will have a positive impact on organizational resistance and employee reluctance, which is another critical factor. Policymakers should take necessary actions and provide support in the development of standards.

TABLE 10.4
Normalized Direct Relation Matrix

	FR1	FR2	FR3	FR4	FR5	FR6	FR7	FR8	FR9	FR10	FR11	FR12	FR13	FR14
FR1	0.000	0.030	0.030	0.045	0.000	0.076	0.015	0.167	0.000	0.030	0.000	0.000	0.015	0.000
FR2	0.076	0.000	0.015	0.030	0.015	0.121	0.030	0.106	0.000	0.030	0.045	0.000	0.015	0.000
FR3	0.076	0.015	0.000	0.015	0.106	0.030	0.000	0.076	0.000	0.076	0.000	0.015	0.000	0.015
FR4	0.000	0.000	0.015	0.000	0.076	0.000	0.000	0.015	0.000	0.061	0.000	0.000	0.000	0.000
FR5	0.000	0.015	0.015	0.091	0.000	0.061	0.000	0.045	0.000	0.045	0.000	0.000	0.000	0.000
FR6	0.152	0.076	0.000	0.000	0.000	0.000	0.000	0.106	0.000	0.000	0.000	0.000	0.000	0.000
FR7	0.076	0.076	0.106	0.106	0.000	0.091	0.000	0.076	0.000	0.106	0.076	0.000	0.030	0.000
FR8	0.000	0.106	0.000	0.030	0.076	0.076	0.015	0.000	0.000	0.061	0.015	0.000	0.091	0.000
FR9	0.030	0.030	0.076	0.076	0.076	0.061	0.000	0.015	0.000	0.182	0.000	0.000	0.000	0.030
FR10	0.106	0.061	0.076	0.106	0.091	0.136	0.000	0.015	0.000	0.000	0.000	0.000	0.000	0.076
FR11	0.000	0.167	0.000	0.000	0.015	0.030	0.015	0.182	0.000	0.000	0.000	0.000	0.061	0.000
FR12	0.015	0.030	0.061	0.015	0.167	0.045	0.000	0.061	0.000	0.030	0.000	0.000	0.000	0.000
FR13	0.000	0.136	0.076	0.061	0.045	0.076	0.182	0.015	0.000	0.106	0.152	0.000	0.000	0.000
FR14	0.182	0.136	0.000	0.015	0.000	0.167	0.000	0.121	0.000	0.061	0.000	0.000	0.000	0.000

TABLE 10.5

Total Relation Matrix

	FR1	FR2	FR3	FR4	FR5	FR6	FR7	FR8	FR9	FR10	FR11	FR12	FR13	FR14
FR1	0.034	0.045	0.044	0.070	0.033	0.117	0.027	0.202	0.000	0.061	0.013	0.001	0.036	0.005
FR2	0.120	0.056	0.034	0.061	0.046	0.172	0.044	0.172	0.000	0.064	0.060	0.001	0.038	0.005
FR3	0.109	0.050	0.018	0.051	0.135	0.083	0.008	0.124	0.000	0.103	0.007	0.015	0.014	0.023
FR4	0.015	0.012	0.023	0.018	0.089	0.022	0.002	0.030	0.000	0.071	0.002	0.000	0.003	0.006
FR5	0.025	0.036	0.024	0.105	0.022	0.085	0.004	0.070	0.000	0.062	0.004	0.000	0.008	0.005
FR6	0.171	0.103	0.012	0.023	0.020	0.045	0.012	0.156	0.000	0.024	0.011	0.000	0.019	0.002
FR7	0.144	0.147	0.134	0.152	0.060	0.173	0.021	0.173	0.000	0.157	0.095	0.002	0.057	0.014
FR8	0.050	0.156	0.026	0.070	0.105	0.138	0.041	0.062	0.000	0.097	0.042	0.000	0.103	0.008
FR9	0.096	0.075	0.102	0.123	0.124	0.133	0.007	0.080	0.000	0.217	0.007	0.002	0.010	0.048
FR10	0.172	0.108	0.092	0.141	0.126	0.203	0.010	0.105	0.000	0.044	0.010	0.001	0.015	0.080
FR11	0.042	0.224	0.021	0.035	0.050	0.101	0.043	0.239	0.000	0.043	0.031	0.000	0.090	0.004
FR12	0.046	0.059	0.073	0.049	0.193	0.089	0.006	0.102	0.000	0.059	0.006	0.001	0.012	0.006
FR13	0.091	0.233	0.123	0.130	0.103	0.185	0.202	0.145	0.000	0.173	0.185	0.002	0.035	0.015
FR14	0.249	0.195	0.024	0.057	0.037	0.248	0.019	0.222	0.000	0.100	0.018	0.000	0.029	0.008

TABLE 10.6

Categorization of the Factors based on $D_i + R_i$ and $D_i - R_i$ Values

Factor	D_i	R_i	$D_i + R_i$	$D_i - R_i$	Category
FR1	0.6880	1.3653	2.0530	−0.6770	Effect
FR2	0.8722	1.5011	2.3733	−0.6290	Effect
FR3	0.7409	0.7485	1.4895	−0.0080	Effect
FR4	0.2928	1.0854	1.3782	−0.7930	Effect
FR5	0.4510	1.1426	1.5939	−0.6910	Effect
FR6	0.5990	1.7936	2.3926	−1.1950	Effect
FR7	1.3292	0.4451	1.7742	0.8841	Cause
FR8	0.8991	1.8806	2.7797	−0.9820	Effect
FR9	1.0234	0.0000	1.0234	1.0234	Cause
FR10	1.1074	1.2748	2.3822	−0.1670	Effect
FR11	0.9218	0.4895	1.4113	0.4323	Cause
FR12	0.7009	0.0265	0.7274	0.6744	Cause
FR13	1.6196	0.4697	2.0893	1.1499	Cause
FR14	1.2060	0.2290	1.4353	0.9771	Cause

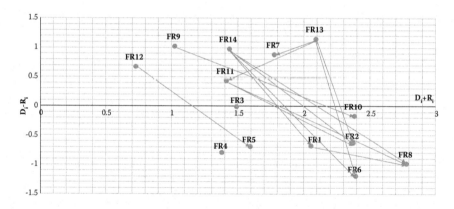

FIGURE 10.1 Impact-Digraph Map for the Causal Relationships among the Factors.

They should also initiate clear legal procedures for industry 4.0 transactions. Improvements in standards and legal procedures will have a positive impact on integration and compatibility issues.

6 CONCLUSIONS

Rapid developments in product technology and consumers' eagerness to buy newer product models have increased the number of products prematurely

TABLE 10.7
Factor Weights

Factor	Weight	Normalized Weight
FR1	2.1621	0.0787
FR2	2.4552	0.0894
FR3	1.4895	0.0542
FR4	1.5899	0.0579
FR5	1.7373	0.0633
FR6	2.6742	0.0974
FR7	1.9823	0.0722
FR8	2.9479	0.1073
FR9	1.4473	0.0527
FR10	2.3881	0.0870
FR11	1.4760	0.0537
FR12	0.9920	0.0361
FR13	2.3848	0.0868
FR14	1.7363	0.0632

disposed of by consumers. In addition, consumer returns are on the rise due to more liberal return policies and a modern way of shopping. Companies try to deal with the increased amount of product returns by forming cost-effective systems for reverse logistics, which comprises all activities required for the collection and recovery or disposal of products returned by consumers for any reason. Industry 4.0 technologies can be used to enhance the effectiveness of a reverse logistics system. Although there are successful applications of sensor technology, the internet of things, and RFID, the use of industry 4.0 technologies in reverse logistics is not widespread. In this chapter, the factors that hinder the widespread use of industry 4.0 technologies in reverse logistics were analyzed using DEMATEL multi-criteria decision-making methodology. Cause and effect factors were determined, and the weights of the factors were calculated. Several suggestions were provided for policymakers and administrators based on the results of DEMATEL analysis.

This study can be extended in several directions. The relationships among the factors can be analyzed using other techniques like interpretive structural modeling (ISM). Comparative analysis of DEMATEL and ISM results can provide interesting managerial insights. The use of fuzzy DEMATEL analysis in considering the linguistic preferences of experts is another interesting future research direction.

REFERENCES

Alqahtani A. Y., Gupta S. M., and Nakashima K. 2019. Warranty and maintenance analysis of sensor embedded products using internet of things in industry 4.0. *International Journal of Production Economics* 208: 483–499.

Alqahtani A. Y., and Gupta S. M. 2019. Warranty cost sharing policies with preventive maintenance strategy for sensor embedded remanufactured products. *International Journal of Industrial and Systems Engineering* 31: 360–394.

Alqahtani A. Y., and Gupta S. M. 2018. Money-back guarantee warranty policy with preventive maintenance strategy for sensor-embedded remanufactured products. *Journal of Industrial Engineering International* 14: 767–782.

Bandyopadhyay S. 2020. *Production and Operations Analysis: Traditional, Latest, and Smart Views*. CRC Press, Boca Raton.

Castelo-Branco I., Cruz-Jesus F., and Oliveira T. 2019. Assessing Industry 4.0 readiness in manufacturing: Evidence for the European Union. *Computers in Industry* 107: 22–32.

Chaudhry S. A., Kim I. L., Rho S., Farash M. S., and Shon T. 2019. An improved anonymous authentication scheme for distributed mobile cloud computing services. *Cluster Computing* 22: 1595–1609.

Dalalah D., Hayajneh M., and Batieha F. 2011. A fuzzy multi-criteria decision making model for supplier selection. *Expert Systems with Applications* 38: 8384–8391.

Dev N. K., Shankar R., and Swami S. 2020. Diffusion of green products in industry 4.0: Reverse logistics issues during design of inventory and production planning system. *International Journal of Production Economics* 223: 107519.

Ding B. 2018. Pharma Industry 4.0: Literature review and research opportunities in sustainable pharmaceutical supply chains. *Process Safety and Environmental Protection* 119: 115–130.

Gabus A., and Fontela E. 1972. *World Problems, an Invitation to Further Thought within the Framework of DEMATEL*. Battelle Geneva Research Center, Geneva, Switzerland.

Gabus A., and Fontela E. 1973. *Perceptions of the World Problematique: Communication Procedure, Communicating with Those Bearing Collective Responsibility*. Battelle Geneva Research Centre, Geneva, Switzerland.

Garrido-Hidalgo C., Olivares T., Ramirez F. J., and Roda-Sanchez L. 2019. An end-to-end Internet of Things solution for Reverse Supply Chain Management in Industry 4.0. *Computers in Industry* 112: 103127.

Garrido-Hidalgo C., Ramirez F. J., Olivares T., and Roda-Sanchez L. 2020. The adoption of Internet of Things in a Circular Supply Chain framework for the recovery of WEEE: The case of Lithium-ion electric vehicle battery packs. *Waste Management* 103: 32–44.

Gu F., Guo J., Hall P., and Gu X. 2019. An integrated architecture for implementing extended producer responsibility in the context of Industry 4.0. *International Journal of Production Research* 57: 1458–1477.

Gupta K., and Sikka V. 2015. Design issues and challenges in wireless sensor networks. *International Journal of Computer Applications* 112: 26–32.

Gupta S. M., and Ilgin M. A. 2018. *Multiple Criteria Decision Making Applications in Environmentally Conscious Manufacturing and Product Recovery*. CRC Press, Boca Raton, FL.

Gu Y., and Liu Q. 2013. Research on the application of the internet of things in reverse logistics information management. *Journal of Industrial Engineering and Management* 6: 963–973.

Horvath D., and Szabo R. Z. 2019. Driving forces and barriers of Industry 4.0: Do multinational and small and medium-sized companies have equal opportunities? *Technological Forecasting and Social Change* 146: 119–132.

Ilgin M. A., and Gupta S. M. 2013. Reverse logistics, in: S. M..Gupta (Ed.) *Reverse Supply Chain: Issues and Analysis*. CRC Press, Boca Raton, FL, pp. 1–60.

Ilgin M. A., and Gupta S. M. 2010. Comparison of economic benefits of sensor embedded products and conventional products in a multi-product disassembly line. *Computers & Industrial Engineering* 59: 748–763.

Ilgin M. A., Gupta S. M., and Nakashima K. 2011. Coping with disassembly yield uncertainty in remanufacturing using sensor embedded products. *Journal of Remanufacturing* 1: 7.

Ilgin M. A., Ondemir O., and Gupta S. M. 2014. An approach to quantify the financial benefit of embedding sensors into products for end-of-life management: A case study. *Production Planning & Control* 25: 26–43.

Ilgin M. A. 2019. A DEMATEL-based disassembly line balancing heuristic. *Journal of Manufacturing Science and Engineering* 141: 021002.

Jabbour A. B. L. d. S, Jabbour C. J. C., Godinho Filho M., and Roubaud D. 2018. Industry 4.0 and the circular economy: A proposed research agenda and original roadmap for sustainable operations. *Annals of Operations Research* 270: 273–286.

Joshi A. D., and Gupta S. M. 2019. Evaluation of design alternatives of End-Of-Life products using internet of things. *International Journal of Production Economics* 208: 281–293.

Kamble S. S., Gunasekaran A., and Sharma R. 2018. Analysis of the driving and dependence power of barriers to adopt industry 4.0 in Indian manufacturing industry. *Computers in Industry* 101: 107–119.

Lin R.-J. 2013. Using fuzzy DEMATEL to evaluate the green supply chain management practices. *Journal of Cleaner Production* 40: 32–39.

Liu S., Zhang G., and Wang L. 2018. IoT-enabled dynamic optimisation for sustainable reverse logistics. *Procedia CIRP* 69: 662–667.

Moreno-Vozmediano R., Montero R. S., and Llorente I. M. 2013. Key challenges in cloud computing: enabling the future internet of services. *IEEE Internet Computing* 17: 18–25.

Ondemir O., Ilgin M. A., and Gupta S. M. 2012. Optimal end-of-life management in closed-loop supply chains using RFID and sensors. *IEEE Transactions on Industrial Informatics* 8: 719–728.

Ondemir O., and Gupta S. M. 2014a. Quality management in product recovery using the Internet of Things: An optimization approach. *Computers in Industry* 65: 491–504.

Ondemir O., and Gupta S. M. 2014b. A multi-criteria decision making model for advanced repair-to-order and disassembly-to-order system. *European Journal of Operational Research* 233: 408–419.

Shah S., Dikgang G., Menon S. 2019. The global perception of industry 4.0 for reverse logistics. *International Journal of Economics and Management Systems* 4: 103–107.

Si S.-L., You X.-Y., Liu H.-C., and Zhang P. 2018. DEMATEL technique: A systematic review of the state-of-the-art literature on methodologies and applications. *Mathematical Problems in Engineering* 2018: 3696457.

Singh R., and Bhanot N. 2020. An integrated DEMATEL-MMDE-ISM based approach for analysing the barriers of IoT implementation in the manufacturing industry. *International Journal of Production Research* 58: 1–23.

Türkeş M. C., Oncioiu I., Aslam H. D., Marin-Pantelescu A., Topor D. I., and Capuşneanu S. 2019. Drivers and barriers in using industry 4.0: A perspective of SMEs in Romania. *Processes* 7: 153.

Thürer M., Pan Y. H., Qu T., Luo H., Li C. D., and Huang G. Q. 2019. Internet of Things (IoT) driven kanban system for reverse logistics: solid waste collection. *Journal of Intelligent Manufacturing* 30: 2621–2630.

11 Considering Service Level in Planning Forward/Reverse Supply Chain Network Under Uncertainties

Murtadha A. Aldoukhi and Surendra M. Gupta
Department of Mechanical and Industrial Engineering,
Northeastern University, 334 Snell Engineering Center, 360
Huntington Avenue, Boston, MA 02115, USA

1 INTRODUCTION

The rapid increase in the amount of industrial waste has a huge negative impact and poses a serious threat to the environment. According to the U.S. Environmental Protection Agency (EPA), the United States produces more than 2 million tons of industrial solid waste annually, and landfills about 50% of that amount. The industrial waste is associated with end of life (EOL) products, which are products that have reached the end of their useful lives. Ilgin and Gupta [1] categorized EOL products to: products returned for upgrading proposes, non-functional returned rental products, products returned due to dissatisfaction or for finding a better deal but found damaged, and unrepairable products.

In addition to waste, emission of carbon dioxide, which is a factor of greenhouse gas (GHG) emissions, causes several environmental issues including climate change and global warming. In 2014, the intergovernmental panel on climate change reported that between the year 2000 and 2010, there has been an increase of 10 billion metric tons of global emissions of GHGs [2]. Thus, the huge environmental impact of carbon emission has influenced researchers to investigate this topic and its impact on supply chain [3]–[8]. To mitigate the impact of carbon emission on the environment, four reduction policies have been utilized widely. These policies include carbon cap, carbon emission tax, cap-and-trade and carbon offset policy.

New government regulations in addition to awareness about the environmental issues have forced supply chain decision makers to take actions to mitigate the environmental risks. Therefore, traditional supply chain (TSC) is reformatted to a closed loop supply chain (CLSC), which consists of forward and reverse flow supply chain. The forward flow supply chain is known as the TSC and the product recovery options pointed as the reverse flow supply chain. However, management of the new supply chain has become more challenging as product recovery is associated with uncertainties in terms of arrival time, number, and condition of the returned products. Today, many industries are employing the concept of CLSC in their practices, e.g. Caterpillar Inc.'s remanufacturing division had over \$2 billion in sales and was the fastest growing division out of all of its other divisions [9]. Different topics under the CLSC have taken the attention of researchers. Ilgin and Gupta [10] divided the topics discussed under the CLSC into: network design, simultaneous consideration of network and product design issues, optimization of transportation.

Design of the CLSC network topic has been a constant concern of researchers, managers and supply chain decision makers since there are many economic and environmental benefits of the CLSC network proven in industry[9]. Therefore, many review papers [10]–[17] were devoted to summarize the discussions on this topic. Although the amount of papers written on this topic was enormous, there is still a need to develop more flexible models to fill gaps found in the literature.

In this paper, we propose a model for designing the CLSC network with three objectives in mind. The first objective is to minimize the total cost of the CLSC network. The second objective is to minimize the carbon emission resulted from the different activities of the CLSC. The third objective is to maximize the level of service of all market zones, which distinguishes our model from the available models in the literature. Besides the level of service considered in our model, our model considers the flexibility of allowing products to be substituted in case of shortage or inability to fulfill customer demand.

The rest of this paper is organized as follows: we show the available literature concerning our topic in section 2. Section 3 explains the product substitution mechanism used in this paper. Section 4 shows the problem description and assumptions. We explain the methodology we used in section 5. Our illustrative example is in section 6 and section 7 presents the result and discussion. Conclusion and suggested future work are illustrated in section 8. Our mathematical model is presented in Appendix A.

2 LITERATURE REVIEW

Two major sections in the literature review are discussed. Section 2.1 examines research conducted in designing the CLSC network with multiple objectives and Section 2.2 illustrates the research with the level of service in designing network problems. Section 2.1 is categorized based on deterministic modeling of the CLSC network design (Section 2.1.1.) and stochastic modeling of the CLSC network design considering multiple objectives (Section 2.1.2.). Section 2.2 is classified by the service level in designing the TSC network (2.2.1.) and the service level in designing the CLSC network (2.2.2.).

2.1 CLSC Network Design under Multiple Objectives

Three important review papers illustrated the research conducted to design the CLSC network with multiple objectives. Ilgin and Gupta, 2010 collected all the research concerned with this topic, published between 1999 and 2010. Reviewing the literature was a part of economically conscious manufacturing and product recovery (ECMPRO). In 2014, Ilgin, Gupta, and Battaïa 2015 published the second review paper that considered multi-criteria decision making generally in the area of ECMPRO, specifically in the CLSC network. The total number of papers reviewed was over 190 between 1996 to 2014. The third review paper was presented in a book by Gupta and Ilgin, 2018. The book extended the review paper by Ilgin, Gupta, and Battaïa, and covered the rest of the papers published since 2014. We divided the topics discussing the issue of considering multiple objectives in designing the CLSC network into deterministic and stochastic models.

2.1.1 Deterministic Models

Pishvaee, Farahani, and Dullaert (2010) developed a bi-objective mixed integer programming to design a CLSC network with two objectives: minimize the total cost of the network and maximize the logistics network responsiveness. Nurjanni, Carvalho, and Costa 2017 presented a model that consisted of two objectives: minimize the total cost and the carbon emission of the CLSC network, respectively. They used the weighted sum method to solve the problem of two conflict objectives.

To minimize the total cost and defect rate of CLSC network design were the objectives of the Amin and Zhang 2014 study. They used the weighted-sum method to solve their proposed model. Shi, Liu, Tang & Xiong 2016 presented a case study from China. Due to the large size of the problem, they used the genetic algorithm method. In this case study, a CLSC network was designed considering three objectives: minimize the overall cost, minimize the carbon emission, and maximize the responsiveness of the CLSC network. Another case study was presented by Mardan et al. (2019) that studied a network design of the wire-and-cable industry. In this study, the network was designed using an accelerated Benders decomposition algorithm. Some objective functions were to minimize the total cost and the carbon emission of the CLSC network. Papen and Amin, 2019 implemented their proposed model for designing a bottled water CLSC network. They considered four objectives in their proposed model—maximize the profit and on-time delivery and minimize the carbon emission and defect rate. They utilized weighted-sums, distance, and ε-constraint methodology and compared the results of the three methodologies. Sadeghi Rad and Nahavandi 2018 discussed a study that encountered the issue of quantity discounts when designing the CLSC network. To solve the proposed problem, they used the Lp-metrics method; the aim of the study was to minimize the total cost of designing and operating the network, to minimize the total amount of carbon emitted, and to maximize customer satisfaction. Pazhani and Ravi Ravindran 2018 proposed a CLSC network design model that optimizes profit and energy usage on the CLSC network. They used bi-criteria mixed integer programming to consider the two proposed objectives.

2.1.2 Stochastic Models

Lee, Bian, and Dong 2007 proposed a model with two objectives: maximize the number of products returned to the collection centers and minimize the total cost of designing the CLSC network. To solve the uncertainty of the objective functions, they used a fuzzy goal programming approach. Amin and Zhang 2013 proposed a model that minimizes the total cost and maximizes the use of environmental-friendly materials and clean technology in the CLSC network. The product demand and number of returned products considered uncertain and the weighted sum and the ε-constraints approaches were implemented in this paper. Paksoy, Pehlivan, and Özceylan 2012 used a fuzzy programming approach to design the CLSC network under the aspiration level of the objective function, the product demand, actor, and truck capacities. The four objectives they considered are minimization of the transportation cost of the forward supply chain, minimization of the transportation cost of the reverse supply chain, carbon emission, and maximization encouraging usage of recyclable products. The goal of the study presented by Özceylan and Paksoy 2014 was to minimize the four costs related to designing the CLSC network—total transportation cost, the total purchasing costs, the total refurbishing costs, and the fixed cost of opening a facility in the CLSC network. To consider the uncertainty of the aspiration level on the objective functions, the capacity of the facilities, product demand, and reverse rate, they proposed fuzzy interactive programming approaches. Jindal and Sangwan 2017 proposed a fuzzy optimization model to design a CLSC network where the uncertainties considered are allocated to different parameters. The proposed model consisted of two objectives: to maximize profit and minimize carbon emission. Another model has been proposed by Zhen, Huang, and Wang 2019 that used the scenario-based method to consider the uncertainty of product demand. Their proposed model consisted of two objectives: an economical objective to minimize the total operating cost and an environmental objective to minimize carbon emission. Govindan et al. (2019) proposed a model that resulted from the integration of fuzzy analytic network process, fuzzy decision-making trial and evaluation laboratory, and multi-objective MILP approaches. The proposed model had two objectives: to minimize the total cost and the shortage cost of the CLSC network. The uncertainties assumed by the proposed model were product demand and other cost parameters on the CLSC network. Another integrated model was proposed by Talaei et al. (2016), where a robust and fuzzy optimization approach was integrated. The objective functions of the proposed model were to minimize the total cost and the carbon emission of the CLSC network while considering the uncertainty of variable costs and product demand on the CLSC network design. Tosarkani and Amin, 2018 discussed the design of the CLSC network under two objectives: to maximize the total profit and green factors of a battery CLSC network. They used a fuzzy analytic network process and ε-constraint approaches to consider the uncertainty of product demand and return, selling prices, and other cost parameters. Jalil et al., (2019) proposed a model using fuzzy programming to consider the uncertainty on the aspiration level of the objective functions. These objective functions were to minimize the total cost of the

CLSC network, the storage cost of raw materials, storage cost of raw materials, and the total defects.

2.2 SERVICE LEVEL IN DESIGNING SUPPLY CHAIN NETWORK

In the literature review, there are different definitions of service level (Farahani et al., 2014). The proportion of the demand satisfied immediately using in-hand inventory is a way of defining service level (Bernstein and Federgruen, 2004, Bernstein and Federgruen, 2007). Other researchers described the consistency of order cycle time, the accuracy of order fulfillment rate, delivery lead time, and flexibility in order quantity as a measurement of the level of service (Selim and Ozkarahan, 2008). Using the service level as a performance measure of competitiveness is another way of using service level (Farahani et al., 2014). This section is classified into the service level in designing the TSC and the service level in designing the CLSC network.

2.2.1 The Service Level in the Tsc Network

Ghavamifar, Makui, and Taleizadeh (Ghavamifar et al., 2018) used the concept of service level as one of the factors to measure the competition between a product manufacturer and resellers in the TSC network. They defined service level as the ratio of demand satisfaction using in-hand inventory without delay. Martino et al. (2017) proposed a model and implemented it in the fashion industry. The level of service was one of the key performance indicators — it is the percentage from the satisfied demand according to the total number of orders. Sabri and Beamon 2000 developed a multi-objective model that incorporated strategic and operational decisions in the TSC network. The level of service was used as one of the objectives the model aimed to achieve, and it was defined as the immediate orders filled proportion. Interactions of the service level to design the TSC network were discussed by Shen and Daskin's (2005) model where the level of service was used to measure to fill a location demand that is located within an exogenously specified distance from the distribution center. Xu, Liu, and Wang (2008) discussed designing a Chinese liquor network where the customer level of service was maximized. The percentage of satisfying the customer demand by-products shipped from the distribution centers within a specified access time represented the customer service level. Selim and Ozkarahan (2008) developed a fuzzy goal programming model for designing the TSC network. They used the concept of maximal covering location problem as a statement for the level of service for the retailers. Baghalian, Rezapour, and Farahani (2013) implemented their proposed model, which was based on robust optimization, on the rice industry in the middle east. In their model, the level of service of each market location was determined by costs related to inventory, salvage, and shortage cost.

2.2.2 The Service Level in the CLSC Network

Subramanian et al. (2013) developed a constructive heuristic based on Vogel's approximation method–total opportunity cost method for a large scale CLSC network design problem with a single objective. The service level was not considered as an objective, but it was calculated according to the distance traveled and the number of distribution centers that were assigned to serve the retailers. In this study, only the number of returned products were considered uncertain. Ramezani, Bashiri, and Tavakkoli-Moghaddam (2013) proposed a multi-objective model to design the CLSC network. The second objective was to maximize the level of service, which was represented by the proper delivery time of products to customers. All the prices, costs, and product demand—as well as the number of returned products—were assumed uncertain. They used the benefit of stochastic programming to solve the assumed uncertainty. However, they neglected the environmental impact of carbon emission. Zarandi and Davari (2011) extended the model proposed by Selim and Ozkarahan (2008). They integrated reverse logistics into the proposed network to form a CLSC network and used the same concept to measure the service level of the retailers and solution approach. It was noticed that the uncertainty considered in the model was the aspiration level of each objective function. Moreover, they used only one demand stream, which did not differentiate between the demand for new and remanufactured products. Jerbia et al. (2018) proposed a single objective model to design the CLSC network, in which they considered the uncertainty of product return rate and quality, revenues, and costs. The service level was not attempted as an objective. However, the service level was calculated as the proportion of the returned products to the retailer which are collected and proceeded. Another study was presented by Pazhani et al. (2013), in which the proposed model aimed to maximize the service level of hybrid facilities (that work as distribution and collection centers) and distributions. The technique for order preference by similarity to an ideal solution (TOPSIS) was implemented to evaluate each hybrid facility and distribution according to different criteria. However, this study neglected the uncertainty impact, except on the aspiration level, and the environmental impact of carbon emission. Table 11.1 summarizes the majority of research in the area of our interest.

Based on the literature review, the following gaps are observed:

1. The majority of proposed models to design the CLSC network are concerned with economic and environmental objectives (Nurjanni et al., 2017), (Shi et al., 2016), (Talaei et al., 2016), (Mardan et al., 2019; Papen and Amin, 2019; Rad and Nahavandi, 2018; Pazhani and Ravi, 2018), (Amin and Zhang, 2013), (Paksoy et al., 2012), (Jindal and Sangwan, 2017), (Zhen et al., 2019).
2. Only a few studies with designing the CLSC network considered the service level as an objective (Subramanian et al., 2013; Ramezani et al., 2013; Zarandi et al., 2011; Jerbia et al., 2018; Pazhani et al., 2013).
3. In the studies that incorporated the service level into the design of the CLSC network, the uncertainty of product demand, and the returned product quantity was rarely assumed (Ramezani et al., 2013).
4. To the best of our knowledge, none of the studies incorporated the service level into the design of the CLSC network considered modeling the problem using robust optimization.

TABLE 11.1

Literature Review Summary.

Article	TSC	CLSC	Econ. Obj.	Carbon Reduction. Obj.	Service Level Obj.	Deterministic Model	Model with Uncertainty	Demand Uncertainty	Return Uncertainty	One Demand Stream	Two Demand Streams
Nurjanni et al., (2017)		•	•	•		•				•	
Shi et al., (2016)		•	•	•		•				•	
Mardan et al., (2019)		•	•	•		•				•	
(Papen and Amin, 2019)		•	•	•		•				•	
Rad and Nahavandi (2018)		•	•	•		•				•	
Paksoy et al., (2012)		•	•	•			•	•		•	
(Özceylan and Paksoy, 2014)		•	•				•	•	•	•	
(Jindal and Sangwan, 2017)		•	•	•			•	•		•	
Zhen et al., (2019)		•	•	•			•	•		•	
Govindan et al., (2019)		•	•				•	•		•	•
Talaei et al., (2016)		•	•	•			•	•		•	

(Continued)

TABLE 11.1 (Continued)

Article	TSC	CLSC	Econ. Obj.	Carbon Reduction. Obj.	Service Level Obj.	Deterministic Model	Model with Uncertainty	Demand Uncertainty	Return Uncertainty	One Demand Stream	Two Demand Streams
Jalil et al., (2019)	•		•				•			•	
(Selim and Ozkarahan, 2008)	•		•		•		•			•	
Ghavamifar et al., (2018)	•		•		•		•	•		•	
Martino et al., (2017)	•		•		•	•				•	
(Sabri and Beamon, 2000)	•		•		•	•				•	
(Shen and Daskin, 2005)	•		•		•	•				•	
Xu et al., (2008)	•		•		•		•	•			
Baghalian et al., (2013)	•		•		•		•	•			
Ramezani et al., (2013)		•	•		•		•	•	•	•	
(Zarandi et al., 2011)		•	•		•		•				
Pazhani et al., (2013)		•	•		•	•				•	
This paper		•	•				•	•	•		•

5. Except (Aldoukhi and Gupta, 2019; Aldoukhi and Gupta, 2018; Ghafarimoghadam et al., 2016; Aldoukhi and Gupta, 2019), none of the previous studies considered the phenomena of product substitution in designing the CLSC network.

In this paper, we attempted to fill the gaps available in literature. We developed a model to design a CLSC network under the uncertainty of product demand and the quantity of returned products. Our model has three objectives: minimize the total cost of the network, minimize carbon emission, and maximize the service levels of the market zone. We integrate robust optimization and goal programming to model our proposed problem. Downward product substitution policy is also integrated into our model.

3 PRODUCT SUBSTITUTION

In general, the phenomena of a product fulfilling the demand of another is called product substitution—partial substitution when a portion of the demand order is satisfied using a substantial product and exclusive substitution when the entire lot size is substituted by a substitutable product (Lang, 2010). In production and inventory management, when a new product substitutes a remanufactured product (but not vice versa), it is called downward substitution as shown in Figure 11.1. Due to the variability of the number of returned products that impact the capacity of remanufacturing, the risk of being out of remanufactured stock is high (Ahiska et al., 2017).

Bayindir, Erkip, and Güllü (2005) studied the benefit of replacing remanufactured products with new products while they are segmented into two different markets. Two years later they extended their study by considering capacity limitations in their model (Bayindir, Erkip, and Güllü, 2007). Li, Chen, and Cai (2006) studied an incapacitated multi-product, multi-period stochastic remanufacturing system with part substitution where no backlog/shortage or disposal was allowed. Piñeyro and Viera (2010) studied the problem of lot-sizing when there are two demand streams for new and remanufactured products, in which new products can substitute remanufactured products. Ahiska, Gocer, and King (2017) compared different inventory control policies of a hybrid manufacturing/

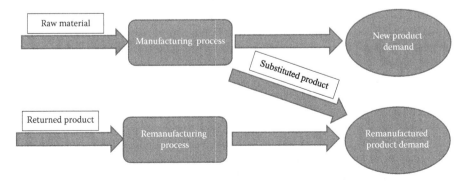

FIGURE 11.1 Downward Product Substitution Mechanism.

remanufacturing system, which allows downward substitution under stochastic demand and products returned. In this paper, we considered a downward product substitution policy where the new product can replace the remanufactured product.

4 PROBLEM DESCRIPTION AND ASSUMPTIONS

A CLSC network is composed of raw material suppliers, hybrid manufacturing/ remanufacturing, distribution, collection, and disposal centers in addition to multiple market zones. The processes involved in the forward flow deals with manufacturing centers that produce new products using the raw materials delivered from the suppliers and remanufacture the returned products. Each hybrid manufacturing/ remanufacturing has its own opening, production costs, and carbon emission rate. Finished products are shipped to different market zones through the distribution center, and each market zone distributes the finished product to different retailers in the same zone (which is out of our scope). In reverse flow, the EOL products returned from the retailers to the main market zone are shipped to collection centers. Due to the different conditions of the returned products, different quality levels are considered. At the collection centers, products are inspected and shipped to be remanufactured or disposed of. In any market zone, to maximize customer satisfaction, new products can substitute the remanufactured products' demand. The first objective function—economic objective—minimizes the cost of opening facilities, cost of purchasing raw material, cost of producing the new product and remanufacturing the returned product, cost allocated to collection facilities, cost of substituting a product, and cost of shipping the product between the CLSC network facilities. The second objective function—environmental objective—deals with minimizing the carbon emission resulted from different processes in the CLSC network. In this objective function, we use the carbon tax policy—an approach used to reduce carbon emission by multiplying each unit of emitted carbon by a tax (Benjaafar et al., 2013), (He et al., 2016). Maximizing the service level of the market zones is the third objective function, where the scores of the service level are calculated using TOPSIS, as illustrated in Section 4.1. The demand for products, new and remanufactured, as well as the number of returned products, are considered uncertain parameters. Our model can find the optimal number of each facility type to open, the quantity of purchased raw materials and their supplier, the quantity of product to produce and ship. It also determines the optimal quantity of new products to substitute remanufactured products and returned product to collect.

The following assumptions are considered:

1. Raw material suppliers, hybrid manufacturing/remanufacturing, distribution, collection centers have capacity restriction.
2. There are predefined market numbers and zones.
3. The cost of purchasing raw materials includes the cost to deliver raw materials from the suppliers to manufacturing facilities.
4. The cost of remanufacturing products includes disassembly and upgrade/repair costs.
5. Returned products are classified into different quality levels.

6. The cost of remanufacturing products depends on the quality level of the returned products. Products returned with major defects cost more to remanufacture.

7. Production at hybrid manufacturing/remanufacturing facilities, transportation of products between facilities in the network, and product disposal activity cause carbon emissions.

8. Production of new product emits more carbon than remanufacturing returned products.

9. We do not consider distributing the finished product to retailers in the market zone.

10. At any market location, we assume it is always acceptable to substitute re-manufactured products with new products, and the substitution can be either partial or exclusive.

11. The substitution ratio is 1:1 — one unit of the new product would substitute one unit of the remanufactured product.

5 METHODOLOGY

Figure 11.2 summarizes the steps adopted to solve the model. We start by selecting the service efficiency scores for the market locations. We use TOPSIS, explained in Section 4.1., to set the service level scores according to four criteria: consumer population, accessibility of the market zone to retailers, number of competitors serving the same product in the zone, and loyalty of the consumer in the zone. Pazhani et al. (2013) used the same concept to find efficiency scores for the hybrid facilities and warehouses.

Then, we use the results from TOPSIS as input to the mathematical model. The mathematical model integrates goal programming (GP) and robust optimization (RO)—explained in Section 4.2. and 4.3., respectively—due to the advantage of GP in solving multiple objectives and RO on solving uncertainties.

5.1 TECHNIQUE FOR ORDER PREFERENCE BY SIMILARITY TO AN IDEAL SOLUTION (TOPSIS)

TOPSIS is a multi-criteria decision technique (MCDM) used to evaluate alternatives based on certain criteria, as shown in Figure 11.3. This procedure ranks the alternatives according to the shortest distance from the ideal solution and the greatest distance from the negative-ideal solution using Euclidean distance (Pochampally et al., 2009). The obtained alternative scores represent the service level score of each market zone in our model. The steps of performing TOPSIS are illustrated in Figure 11.4.

Where i represents the alternative, j represents the criteria, r_{ij} is an element of alternative I, and criteria j in the normalized matrix. z_{ij} is a rate of alternative i and j criteria, v_{ij} is a weighted element in the normalized matrix, A^* is the ideal solution, and A^- is the negative-ideal solution. S_{i_*} represents the rating separation from alternative i and the ideal solution, while S_{i_-} represents the rating separation from alternative i and the negative-ideal solution.

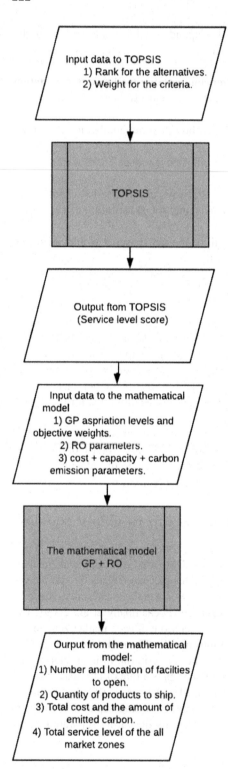

FIGURE 11.2 The Followed Procedures of the Proposed Model.

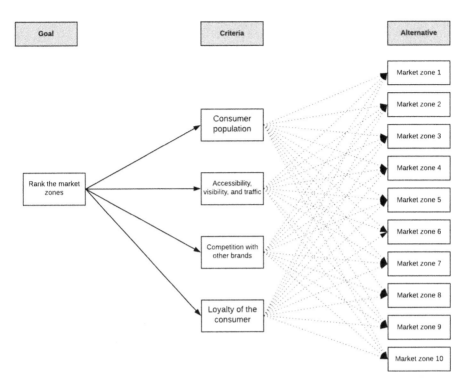

FIGURE 11.3 Hierarchy Structure for TOPSIS.

5.2 GOAL PROGRAMMING

Goal programming is an approach that can solve problems with conflict objectives. A book by Jones and Tamiz (2010) on goal programming can be referred to for more detailed information. In non-preemptive goal programming (NPGP), the decision-maker sets aspiration levels that represent a desired goal for each objective. It also requires the decision-maker to assign weights to each objective, which refers to the importance of achieving the goals. It is formulated as:

$$\text{MIN } \sum_i W_i(d_i^+ + d_i^-) \tag{11.1}$$

$$s.t. \ f_i(X) - d_i^+ + d_i^- = T_i \tag{11.2}$$

$$d_i^+, \ d_i^- \geq 0 \tag{11.3}$$

Equation (11.1) minimizes over-achievement d_i^+ and under-achievement d_i^- variables of each goal i, where W_i represents the weight of each goal i. The objective functions $f_i(X)$ becomes a constraint as shown in equation (11.2). Equation (11.3) ensures the deviational variables are positive numbers.

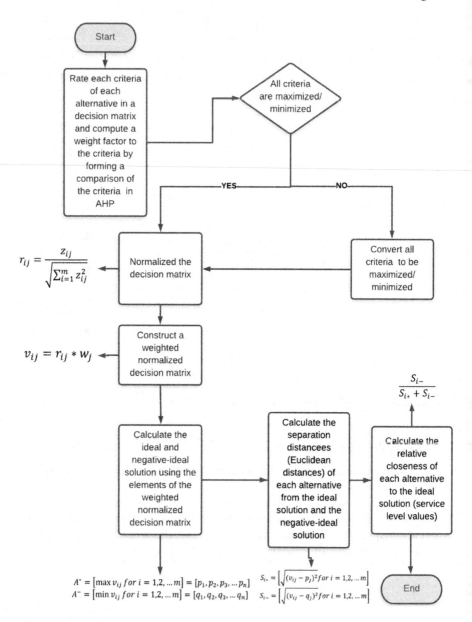

FIGURE 11.4   The Followed Procedures of TOPSIS.

5.3   Robust Optimization

Among all approaches that consider problems with uncertainties, the RO does not require having an exact probability distribution of the uncertainty parameter. Mulvey, Vanderbei, and Zenios (1995) developed an RO approach in terms of two types of robustness: solution robustness and model robustness. Solution robustness is to ensure that any realization of the scenario is a near-optimal solution, while model robustness is when the realization of any scenario is feasible. In this approach, the uncertain parameters are represented as a set of scenarios (SC) and a discrete probability of occurrence of each scenario. Efforts by Li (1996) and Yu and Li (2000) to develop and modify this approach showed its impacts in reducing computation time. The linearized format of the RO approach is represented in (4), (5), (6), (7), and (8).

$$\text{MIN } Z = \sum_{sc} \rho_{sc} \, \xi_{sc} + \lambda \sum_{sc} \rho_{sc} [(\xi_{sc} - \sum_{sc'} \rho_{sc'} \, \xi_{sc'}) + 2\theta_{sc}] + \omega \sum_{sc} \rho_{sc} \, \delta_{sc} \quad (11.4)$$

$$s.\, t \; Ax = b \quad (11.5)$$

$$B_{sc}x + C_{sc}y_{sc} + \delta_{sc} = e_{sc} \quad (11.6)$$

$$\xi_{sc} - \sum_{sc} \rho_{sc} \, \xi_{sc} + \theta_{sc} \geq 0 \quad (11.7)$$

$$x, \, y_{sc}, \, \delta_{sc} \geq 0 \quad (11.8)$$

Equation (11.4) is the objective function, which consists of solution robustness (part 1 and part 2) and model robustness (part 3) of the function. λ controls the solution robustness, ω controls the model robustness. In the above model, x is the design variable, y_{sc} is the control variable, and $\xi_{sc} = x + y_{sc}$. Equations (11.5) and (11.6) are the design and control constraints, respectively. In equation (11.6), δ_{sc} is the violation variable that occurs in case of infeasibility of any scenario realization, and it is penalized in the objective function to ensure model robustness (4). Equation (11.7) is used as an auxiliary constraint to linearize the quadratic format of the model proposed by Mulvey et al. (1995). Equation (11.8) is the non-negativity constraint.

We developed a model for designing CLSC network, which incorporates NPGP and RO approach proposed by Mulvey et al. (1995) and Yu and Li (2000). The developed formulation is as follow:

$$\text{MIN } Z = \sum_{i} W_i \, (d_i^+ + d_i^-) \quad (11.9)$$

$$s.\, t. \; f_i (X) - d_i^+ + d_i^- = T_i \quad (11.10)$$

$$Ax = b \quad (11.11)$$

$$B_{sc}x + C_{sc}y_{sc} + \delta_{sc} = e_{sc} \quad (11.12)$$

$$\xi_{sc} - \sum_{sc} \rho_{sc} \, \xi_{sc} + \theta_{sc} \geq 0 \quad (11.13)$$

$$x, y_{sc}, \delta_{sc}, d_i^+, d_i^- \geq 0 \qquad\qquad (11.14)$$

Where $f_i(X)$ represents equation (11.4).
Model notations and formulation are available in appendix A.

6 NUMERICAL EXAMPLE

To illustrate the usability of our proposed model, we used available data for the proposed project by the National Industrial Clusters Development Program (NICDP) to start a tire manufacturing industry in Saudi Arabia. The government in Saudi Arabia has worked towards its vision (Saudi Vision 2030) to diversify its economic sources and grow the manufacturing portion to Gross Domestic Product (GDP). Besides the incentives and support that the Saudi government is willing to provide, Saudi Arabia believes that it is a perfect place to invest in due to its strategic location at the crossroads of three continents: Asia, Africa, and Europe. In addition, Saudi Arabia established about 35 industrial cities all around the country, with a total area exceeding 198 million square meters—the available number of production facilities being around 3,474 (2016), (2020). Among a variety of investment opportunities that has been announced by NICDP (2016), the tire manufacturing plant is supposed to attract investors to start a business that will be the first of its kind in the region. According to Alpen Capital report (2016), in Gulf Cooperation Council (GCC) countries—Saudi Arabia, United Arab Emirates (UAE), Qatar, Oman, Kuwait, and Bahrain—there were about 10.3M passenger cars in use in 2015. This number was expected to increase and reach 13.3M in 2020 with a Compound Annual Growth Rate (CAGR) of 5%. Moreover, they forecasted the number of new passenger cars sold in this area to hit 1.4M in 2020 compared to the 1.2M sold in 2015. This automatically increases the consumption of tires, on top of the extreme weather condition in GCC areas which shortens the tire life cycle and requires tires to be replaced more often (Smith, 2017). According to NICDP (2016), the demand for tires is projected to be at 41M by 2020; 62% goes to the market of Saudi Arabia. Figure 11.5 shows the potential facility locations in the network. In our example, we use the 215/60R16 tire size which weighs about 10 kg. This is a standard size for many cars in that region, more specifically Toyota Hilux and Camry—the top sold cars in the last few years (2017). The number and location of facilities in the CLSC network considered in this example are shown in Table 11.2. We used Riyadh as a possible location for the hybrid manufacturing/remanufacturing facilities as suggested by NICDP (2016). Since there is no information about other possible locations of the other facility types, we used Analytic Hierarchy Process (AHP) to evaluate the 35 industrial cities available and picked the top three as candidate location facilities in our proposed mathematical model according to three criteria: land space available at each location, accessibility, and distance from the market zone. For more details about AHP, review the paper published by Saaty (Saaty, 2008). We used Google maps to calculate the distance between facilities and embedded them in the model. The data on carbon emission is estimated based on (Lin et al., 2017). In Table 11.3, we summarized

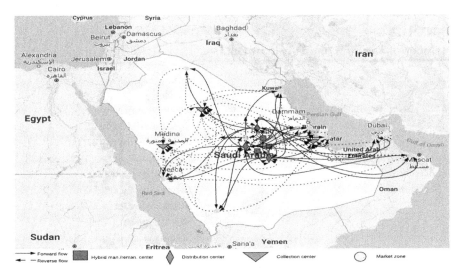

FIGURE 11.5   Potential Facility Locations in the Network.

some of the data that are hypnotized. There are three scenarios of the new and remanufactured tire demand and the number of returned tires as illustrated in Table 11.4. We generated 27 different scenarios for our example using the decision tree analysis shown in Figure 11.6, where the probability of each scenario is calculated as illustrated in Table 11.5. According to Lang (2010), the incurred cost resulted from the substitution between two products should include the difference between the price of selling both products if the objective does not incorporate the selling price of both products. Therefore, we estimated the substitution cost according to this pattern.

7   RESULTS AND DISCUSSION

In this part, we show the results obtained using the numerical example with a carbon tax policy. We used XLSTAT — a special package in Microsoft Office Excel — for criteria evaluation to select the top three candidate locations of distribution and collection facilities using AHP. We also used Microsoft Office Excel to find the service level score of each market zone using TOPSIS and then inputted the results to Lingo 18.0 to solve the mathematical model. Appendix B shows the calculation for the service level score. To set the aspiration levels, we ran the model for each objective with its corresponding constraints separately. We set $\omega = 10000$, $\lambda = 1$, for all three objective functions. The weight of the first objective is .5, the second objective is .3 and the third objective is .2.

The results show that the hybrid manufacturing/remanufacturing facility should be located at Riyadh industrial city 1, the distribution facility should be located at Durma industrial city and the collection facility should be located at Alahsa industrial city 2/Salwa. We selected the raw material supplier 1 and 2. Table 11.6 demonstrates more details about the optimal values of each scenario for each

TABLE 11.2

Number of Facilities and Locations Data

Facility	Number	Location
.	3	Riyadh
Hybrid man./reman. Facility	1	Riyadh industrial city 1
	1	Riyadh industrial city 2
	1	Riyadh industrial city 3
Distribution facility	1	Suair industrial city
Collection facility	1	Alkharj industrial city
Market zone	1	Durma industrial city
	1	Alahsa industrial city 2/ Salwa
	1	Hail industrial city 2
	1	Madinah industrial city
	5	Saudi Arabia $\begin{cases} \text{Central region (SACR)} \\ \text{Eastern region (SAER)} \\ \text{Western region (SAWR)} \\ \text{Western region (SAWR)} \\ \text{Southern region (SASR)} \end{cases}$
	1	Bahrain – Sitra industrial city (BHR)
	1	Oman – Rusayl industrial city (OMN)
	1	Qatar – Alrayyan industrial city (QAT)
	1	Kuwait – Shuwailkh industrial city. (KT)
	1	Alquiz industrial area 4 – (UAE)

objective, the robust solution of each objective, the amount of carbon emitted of each scenario, and the total quantity of new, remanufactured, and substitute products to ship in all market zones. Figure 11.7 displays the impact of increasing the value of λ on the robustness of the first objective function. Figure 11.8 shows the variation of λ on the robustness of the second objective function. Figure 11.9

TABLE 11.3

Hypnotized Data

Parameters	Values
$CAPM_m$	Unit (20000, 33000)
B_k	.5, .3, .1, .1.15
α	$ 20 per Kg
TAX	$ (10, 14)
RCC_k	$ 20
PC_m	

TABLE 11.4
New Tire Demand, Remanufactured Tire Demand and Number of Returned Tire Data

	New Tire Demand			Remanufactured Tire Demand			Returned Tire		
	Low	Mid	High	Low	Mid	High	Low	Mid	High
SACR	(2500–f3000)	(3000–3700)	(3700–4500)	(1200–1480)	(1500–1750)	(1800–2220)	(750–925)	(1500–1750)	(1950–2404)
SAER	(2000–2300)	(2300–2600)	(2600–3500)	(920–1040)	(1150–1300)	(1380–1560)	(575–650)	(1150–1300)	(1495–1690)
SAWR	(2500–2800)	(2800–3100)	(3100–3350)	(1120–1240)	(1400–1550)	(1680–1860)	(700–775)	(1400–1550)	(1820–2015)
SANR	(1000–1150)	(1150–1300)	(1300–1600)	(460–760)	(575–650)	(690–1140)	(288–475)	(575–650)	(758–1235)
SASR	(1450–1700)	(1700–1900)	(1900–2000)	(680–760)	(850–950)	(1020–1140)	(425–475)	(850–950)	(1105–1235)
BHR	(350–500)	(500–600)	(600–700)	(200–240)	(250–300)	(300–360)	(125–150)	(250–300)	(325–390)
OMN	(1350–1450)	(1450–1600)	(1600–1700)	(580–640)	(725–800)	(870–960)	(363–400)	(725–800)	(943–1040)
QAT	(850–1000)	(1000–1200)	(1200–1300)	(400–480)	(500–600)	(600–720)	(250–300)	(500–600)	(650–780)
KT	(1350–1450)	(1450–1600)	(1600–1700)	(580–640)	(725–800)	(870–960)	(363–400)	(725–800)	(943–1040)
UAE	(3000–3500)	(3500–3700)	(3700–3900)	(1400–1480)	(1750–1850)	(2100–2220)	(875–925)	(1750–1850)	(2275–2405)

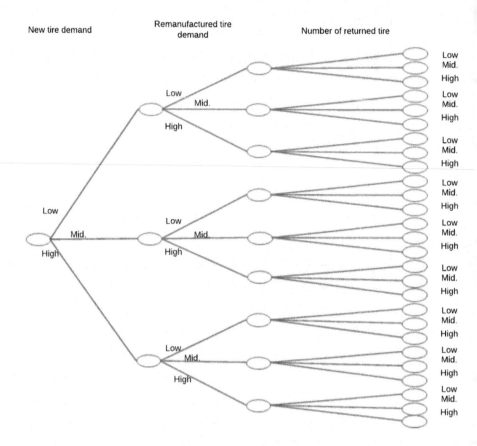

FIGURE 11.6   Decision Tree Analysis.

displays how the robustness of the third objective function varies by increasing the value of λ.

In Figure 11.7–11.9, it is clear that as the value of λ increases, the solution of each objective becomes more robust. This is due to the reduction in the variance between the scenarios of each objective. Hence, at λ = 10,000, the solution of all scenarios is the same for each objective function.

8 CONCLUSION

CLSC management has been a hot research area for the last few years due to the increase in market competition and environmental awareness. CLSC extends its benefit and has proven its capability on the economic ambit, as proven by companies that shifted from TSC to CLSC. Considering flexibility and robustness in designing the CLSC network improves its efficiency. This paper differs from other researches in the area of CLSC network design by three objectives — economic, environmental, and service level of the market zones. Distinguishing the demand of the new and

TABLE 11.5
Probability of Occurrence of Each Scenario

	Scenario (New Tire Demand. Reman. Tire Demand. Returned Tire)	Probability of Each Secnario Low = .2, Mid. = .5, High = .3
SC1	low.low.low	0.008
SC2	low.low.mid	0.02
SC3	low.low.high	0.012
SC4	low.mid.low	0.02
SC5	low.mid.mid	0.05
SC6	low.mid.high	0.03
SC7	low.high.low	0.012
SC8	low.high.mid	0.03
SC9	low.high.high	0.018
SC10	mid.low.low	0.02
SC11	mid.low.mid	0.05
SC12	mid.low.high	0.03
SC13	mid.mid.lxow	0.05
SC14	mid.mid.mid	0.125
SC15	mid.mid.high	0.075
SC16	mid.high.low	0.03
SC17	mid.high.mid	0.075
SC18	mid.high.high	0.045
SC19	high.low.low	0.012
SC20	high.low.mid	0.03
SC21	high.low.high	0.018
SC22	high.mid.low	0.03
SC23	high.mid.mid	0.075
SC24	high.mid.high	0.045
SC25	high.high.low	0.018
SC26	high.high.mid	0.045
SC27	high.high.high	0.027
		$\Sigma = 1$

remanufactured product, allowing product substitution, and considering the uncertainty of the product demand and number of the returned product are further features toward the uniqueness of the paper. The CLSC network in this paper covers raw material suppliers, hybrid plants (which manufacture new products and remanufacture the re-turned` product), distribution centers, collection centers, and multiple market zones. We presented a demonstrative example of a tire CLSC network to study the NICDP proposal to establish a tire manufacturing business in Saudi Arabia. To reach the optimal value of the three proposed objectives, our results recommend the following:

TABLE 11.6

Results Summary of Running the Mathematical Mod

		Objective Function			Robust Objective Function			Amount of Carbon Emitted	Quantity of New Product to Ship	Quantity of Substituted Product to Ship	Quantity of Remanufactured Product to Ship
		f1	f2	f3	G1	G2	G3				
					40,444,170	43,577,020	16,110				
SC1	low.low.low	28,737,693	26,179,420	14,292				1,308,971	17,561	4,301	3,761
SC2	low.low.mid	28,668,139	25,190,557	14,439				1,259,528	17,561	8,158	-
SC3	low.low.high	28,672,285	25,330,901	14,435				1,266,545	17,806	7,895	-
SC4	low.mid.low	28,805,095	28,190,299	15,403				1,409,515	17,834	4,465	5,290
SC5	low.mid.mid	28,732,363	27,125,364	15,448				1,356,268	17,578	8,492	1,521
SC6	low.mid.high	28,709,830	26,930,819	15,444				1,346,541	17,773	10,005	-
SC7	low.high.low	28,891,411	30,602,613	16,685				1,530,131	17,676	4,369	7,753
SC8	low.high.mid	28,848,214	30,430,478	17,069				1,521,524	17,873	8,281	4,506
SC9	low.high.high	28,761,067	28,628,128	16,560				1,431,406	17,333	11,370	1,065
SC10	mid.low.low	28,803,566	28,551,847	15,480				1,427,592	19,807	4,297	3,783
SC11	mid.low.mid	28,751,293	28,190,071	15,894				1,409,504	20,476	8,032	-
SC12	mid.low.high	28,748,280	28,121,211	15,810				1,406,061	20,190	8,251	-
SC13	mid.mid.low	28,874,187	30,556,955	16,627				1,527,848	19,717	4,334	5,721
SC14	mid.mid.mid	28,814,662	30,131,069	17,016				1,506,553	20,514	8,531	1,423
SC15	mid.mid.high	28,769,983	29,104,110	16,678				1,455,205	19,862	9,947	-
SC16	mid.high.low	28,949,817	32,753,659	17,621				1,637,683	19,837	4,367	7,627
SC17	mid.high.mid	28,892,362	32,277,379	18,078				1,613,869	20,182	8,543	3,752
SC18	mid.high.high	28,827,852	31,033,719	17,770				1,551,686	19,631	11,259	1,091
SC19	high.low.low	28,877,774	31,210,026	16,799				1,560,501	22,198	4,315	3,890
SC20	high.low.mid	28,806,227	30,170,824	16,735				1,508,541	22,416	7,907	-

(*Continued*)

TABLE 11.6 (Continued)

		Objective Function			Robust Objective Function			Amount of Carbon Emitted	Quantity of New Product to Ship	Quantity of Substituted Product to Ship	Quantity of Remanufactured Product to Ship
		f1	f2	f3	G1	G2	G3				
SC21	high.low.high	28,822,328	30,746,898	17,495				1,537,345	22,916	8,164	-
SC22	high.mid.low	28,950,001	33,410,989	17,929				1,670,549	22,712	4,423	5,370
SC23	high.mid.mid	28,890,359	32,797,494	18,398				1,639,875	22,914	8,450	1,627
SC24	high.mid.high	28,843,131	31,727,122	18,083				1,586,356	22,446	9,934	-
SC25	high.high.low	28,975,788	33,914,461	18,441				1,695,723	21,896	4,327	6,778
SC26	high.high.mid	28,898,981	32,708,928	18,503				1,635,446	21,486	8,532	2,982
SC27	high.high.high	28,848,605	32,008,075	18,566				1,600,404	21,769	11,097	134

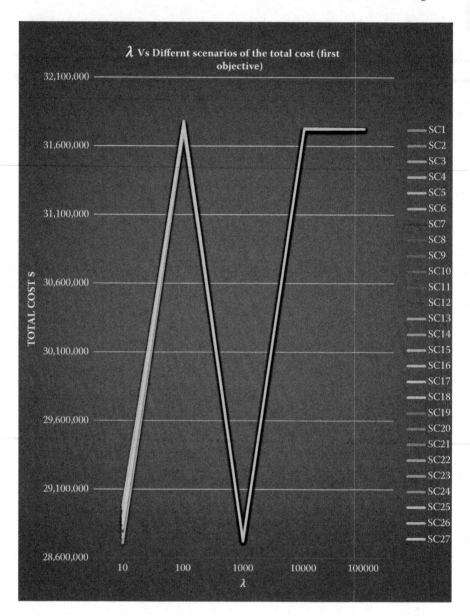

FIGURE 11.7 λ Vs the First Objective.

- Select raw material supplier 1 and 2.
- Locate the hybrid manufacturing/remanufacturing facility at Riyadh industrial city.
- Locate the distribution facility at Durma industrial city.
- Locate the collection facility at Alahsa industrial city 2/ Salwa.

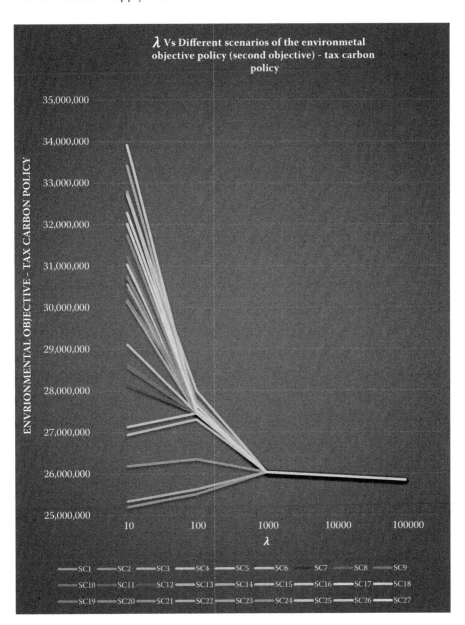

FIGURE 11.8 λ Vs the Second Objective.

Additionally, our results determine the quantity of the new, remanufactured, and substituted products to deliver to the market zone through the distribution centers, as well as the number of returned products to be collected. Besides all the benefits obtained from our model, it also helps supply chain decision-makers take more flexible decisions in determining the number of products that can be substituted to satisfy the market demand under its uncertainty.

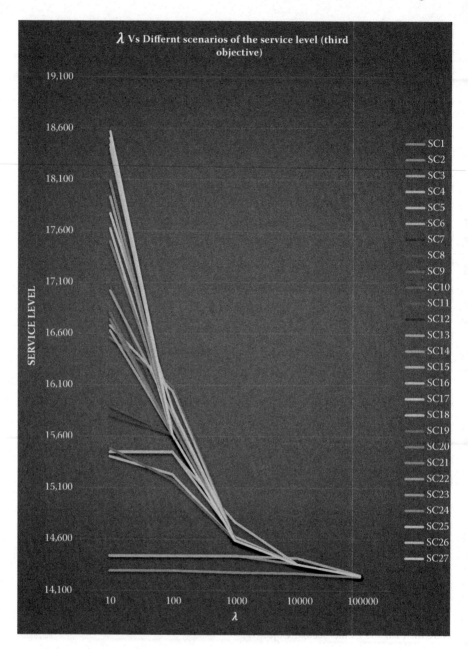

FIGURE 11.9 λ Vs the Third Objective.

For future study, we are working to incorporate different product substitution policies and use varying approaches to solve the problem stated in this paper. Using heuristics models can be an extension to the proposed model to solve a larger problem size. It is also adaptable to use this mathematical model in different industries where product substitution is applicable.

REFERENCES

Ahiska, S. S., Gocer, F., & King, R. E. (2017). Heuristic inventory policies for a hybrid manufacturing/remanufacturing system with product substitution. *Computers & Industrial Engineering, 114*, 206–222. https://doi.org/10.1016/j.cie.2017.10.014

Aldoukhi, M. A., & Gupta, S. M. (2018). Use of a roubst optimization in design a closed-loop supply chain network. *16th International Logistics and Supply Chain Congress*, 87–96. Denizli, Turkey.

Aldoukhi, M. A., & Gupta, S. M. (2019a). A Robust Closed Loop Supply Chain Network Design Under Different Carbon Emission Policies. *Pamukkale University Journal of Engineering Sciences, 25*(9), 1020–1032. https://doi.org/10.5505/pajes.2019.51460

Aldoukhi, M. A., & Gupta, S. M. (2019b). Use of goal programming in designing a closed-loop supply chain network under uncertainty. *Northeast Decision Sciences Institute 2019 Annual Conference*, 883–890. Philadelphia, PA.

Amin, S. H., & Zhang, G. (2013). A multi-objective facility location model for closed-loop supply chain network under uncertain demand and return. *Applied Mathematical Modelling, 37*(6), 4165–4176. https://doi.org/10.1016/j.apm.2012.09.039

Amin, S. H., & Zhang, G. (2014). Closed-loop supply chain network configuration by a multi-objective mathematical model. *International Journal of Business Performance and Supply Chain Modelling, 6*(1), 1–15. https://doi.org/10.1504/IJBPSCM.2014.058890

Aravendan, M., & Panneerselvam, R. (2014). Literature Review on Network Design Problems in Closed Loop and Reverse Supply Chains. *Intelligent Information Management, 06*(03), 104–117. https://doi.org/10.4236/iim.2014.63012

Baghalian, A., Rezapour, S., & Farahani, R. Z. (2013). Robust supply chain network design with service level against disruptions and demand uncertainties: A real-life case. *European Journal of Operational Research, 227*(1), 199–215. https://doi.org/10.1016/j.ejor.2012.12.017

Bayindir, Z. P., Erkip, N., & Güllü, R. (2005). Assessing the benefits of remanufacturing option under one-way substitution. *Journal of the Operational Research Society, 56*(3), 286–296.

Bayindir, Z. P., Erkip, N., & Güllü, R. (2007). Assessing the benefits of remanufacturing option under one-way substitution and capacity constraint. *Computers and Operations Research, 34*(2), 487–514. https://doi.org/10.1016/j.cor.2005.03.010

Benjaafar, S., Li, Y., & Daskin, M. (2013). Carbon footprint and the management of supply chains: Insights from simple models. *IEEE Transactions on Automation Science and Engineering, 10*(1), 99–116. https://doi.org/10.1109/TASE.2012.2203304

Bernstein, F., & Federgruen, A. (2004). A General Equilibrium Model for Industries with Price and Service Competition. *Operations Research, 52*(6), 868–886. https://doi.org/10.1287/opre.1040.0149

Bernstein, F., & Federgruen, A. (2007). Coordination Mechanisms for Supply Chains Under Price and Service Competition. *Manufacturing & Service Operations Management, 9*(3), 242–262. https://doi.org/10.1287/msom.1070.0159

Diabat, A., Abdallah, T., Al-refaie, A., Svetinovic, D., & Govindan, K. (2013). Strategic Closed-Loop Facility Location Problem. *IEEE Transactions on Engineering Management, 60*(2), 398–408. https://doi.org/10.1109/TEM.2012.2211105

Du, S., Hu, L., & Song, M. (2016). Production optimization considering environmental performance and preference in the cap-and-trade system. *Journal of Cleaner Production, 112*, 1600–1607. https://doi.org/10.1016/j.jclepro.2014.08.086

Farahani, R. Z., Rezapour, S., Drezner, T., & Fallah, S. (2014). Competitive supply chain network design: An overview of classifications, models, solution techniques and applications. *Omega (United Kingdom)*, *45*, 92–118. https://doi.org/10.1016/j.omega.2013.08.006

Fareeduddin, M., Hassan, A., Syed, M. N., & Selim, S. Z. (2015). The impact of carbon policies on closed-loop supply chain network design. *Procedia CIRP*, *26*, 335–340. https://doi.org/10.1016/j.procir.2014.07.042

Ferguson, M. (2009). Strategic and Tactical Aspects of Closed-Loop Supply Chains. *Foundations and Trends® in Technology, Information and Operations Management*, *3*(2), 101–200. https://doi.org/10.1561/0200000019

GCC Automobile Industry Report. (2016). Retrieved from http://www.alpencapital.com/downloads/reports/2016/GCC-Automobile-Industry-Report-December-2016.pdf

GCC Best Selling Cars. The Top 50 in 2017. (2017). Retrieved from https://focus2move.com/gcc-best-selling-cars/

Ghafarimoghadam, A., Karimi, A., Mousazadeh, M., & Pishvaee, M. S. (2016). A robust optimisation model for remanufacturing network design problem with one-way substitution. *International Journal of Services and Operations Management*, *24*(4), 484–503. https://doi.org/10.1504/IJSOM.2016.077785

Ghavamifar, A., Makui, A., & Taleizadeh, A. A. (2018). Designing a resilient competitive supply chain network under disruption risks: A real-world application. *Transportation Research Part E: Logistics and Transportation Review*, *115*, 87–109. https://doi.org/10.1016/j.tre.2018.04.014

Govindan, K., Fattahi, M., & Keyvanshokooh, E. (2017). Supply chain network design under uncertainty: A comprehensive review and future research directions. *European Journal of Operational Research*, *263*(1), 108–141. https://doi.org/10.1016/j.ejor.2017.04.009

Govindan, K., Mina, H., Esmaeili, A., & Gholami-Zanjani, S. M. (2019). An Integrated Hybrid Approach for Designing a Green Closed-loop Supply Chain Network under Uncertainty. *Journal of Cleaner Production*, *242*, 118317. https://doi.org/10.1016/j.jclepro.2019.118317

Govindan, K., & Soleimani, H. (2017). A review of reverse logistics and closed-loop supply chains: a Journal of Cleaner Production focus. *Journal of Cleaner Production*, *142*, 371–384. https://doi.org/10.1016/j.jclepro.2016.03.126

Govindan, K., Soleimani, H., & Kannan, D. (2015). Reverse logistics and closed-loop supply chain: A comprehensive review to explore the future. *European Journal of Operational Research*, *240*(3), 603–626. https://doi.org/10.1016/j.ejor.2014.07.012

Guide Jr, V. D. R., & Van Wassenhove, L. N. (2010). The Evolution of Closed-Loop Supply Chain Research. *Independent Review*, *14*(3), 363–375. https://doi.org/10.1287/opre.1080.0628

Gupta, S. M., & Ilgin, M. A. (2018). Multiple Criteria Decision Making Applications in Environmentally Conscious Manufacturing and Product Recovery. In *CRC Press*. Retrieved from https://www.crcpress.com/Multiple-Criteria-Decision-Making-Applications-in-Environmentally-Conscious/Gupta-Ilgin/p/book/9781498700658

He, L., Hu, C., Zhao, D., Lu, H., Fu, X., & Li, Y. (2016). Carbon emission mitigation through regulatory policies and operations adaptation in supply chains: theoretic developments and extensions. *Natural Hazards*, *84*(1), 179–207. https://doi.org/10.1007/s11069-016-2273-5

Ilgin, M. A., & Gupta, S. M. (2010). Environmentally conscious manufacturing and product recovery (ECMPRO): A review of the state of the art. *Journal of Environmental Management*, *91*(3), 563–591. https://doi.org/10.1016/j.jenvman.2009.09.037

Ilgin, M. A., & Gupta, S. M. (2013). Reverse Logistics. In S. M. Gupta (Ed.), *Reverse Supply Chains: Issues and Analysis*. Boca Raton, Florida: CRC Press.

Ilgin, M. A., Gupta, S. M., & Battaïa, O. (2015). Use of MCDM techniques in environmentally conscious manufacturing and product recovery: State of the art. *Journal of Manufacturing Systems*, *37*, 746–758. https://doi.org/10.1016/j.jmsy.2015.04.010

Jalil, S. A., Hashmi, N., Asim, Z., & Javaid, S. (2019). A de-centralized bi-level multi-objective model for integrated production and transportation problems in closed-loop supply chain networks. *International Journal of Management Science and Engineering Management*, *14*(3), 206–217. https://doi.org/10.1080/17509653.2018.1545607

Jerbia, R., Kchaou Boujelben, M., Sehli, M. A., & Jemai, Z. (2018). A Stochastic Closed-Loop Supply Chain Network Design Problem with Multiple Recovery Options. *Computers and Industrial Engineering*, *118*, 23–32. https://doi.org/10.1016/j.cie.2018.02.011

Jin, M., Granda-Marulanda, N. A., & Down, I. (2014). The impact of carbon policies on supply chain design and logistics of a major retailer. *Journal of Cleaner Production*, *85*, 453–461. https://doi.org/10.1016/j.jclepro.2013.08.042

Jindal, A., & Sangwan, K. S. (2017). Multi-objective fuzzy mathematical modelling of closed-loop supply chain considering economical and environmental factors. *Annals of Operations Research*, *257*(1–2), 95–120. https://doi.org/10.1007/s10479-016-2219-z

Jones, D., & Tamiz, M. (2010). *Practical Goal Programming*. https://doi.org/10.1007/978-1-4419-5771-9

Krikke, H., Hofenk, D., & Wang, Y. (2013). Revealing an invisible giant: A comprehensive survey into return practices within original (closed-loop) supply chains. *Resources, Conservation and Recycling*, *73*, 239–250. https://doi.org/10.1016/j.resconrec.2013.02.009

Lang, J. C. (2010). Production and inventory management with substitutions. In *Springer Science & Business Media* (Vol. 636). https://doi.org/10.1007/978-3-642-04247-8_1

Lee, D. H., Bian, W., & Dong, M. (2007). Multiobjective model and solution method for integrated forward and reverse logistics network design for third-party logistics providers. *Transportation Research Record*, (2032), 43–52. https://doi.org/10.3141/2032-06

Li H. L. (1996). An efficient method for solving linear goal programming problems. *J Optim Theory*, *90*(2), 465–469.

Li, Y., Chen, J., & Cai, X. (2006). Uncapacitated production planning with multiple product types, returned product remanufacturing, and demand substitution. *OR Spectrum*, *28*(1), 101–125. https://doi.org/10.1007/s00291-005-0012-5

Lin, T. H., Chien, Y. S., & Chiu, W. M. (2017). Rubber tire life cycle assessment and the effect of reducing carbon footprint by replacing carbon black with graphene. *International Journal of Green Energy*, *14*(1), 97–104. https://doi.org/10.1080/15435075.2016.1253575

Mardan, E., Govindan, K., Mina, H., & Gholami-Zanjani, S. M. (2019). An accelerated benders decomposition algorithm for a bi-objective green closed loop supply chain network design problem. *Journal of Cleaner Production*, *235*, 1499–1514. https://doi.org/10.1016/j.jclepro.2019.06.187

Martino, G., Iannnone, R., Fera, M., Miranda, S., & Riemma, S. (2017). Fashion retailing: A framework for supply chain optimization. *Uncertain Supply Chain Management*, *5*, 243–272. https://doi.org/10.5267/j.uscm.2016.12.002

Mohajeri, A., & Fallah, M. (2016). A carbon footprint-based closed-loop supply chain model under uncertainty with risk analysis: A case study. *Transportation Research Part D: Transport and Environment*, *48*, 425–450. https://doi.org/10.1016/j.trd.2015.09.001

Mulvey, J. M., Vanderbei, R. J., & Zenios, S. A. (1995). Robust Optimization of Large-Scale Systems. *Operations Research*, *43*(2), 264–281.

National Industrial Clusters Development Program (NICDP) in Saudi Arabia. (2016).

Nurjanni, K. P., Carvalho, M. S., & Costa, L. (2017). Green supply chain design: A mathematical modeling approach based on a multi-objective optimization model. *International Journal of Production Economics*, *183*, 421–432. https://doi.org/10.1016/j.ijpe.2016.08.028

Özceylan, E., & Paksoy, T. (2014). Interactive fuzzy programming approaches to the strategic and tactical planning of a closed-loop supply chain under uncertainty. *International Journal of Production Research*, *52*(8), 2363–2387. https://doi.org/10.1080/00207543.2013.865852

Paksoy, T., Pehlivan, N. Y., & Özceylan, E. (2012). Fuzzy Multi-Objective Optimization of a Green Supply Chain Network with Risk Management that Includes Environmental Hazards. *Human and Ecological Risk Assessment, 18*(5), 1120–1151. https://doi.org/10.1080/10807039.2012.707940

Papen, P., & Amin, S. H. (2019). Network configuration of a bottled water closed-loop supply chain with green supplier selection. *Journal of Remanufacturing, 9*(2), 109–127. https://doi.org/10.1007/s13243-018-0061-y

Pazhani, S., Ramkumar, N., Narendran, T. T., & Ganesh, K. (2013). A bi-objective network design model for multi-period, multi-product closed-loop supply chain. *Journal of Industrial and Production Engineering, 30*(4), 264–280. https://doi.org/10.1080/21681015.2013.830648

Pazhani, S., & Ravi Ravindran, A. (2018). A bi-criteria model for closed loop supply chain network design. *International Journal of Operational Research, 31*(3), 330–356. https://doi.org/10.1504/IJOR.2018.089735

Piñeyro, P., & Viera, O. (2010). The economic lot-sizing problem with remanufacturing and one-way substitution. *International Journal of Production Economics, 156*, 167–168. https://doi.org/10.1016/j.ijpe.2014.06.003

Pishvaee, M. S., Farahani, R. Z., & Dullaert, W. (2010). A memetic algorithm for bi-objective integrated forward/reverse logistics network design. *Computers and Operations Research, 37*(6), 1100–1112. https://doi.org/10.1016/j.cor.2009.09.018

Pochampally, K. K., Nukala, S., & Gupta, S. M. (2009). Strategic planning models for reverse and closed-loop supply chains. In *CRC Press* (Vol. 47). https://doi.org/10.1080/00207540902808233

Ramezani, M., Bashiri, M., & Tavakkoli-Moghaddam, R. (2013). A new multi-objective stochastic model for a forward/reverse logistic network design with responsiveness and quality level. *Applied Mathematical Modelling, 37*(1–2), 328–344. https://doi.org/10.1016/j.apm.2012.02.032

Saaty, T. L. (2008). Decision making with the analytic hierarchy process. *International Journal of Services Sciences, 1*(1), 83–98. https://doi.org/10.1016/0305-0483(87)90016-8

Sabri, E. H., & Beamon, B. M. (2000). A Multi-objective Approach to Simultaneous Strategic and Operational Planning in Supply Chain Design. *Omega, 28*(5), 581–598. https://doi.org/10.1016/S0305-0483(99)00080-8

Sadeghi Rad, R., & Nahavandi, N. (2018). A novel multi-objective optimization model for integrated problem of green closed loop supply chain network design and quantity discount. *Journal of Cleaner Production, 196*, 1549–1565. https://doi.org/10.1016/j.jclepro.2018.06.034

Saudi Authority for Industrial Cities and Technology Zones (MODON). (2020).

Selim, H., & Ozkarahan, I. (2008). A supply chain distribution network design model: An interactive fuzzy goal programming-based solution approach. *International Journal of Advanced Manufacturing Technology, 36*(3–4), 401–418. https://doi.org/10.1007/s00170-006-0842-6

Shen, Z.-J. M., & Daskin, M. S. (2005). Trade-offs Between Customer Service and Cost in Integrated Supply Chain Design. *Manufacturing & Service Operations Management, 7*(3), 188–207. https://doi.org/10.1287/msom.1050.0083

Shi, J., Liu, Z., Tang, L., & Xiong, J. (2016). Multi-objective optimization for a closed-loop network design problem using an improved genetic algorithm. *Applied Mathematical Modelling, 45*, 14–30. https://doi.org/10.1016/j.apm.2016.11.004

Smith, S. (2017). *Saudi Arabia Tire Market Forecast & Opportunities*.

Stindt, D., & Sahamie, R. (2014). Review of research on closed loop supply chain management in the process industry. *Flexible Services and Manufacturing Journal, 26*(1–2), 268–293. https://doi.org/10.1007/s10696-012-9137-4

Subramanian, P., Ramkumar, N., Narendran, T. T., & Ganesh, K. (2013). PRISM: PRIority based SiMulated annealing for a closed loop supply chain network design problem. *Applied Soft Computing Journal, 13*(2), 1121–1135. https://doi.org/10.1016/j.asoc.2012.10.004

Talaei, M., Farhang Moghaddam, B., Pishvaee, M. S., Bozorgi-Amiri, A., & Gholamnejad, S. (2016). A robust fuzzy optimization model for carbon-efficient closed-loop supply chain network design problem: A numerical illustration in electronics industry. *Journal of Cleaner Production, 113,* 662–673. https://doi.org/10.1016/j.jclepro.2015.10.074

Tosarkani, B. M., & Amin, S. H. (2018). A possibilistic solution to configure a battery closed-loop supply chain: Multi-objective approach. *Expert Systems with Applications, 92,* 12–26. https://doi.org/10.1016/j.eswa.2017.09.039

Xu, J., Liu, Q., & Wang, R. (2008). A Class of Multi-objective Supply Chain Networks Optimal Model Under Random Fuzzy Environment and Its Application to the Industry of Chinese Liquor. *Information Sciences, 178*(8), 2022–2043. https://doi.org/10.1016/j.ins.2007.11.025

Yu, C., & Li, H. (2000). A robust optimization model for stochastic logistic problems. *International Journal of Production Economics, 64*(1–3), 385–397.

Zarandi, M. H. F., Sisakht, A. H., & Davari, S. (2011). Design of a closed-loop supply chain (CLSC) model using an interactive fuzzy goal programming. *International Journal of Advanced Manufacturing Technology, 56*(5–8), 809–821. https://doi.org/10.1007/s00170-011-3212-y

Zhen, L., Huang, L., & Wang, W. (2019). Green and sustainable closed-loop supply chain network design under uncertainty. *Journal of Cleaner Production, 227,* 1195–1209. https://doi.org/10.1016/j.jclepro.2019.04.098

Appendix A

NOTATIONS

GENERAL PARAMETERS

$DM^{sc}_{pr,c}$: demand for product pr by the market zone c under scenario sc.

R^{sc}_{c}: returned product to the market location c under scenario sc.

B_k: fraction of quality level k of return product.

CO: number of raw material required to produce a unit of product.

$FRAC_{c,pr}$: substitution fraction at the market location c of product pr.

α: fraction of returned product that is disposed of

λ: weighing factor to measure solution robustness.

ω: weighting factor to measure model robustness.

ρ_{sc}: probability of scenario sc.

FIXED COSTS

FM_m: fixed cost of constructing the hybrid manufacturing/remanufacturing center in location m.

FW_w: fixed cost of constructing the distribution center in location w.

FH_h: fixed cost of constructing the collection center in location h.

UNIT COSTS

PC_m: unit production cost of new product at the hybrid manufacturing /re-manufacturing center in location m.

$RCC_{k,m}$: unit remanufacturing cost of returned product of quality level k at the hybrid manufacturing /remanufacturing center in location m.

INS_h: unit inspection and sorting cost of returned product at the collection center in location h.

$P_{s,m}$: component purchasing cost from supplier s by the hybrid manufacturing /remanufacturing center in location m.

$CRMSUB_{pr,px}$: cost of substituting product pr by product px.

$Scof_c$: the service level score of the market zone c.

SHIPPING COSTS

$TRMW_{m,w}$: cost of shipping a unit of product from hybrid manufacturing /re-manufacturing center location m to the distribution center in location w.

$TRWC_{w,c}$: cost of shipping a unit of product from the distribution center in location w to the market zone c.

$TRCH_{c,h}$: cost of shipping a unit of returned product from the market location c to the collection center in location h.

$TRHM_{h,m}$: cost of shipping a unit of returned product from the collection center in location h to the hybrid manufacturing/remanufacturing center m in location m.

CAPACITY PARAMETERS

$CAPM_m$: production capacity at the hybrid manufacturing/remanufacturing center in location m, in units.

$CAPW_w$: capacity of the distribution center in location w, in units.

$CAPH_h$: capacity of the collection center in location h, in units.

$CAPS_s$: capacity of the supplier s, in units.

CARBON EMISSION PARAMETERS

$EP_{m,pr}$: carbon emission in kg at the hybrid manufacturing/remanufacturing center in location m due to production a unit of product pr.

$ETMW_{m,w}$: carbon emission in kg due to shipping a unit of product from the hybrid manufacturing/remanufacturing center in location m to the distribution center in location w.

$ETWC_{w,c}$: carbon emission in kg due to shipping a unit of product from the distribution in location w to the market zone c.

$ETCH_{c,h}$: carbon emission in kg due to shipping a unit of EOL product from the market zone c to the collection center in location h.

$ETHM_{h,m}$: carbon emission in kg due to shipping a unit of returned product from the collection center in location h to hybrid manufacturing/remanufacturing center in location m in location m. *ETDIP*: carbon emission in kg due to disposal process. *Tax*: penalty per kg of carbon emitted.

DESIGN VARIABLES

YM_m: 1 if the hybrid manufacturing /remanufacturing center is constructed in location m, 0 otherwise.

YW_w: 1 if the distribution is constructed in location w, 0 otherwise. YC_h: 1 if the collection center is constructed in location h, 0 otherwise.

CONTROL VARIABLES

$XP^{sc}_{pr,m}$: quantity of product pr produced in the hybrid manufacturing/remanufacturing center in location m under scenario sc.

$XQSM^{sc}_{s,m}$: quantity of raw material shipped from the supplier s to the hybrid manufacturing /remanufacturing center in location m using mod transportation under scenario sc.

$XQMW^{sc}_{pr,m,w}$: quantity of product pr shipped from the hybrid manufacturing/remanufacturing center in location m to the distribution center in location w under scenario sc.

$XQWC^{sc}_{pr,w,c}$: quantity of products pr shipped from the distribution center in location w to the market zone c under scenario sc.

$XRQCH^{sc}_{c,h}$: quantity of returned products shipped from the market zone c to the collection center in location h under scenario sc.

$XRQHM^{sc}_{k,h,m}$: quantity of returned products with quality level k shipped from the collection center in location h to the hybrid manufacturing /remanufacturing center in location m under scenario sc.

$DISP^{sc}_h$: quantity of returned products disposed of from the collection center in location h under scenario sc.

$U^{sc}_{pr,c}$: amount of unsatisfied product pr which can be substituted at the market zone c under scenario sc.

$\delta DP^{sc}_{c,pr}$: violation of demand constraint, amount of not meeting the market zone c demanding product pr under scenario sc.

δRE^{sc}_c: violation of number of returned product constraint, amount of returned product not collected at the market zone c under scenario sc.

θI_{sc}: deviation for violation of the mean under scenario sc of the first objective function.

θII_{sc}: deviation for violation of the mean under scenario sc of the second objective function.

θIII_{sc}: deviation for violation of the mean under scenario sc of the third objective function.

MODEL FORMULATION

We use the format introduced in equations (11.9)–(11.14).

OBJECTIVE FUNCTIONS

The first objective function is to minimize the total cost of the CLSC network (*TECON*).

$$f_1(X) - d_i^+ + d_i^- = TECON*$$ (11.15)

This includes the total fixed cost (TFC), the total cost of purchasing raw material (TRM_{sc}), the total production cost (TPC_{sc}), the total transportation cost (TTC_{sc}), the total product substitution cost ($TSUBC_{sc}$) and the total collection cost (TCC_{sc}).

$$TECON_{sc} = TFC + TRM_{sc} + TPC_{sc} + TTC_{sc} + TCC_{sc} + TSUBC_{sc} \quad \forall \, sc \quad (11.16)$$

The total fix cost contains all cost associated with opening a new facility in the CLSC; opening hybrid manufacturing/remanufacturing center, distribution and collection center.

$$TFC = \sum_m (FM_m * YM_m) + \sum_w (FW_w * YW_w) + \sum_H (Fh_H * Yh_{hw}) \quad (11.17)$$

The total cost of purchasing raw material (TRM_{sc}) includes also the cost to deliver raw materials from suppliers to hybrid manufacturing/remanufacturing center.

$$TRM_{sc} = \sum_s \sum_m P_{s,m} * XQSM_{s,m}^{sc} \quad \forall \, sc \quad (11.18)$$

The total production cost (TPC_{sc}) includes the cost to produce new product ($TPCI_{sc}$) and the cost to remanufactured returned product ($TPCII_{sc}$).

$$TPC_{sc} = TPCI_{sc} + TPCII_{sc} \quad \forall \, sc \quad (11.19)$$

where

$$TPCI_{sc} = \sum_m PC_m * XP_{pr,m}^{sc} \quad \forall \, sc, \, pr = new \quad (11.20)$$

$$TPCII_{sc} = \sum_k \sum_h \sum_m RCC_{k,m} * XRQHM_{k,h,m}^{sc} \quad \forall \, sc \quad (11.21)$$

The total transportation cost (TTC_{sc}) is the cost of shipping a unit of product among all facilities in the CLSC network.

$$
\begin{aligned}
TTC_{sc} = \; & \Sigma_m \Sigma_w \Sigma_{pr} \, TRMW_{m,w} * XQMW_{pr,m}^{sc} \\
& + \Sigma_w \Sigma_c \Sigma_{pr} \, TRWC_{w,c} * XQWC_{pr,w,c}^{sc} \\
& + \Sigma_c \Sigma_h \, TRCH_{c,h} * XRQCH_{c,h}^{sc} \\
& + \Sigma_k \Sigma_h \Sigma_m \, TRHM_{h,m} * XRQHM_{k,h,m}^{sc}
\end{aligned}
\quad \forall \, sc \quad (11.22)
$$

The total collection cost (TCC_{sc}) related to inspection and sorting of the EOL product at the collection centers.

$$TCC_{sc} = \sum_c \sum_h INS_h * XRQCH_{c,h}^{sc} \quad \forall \ sc \quad (11.23)$$

The total product substitution cost ($TSUBC_{sc}$) occurs when the new product substitute the remanufactured product.

$$TSUBC_{sc} = \sum_w \sum_c \sum_{pr} \sum_{px} CRMSUB_{pr,px} * XQWC_{pr,px,w,c}^{sc} \quad \forall \ sc \quad (11.24)$$

The second objective is to minimize the carbon emission in the CLSC network ($TENVI_{sc}$)

$$f_2(X) - d_i^+ + d_i^- = TENVI* \quad (11.25)$$

$$TENVI_{sc} = Tax * ENVI_{sc}$$

In this objective, the carbon emission is resulted from production, transportation and disposal activities. We use the carbon tax policy as described in Section 5.

$$
\begin{aligned}
ENVI_{sc} = \ & \Sigma_m \Sigma_{pr} EP_m * XP_{pr,m}^{sc} + \Sigma_m \Sigma_w \Sigma_{pr} ETMW_{m,w} * XQMW_{pr,m,w}^{sc} \\
& + \Sigma_w \Sigma_c \Sigma_{pr} \Sigma_{px} ETWC_{w,c} * XQWC_{pr,px,w,c}^{sc} \\
& + \Sigma_c \Sigma_h ETCH_{c,h} * XRQCH_{c,h}^{sc} \quad\quad\quad \forall \ sc \\
& + \Sigma_k \Sigma_h \Sigma_m ETHM_{h,m} * XRQHM_{k,h,m}^{sc} \\
& + \Sigma_h ETDIP * DISP_h^{sc}
\end{aligned}
$$

$$(11.26)$$

The third objective is to determine the inventory level of each market location (SL).

$$f_3(X) - d_i^+ + d_i^- = SL* \quad (11.27)$$

In this objective, the result obtained from TOPSIS, where the market zones are evaluated, represent the service score of the market zone.

$$SL_{sc} = \sum_w \sum_c \sum_{pr} \sum_{px} Scof_c * XQWC_{pr,px,w,c}^{sc} \quad \forall \ sc \quad (11.28)$$

where $f_i(X) \ i = 1, \ 2, \ 3.$ formulated based on Mulvey et al. (1995).

CONSTRAINTS

The New product are produced using the raw material delivered from all suppliers.

$$\sum_s XQSM_{s,m}^{sc} = CO_* XP_{pr,m}^{sc} \quad \forall \; m, \, sc, \; pr = new \qquad (11.29)$$

The remanufactured product are produced using all quality levels of returned products delivered from all collection centers.

$$XP_{pr,m}^{sc} = \sum_k \sum_h XRQHM_{k,h,m}^{sc} \quad \forall \; m, \, sc, \; pr = remanufacture \qquad (11.30)$$

All products produced at any hybrid manufacturing/remanufacturing center are shipped to all distribution centers.

$$XP_{pr,m}^{sc} = \sum_w XQMW_{pr,m,w}^{sc} \quad \forall \; m, \, sc, \, pr \qquad (11.31)$$

Any product shipped from all hybrid manufacturing/remanufacturing centers to any distribution center are shipped to all market zones.

$$\sum_m XQMW_{pr,m,w}^{sc} = \sum_c \sum_{px \neq pr} XQWC_{pr,px,w,c}^{sc} \quad \forall \; w, \, sc, \, pr \qquad (11.32)$$

The returned product are either collected and shipped to the collection centers or not collected.

$$\sum_h XRQCH_{c,h}^{sc} + \delta RE_c^{sc} = RE_c^{sc} \quad \forall \; c, \, sc \qquad (11.33)$$

Part of returned products shipped from all market zones to any collection center is disposed of.

$$DISP_h^{sc} = \alpha_* \sum_c XRQCH_{c,h}^{sc} \quad \forall \; h, \, sc \qquad (11.34)$$

The rest of product are shipped to all hybrid manufacturing/remanufacturing centers for remanufacturing with different quality levels.

$$\sum_m XRQHM_{k,h,m}^{sc} = B_{k*}((1 - \alpha)_* \sum_c XRQCH_{c,h}^{sc}) \quad \forall \; k, \, h, \, sc \qquad (11.35)$$

Demand of product can be satisfied either directly or indirectly. Otherwise, violation of not meeting the demand (shortage).

$$\sum_W XQWC^{sc}_{pr,pr,w,c} + U^{sc}_{c,pr} + \delta DP^{sc}_{c,pr} = DM^{sc}_{c,pr} \quad \forall \, sc, c, pr \qquad (11.36)$$

At any market zone, satisfying the demand of any product indirectly can be done by substitution of the product

$$U^{sc}_{c,pr} = \sum_W \sum_{px \neq pr} XQWC^{sc}_{px,pr,w,c} \quad \forall \, sc, c, pr \qquad (11.37)$$

At any market zone, the fraction shows the acceptance of substituting product.

$$\sum_W \sum_{px \neq pr} XQWC^{sc}_{px,pr,w,c} \leq FRAC_{c,px,pr} * U^{sc}_{c,pr} \quad \forall \, sc, c, pr, px \qquad (11.38)$$

Capacity constraints

$$\sum_m XQSM^{sc}_{s,m} \leq CAPS_s \quad \forall \, s, sc \qquad (11.39)$$

$$\sum_{pr} XP^{sc}_{pr,m} \leq CAPM_m * YM_m \quad \forall \, m, sc \qquad (11.40)$$

$$\sum_m \sum_{pr} XQMW^{sc}_{pr,m,w} \leq CAPW_w * YW_w \quad \forall \, w, sc \qquad (11.41)$$

$$\sum_c XRQCH^{sc}_{c,h} \leq CAPH_h * YC_h \quad \forall \, h, sc \qquad (11.42)$$

Auxiliary constraint for linearization

$$TECON_{sc} - \sum_{sc} \rho_{sc} * TECON_{sc} + \theta I_{sc} \geq 0 \quad \forall \, sc \qquad (11.43)$$

$$TENVI_{sc} - \sum_{sc} \rho_{sc} * TENVI_{sc} + \theta II_{sc} \geq 0 \quad \forall \, sc \qquad (11.44)$$

$$SL_{sc} - \sum_{sc} \rho_{sc} * SL_{sc} + \theta III_{sc} \geq 0 \quad \forall \, sc \qquad (11.45)$$

Appendix B

AHP comparison matrix to weight the evaluation criteria – step 1

	Customer Population	Accessibility, Visibility, and Traffic	Competition and Neighbors	Loyalty
Customer population	1.00	11.11	6.67	0.50
Accessibility, visibility, and traffic	0.09	1.00	10.00	3.33
Competition and neighbors	0.15	0.10	1.00	2.00
Loyalty	2.00	0.30	0.50	1.00

Step 2

	Customer Population	Accessibility, Visibility, and Traffic	Competition and Neighbors	Loyalty	Weights
Customer population	0.31	0.89	0.37	0.07	0.40
Accessibility, visibility, and traffic	0.03	0.08	0.55	0.49	0.30
Competition and neighbors	0.05	0.01	0.06	0.29	0.10
Loyalty	0.62	0.02	0.03	0.15	0.20

Market zones rating in terms of jth criteria for TOPSIS

	Customer Population	Accessibility, Visibility, and Traffic	Competition and Neighbors	Loyalty
SACR	10	8	9	6
SAER	4	8	9	6
SAWR	8	8	9	6
SANR	3	7	7	7
SASR	6	6	6	7
BHR	1	6	5	8
OMN	5	5	5	7
QAT	2	5	8	6
KT	5	7	7	7
UAE	9	7	8	6

(Continued)

	Customer Population	Accessibility, Visibility, and Traffic	Competition and Neighbors	Loyalty
	max	Max	min	max
Weights	0.40	0.30	0.10	0.20
Ideal	10	8	5	8
Worst	1	5	9	6

All criterions to be maximized

	Customer Population	Accessibility, Visibility, and Traffic	Competition and Neighbors	Loyalty
SACR	10	8	0	6
SAER	4	8	0	6
SAWR	8	8	0	6
SANR	3	7	2	7
SASR	6	6	3	7
BHR	1	6	4	8
OMN	5	5	4	7
QAT	2	5	1	6
KT	5	7	2	7
UAE	9	7	1	6
$\sqrt{\Sigma_m z_{ij}^2}$	19	21.47	7.14	20.98

Normalized matrix

	Customer Population	Accessibility, Visibility, and Traffic	Competition and Neighbors	Loyalty
SACR	0.53	0.37	0.00	0.29
SAER	0.21	0.37	0.00	0.29
SAWR	0.42	0.37	0.00	0.29
SANR	0.16	0.33	0.28	0.33
SASR	0.32	0.28	0.42	0.33
BHR	0.05	0.28	0.56	0.38
OMN	0.26	0.23	0.56	0.33
QAT	0.11	0.23	0.14	0.29
KT	0.26	0.33	0.28	0.33
UAE	0.47	0.33	0.14	0.29

Weighted normalized matrix

	Customer Population	Accessibility, Visibility, and Traffic	Competition and Neighbors	Loyalty
SACR	0.21	0.11	0.00	0.06
SAER	0.08	0.11	0.00	0.06
SAWR	0.17	0.11	0.00	0.06
SANR	0.06	0.10	0.03	0.07
SASR	0.13	0.08	0.04	0.07
BHR	0.02	0.08	0.06	0.08
OMN	0.11	0.07	0.06	0.07
QAT	0.04	0.07	0.01	0.06
KT	0.11	0.10	0.03	0.07
UAE	0.19	0.10	0.01	0.06
Ideal$A$*	0.21	0.11	0.06	0.08
Negative-idealA^-	0.02	0.07	0.00	0.06

The separation distance

	S_{i_*}	S_{i-}
SACR	0.06	0.19
SAER	0.14	0.08
SAWR	0.07	0.15
SANR	0.15	0.06
SASR	0.09	0.11
BHR	0.19	0.06
OMN	0.11	0.10
QAT	0.18	0.03
KT	0.11	0.09
UAE	0.05	0.17

The service level score of each market zone

SACR	SAER	SAWR	SANR	SASR	BHR	OMN	QAT	KT	UAE
0.77	0.35	0.68	0.28	0.56	0.24	0.47	0.12	0.46	0.77

12 Two-Stage Repair–Full Refund Warranty Policy Analysis for Remanufactured Products in Reverse Supply Chain

Ammar Y. Alqahtani[1], Albraa A. Rajkhan[1], and Surendra M. Gupta[2]

[1]Faculty of Engineering, Department of Industrial Engineering, King Abdulaziz University, Jeddah, 22254, Saudi Arabia

[2]Northeastern University, Department of Mechanical and Industrial Engineering, 334 Snell Engineering Center, 360 Huntington Avenue, Boston, MA 02115, USA

1 INTRODUCTION

In recent years, it has become challenging to define marketing due to its complexity and applications. Many different points of view rose in its role (Szabo & Webster, 2020). Yet, marketing creates and manages the trade by studying places of behavioral interaction between consumers and all types of products, including remanufactured products. Furthermore, one of the most significant approaches to creating and managing the market is offering product warranties, as it addresses the consumer's doubt of the product's performance or useful life.

Warranties are a valuable tool to utilize in marketing. It's a way to communicate with consumers as a powerful marketing tool in either promotion or preservation. In terms of promotional, it boosts the quality and reliability of a product; in terms of preservation, it guarantees performance satisfaction against product failure over warranty duration. Indeed, marketing warranties successfully reduces the risk associated with product purchasing and increase the value of a product.

Warranties are well-known to reduce the risk of purchasing products by protecting against defect of products within the warranty duration. On the other side, warranties also reduce the financial risk for consumers, as manufacturers guarantee repair costs that fall into the scope of warranty to ensure product quality and performance. Nonetheless, providing warranties are additive costs to manufacturers that led them to invest in the quality and reliability of products that are offered with warranties. Thus, consumers generally perceive warranties positively by their association with the product's quality and reliability.

The scope of this chapter is limited to specific factors as product manufacturing processes became more complex and uncertain in recent years. The end-of-life products (EOLPs) that are received at the manufacturing facility along with the demanded components are based on Poisson distribution. Accordingly, the disassembly and remanufacturing times are exponentially assigned to each station. The costs related to backorders are estimated based on the duration of the above-mentioned. Also, a stringent disposal policy applies to the excessive and nonessential EOL products and components. In this chapter, a pull control production mechanism is implemented in all disassembly line settings considered and reviewed. Different warranty policies are also reviewed in terms of costs and durations.

The primary contribution offered by this chapter is a quantitative assessment of warranty offers on remanufactured items from a manufacturer's perspective; it proposes an appealing price in the eyes of the consumer. While there are developmental studies on warranty policies for brand new and secondhand products, there are no studies that evaluate the potential benefits of Two-Stage Repair–Full Refund Warranty on remanufactured products in a quantitative and comprehensive manner. In these related studies, profit improvements are achieved through warranty offers for different policies to determine the range of money that can be invested in a warranty while keeping it profitable overall.

The rest of the chapter is organized as follows: Section 2 lists all the related work from the literature review. System descriptions and Two-Stage Repair–Full Refund Warranty are presented in Section 3 and Section 4, respectively. Assumptions and notations are given in Section 5. Section 6 describes the warranty cost. Finally, results and conclusions are given in Section 7 and Section 8, respectively.

2 LITERATURE REVIEW

2.1 Environmentally-Conscious Manufacturing and Product Recovery

The rise of environmental concerns in product manufacturing has been widespread across the globe. These concerns have impacted the manufacturing process as well as product recovery due to government regulations and environmental activists (see Pazoki & Zaccour, 2019; Cai et al., 2019). From a business point of view, environmental concerns have influenced decision-making on many levels—from choosing facility locations to reverse logistics to support product recycling (see Ilgin & Gupta, 2010; Govindan et al., 2016; Meng et al., 2017). From another perspective, consumers become more aware of environmental and social concerns when manufacturers responded through applying strict environmental rules and guidelines that minimize products waste by the recovery of end-of-life products (EOLPs) as well as its material and components (see Agrawal et al., 2017; Feng et al., 2018).

Meanwhile, many researchers have addressed related issues that affect the manufacturing industry, especially in logistical issues—the rise in product manufacturing and product recovery along with environmental concerns. Accordingly, businesses have modified their overall disassembly lines and logistics to adapt to the challenges brought by the significant environmental role of the industries (See Ilgin et al., 2015; Sharma et al., 2016; Govindan et al., 2016; Liu & Wang, 2017; Alqahtani et al., 2019a).

2.2 WARRANTY ANALYSIS

Warranty is a liability on sold products against any failure within its scope, fulfilled by the manufacturer. Moreover, warranties guarantee that a product's performance meets consumer expectations, and the buyer has the right to seek compensations as defined in the warranty if the product fails to meet them (Blischke & Murthy, 1993). In other words, warranties transfer the risk of product failure from the buyer to the manufacturer, as insurance and protection are their focal aspects (Heal, 1977). Warranties may likewise increase the confidence of the buyers as they associate it with product quality. Meanwhile, vendors may obtain additional profits by providing warranties (See Spence, 1977; Gal-Or, 1989; Lutz & Padmanabhan, 1995; Blischke, 1995; Soberman, 2003).

Still, many of the research efforts focus on new product warranties while less attention is given to remanufactured products. In recent years, advanced technologies have been exponentially improving, and manufacturers and consumers are beginning to recognize remanufactured or second-hand products worldwide. However, different approaches have been introduced for remanufactured products to address key issues. A game theory model was implemented by Jin and Zhou (2020) to determine if remanufactured products should have the same premium warranty policy as new products. While consumers are influenced by the warranty policy of the remanufactured product, the ideal warranty policy depends on the cost structure from the manufacturer's perspective. Warranties may also affect the sales of both new and remanufactured products, as the outcomes of Tang et al., (2020) represent after analyzing the decision of pricing and warranty in a two-period closed-loop supply chain for new and remanufactured products using the Stackelberg game. By offering a warranty to remanufactured products, the price of the new products would not be affected in period 2, but it would be affected in period 1 as well as remanufactured products in period 2.

Evaluating different warranty policies and the condition of the remanufacturing products is fundamental to warranty costs analysis for the manufacturer. The age limit of receiving products is the key issue addressed by Yazdian et al. (2016). From another point of view, Liao et al. (2015) examined the effect of warranty policies on consumer behavior. Furthermore, offering warranties would be a powerful marketing strategy to enhance consumer purchase of remanufactured products (Alqahtani & Gupta, 2017a). Thus, different warranty policies have been evaluated for sensor-embedded remanufactured products to find the best profitable practice from the manufacturer's perspective. Renewing and non-renewing one-dimensional warranty were examined using Free Replacement Warranty (FRW), Pro-Rata

Warranty (PRW) policy, and a combination of different warranty periods for sensor-embedded remanufactured products and disassembled components to minimize warranty costs. (Alqahtani & Gupta, 2017b; Alqahtani et al. 2017a). Likewise, the impact of two-dimensional warranty was evaluated on remanufactured products using the information from the sensors on the product's age and usage to minimize the manufacturer's cost of warranties and maximize profit. (Alqahtani & Gupta, 2017c, 2017d; Alqahtani et al. 2017b;).

However, preventive maintenance (PM) activates at the pre-specified value of remaining life found to be economically viable in reducing the costs of warranty servicing and increasing the profitability of the manufacturer (See Alqahtani & Gupta, 2018a; Alqahtani & Gupta, 2018b, 2019a). In fact, advanced technology such as the Internet of Things (IoT) has been adapted along with SEP as it plays a key role in monitoring EOLPs in real-time. Identifying the condition of products helps the manufacturer in evaluating different warranties, the reselling price of SEPs, as well as planning for maintenance activities with minimal costs (see Alqahtani et al., 2019b, 2019c; Zong et al., 2020). The use of technology is particularly important, as different warranty policies are being introduced regularly into the market—such as the two-stage repair-or-full-refund policy that been considered by Park et al. (2020) — to determine the ideal length of the warranty period. This policy proposes that products would be minimally repaired or a full refund would be occurred to maximize the profitability of the dealer.

3   SYSTEM DESCRIPTION

To recognize the ultimate implementation of a Two-Stage Repair–Full Refund Warranty Policy in remanufactured items, a discrete-event simulation was applied—a recovery system called Advanced Remanufacturing-to-Order (ARTO). Additionally, Taguchi's Orthogonal Arrays was constructed in this experiment as the foundation design that represents the whole recovery system to enhance the detection of system behavior in different conditions. Furthermore, several scenarios were evaluated through t-tests, Tukey pairwise comparison tests, as well as one-way (ANOVA) for each scenario, to identify the ideal strategy that should be offered by the manufacturer.

The ARTO system is considered as a type of product recovery system where a series of operations are performed in different condition levels, throughout a case of sensor-embedded Dishwasher Machine (DW). The reusable parts may play a significant role to fulfill product requirements of refurbishing and repairing, which satisfy the need for both internal and external components across the recovered disassembly process. However, the ARTO system might handle three various classes of products: end-of-life products to go through the recovery process, failed SEPs to go through refurbished process, or SEPs to go through the maintenance process.

The first step is that end-of-life dishwashers proceed through the ARTO system that is equipped with a radio frequency data reader to generate data that will be stored in the facility's database. Next, to ensure that every component of the DWs is well extracted, a six-station disassembly line is fully performed. The priority relationship between DW components is listed in Table 12.1, while Figure 12.1 shows

TABLE 12.1
DW Components and Precedence Relationship

Component Name	Station	Code	Preceding Component
Metal cover	1	A	–
Door latch	2	B	–
Spray arm	3	C	A, B
Float switch	3	D	A, B, C
Motor	4	E	A, B, C, D
Pump	5	F	A, B, C, D, F
Drain hoses	5	G	F
Gasket	6	H	D, F, H
Filtration system	6	I	H

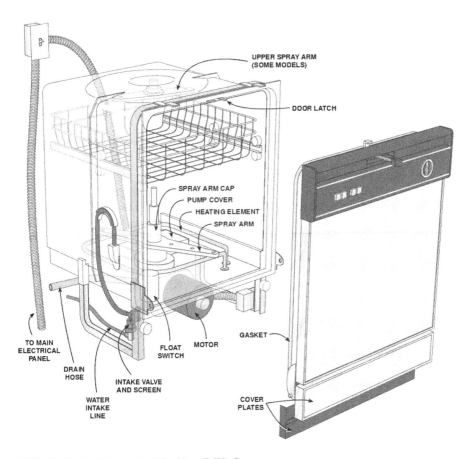

FIGURE 12.1   Dishwasher Machine (DW) Components.

the nine components of the DW—Metal Cover, Door Latch, Spray Arm, Float Switch, Motor, Pump, Drain Hoses, Gasket, and Filtration System. The exponential distribution is used to estimate the times of station disassembly, with the inter-arrival times for each component demand and the inter-arrival times for EOL DWs. After generating the data, all EOLPs are sent to station 1 for disassembly or other relevant stations, if only specific parts need to be repaired. The parameters, factors, and factor levels are given in Table 12.2.

There are two disassembly options: non-destructive or destructive, depending on the components' condition. Non-functional disassembled components, regardless of whether broken or its life cycle is over, undergoes the destructive option to check the functionality of other components. Thus, the cost of a unit disassembly for a functional component is more than for non-functional parts. After disassembly, further testing of components is pointless as data is accessible through the embedded sensor. Yet, two assumptions are made—either EOLP information of demand and life cycle are known or the cost of generating data from sensors is less than testing and inspection operations.

The implementation of recovery options depends on the remaining life and general condition of each sensor-embedded product. Spare parts are supplied by recovered components, whereas material demands are supplied by reused components. The recovered components and products are then classified and grouped based on their remaining life, to be used in recovery operations. From an economic perspective, it is particularly important to determine the remaining life of the product and components since any group of low remaining life loses the value of higher life. Any inventory level exceeding the product, component, or material of a specific class is marked as excess and is either discarded or used to fulfill needs.

Refurbish and repair options must be chosen carefully to fulfill product demands. Indeed, there is some limitation to the EOLP because of nonfictional components that will be replaced or replenished throughout the process of repair and refurbish to line up with specific requirements of remaining life. Moreover, a lower remaining life could be composed of EOLPs that need replacement. If any SEP failure has been detected within the warranty period, the DWs should arrive at the ARTO system for data generation and analysis, and then recovery operations take place for EOLPs.

The last step of risk mitigation is to perform PM actions within the warranty period. In this chapter, if the remanufactured DW reaches a specific remaining life that has been early defined, the SEPs are placed into the ARTO system for data generation. Next, four maintenance actions take place for the SEPs depending on the generated data—adjustment, cleaning, measuring, and component replacement. The PM actions are executed with degree δ and the remanufactured DW's remaining life to δ units of time before starting the process. Consumers are not charged for any failure between two consecutive PM actions within the warranty period.

4   TWO-STAGE REPAIR–FULL REFUND WARRANTY

Many factors influence a buyer's decision. Consumers tend to compare various products based on their features and brand due to competition of functionality

TABLE 12.2
ARTO System Parameters

Parameters	Unit	Value	Parameters	Unit	Value
Backorder cost rate	%	40	Sales price for a 3-year float switch	$	15
Holding cost rate	$/hour	10	Sales price for a 3-year motor	$	60
Remanufacturing cost	$	1.5	Sales price for a 3-year pump	$	25
Disassembly cost per minute	$	1	Sales price for a 3-year drain hoses	$	20
Sales price for a 1-year metal cover	$	10	Sales price for a3-year gasket	$	20
Sales price for a 1-year door latch	$	20	Sales price for a 3-year filtration system	$	65
Sales price for a 1-year spray arm	$	5	Weight for metal cover	kg	3.63
Sales price for a 1-year float switch	$	5	Weight for door latch	kg	1.81
Sales price for a 1-year motor	$	45	Weight for spray arm	kg	0.91
Sales price for a 1-year pump	$	15	Weight for float switch	kg	0.91
Sales price for a 1-year drain hoses	$	15	Weight for motor	kg	2.72
Sales price for a 1-year gasket	$	15	Weight for pump	kg	5.44
Sales price for a 1-year filtration system	$	50	Weight for drain hoses	kg	1.36
Sales price for a 2-year metal cover	$	15	Weight for gasket	kg	1.36
Sales price fora 2-year door latch	$	30	Weight for filtration system	kg	2.72
Sales price for a 2-year spray arm	$	12	Unit copper scrap revenue	$/kg	0.6
Sales price for a 2-year float switch	$	12	Unit Fiberglass scrap revenue	$/kg	0.9
Sales price for a 2-year motor	$	55	Unit steel scrap revenue	$/kg	0.2
Sales price for a 2-year pump	$	18	Unit disposal cost	$/kg	0.3
Sales price for a 2-year drain hoses	$	18	Unit copper scrap cost	$/kg	0.3
Sales price for a 2-year gasket	$	20	Unit fiberglass scrap cost	$/kg	0.45
Sales price for a 2-year filtration system	$	60	Unit steel scrap cost	$/kg	0.1
Sales price for a 3-year metal cover	$	20	Sales price of a 1-year DW	$	180
Sales price for a 3-year door latch	$	35	Sales price of a 2-year DW	$	240
Sales price for a 3-year spray arm	$	15	Sales price of a 3-year DW	$	275
Cost of supplying for metal cover	$	4	Cost of supplying for pump	$	1.66
Cost of supplying for door latch	$	4	Cost of supplying for drain hoses	$	2.34
Cost of supplying for spray arm	$	2.8	Cost of supplying for gasket	$	0.6
Cost of supplying for float switch	$	1.2	Cost of supplying for filtration system	$	3.4
Cost of supplying for motor	$	4	Cost of supplying for DW	$	55

products that include costs, features and characteristics, quality, and warranty. Even after-sale services have become more important in terms of warranties and the availability of part and repair services. Therefore, all these elements play a significant role in purchasing decisions.

Warranties represent an agreement between the manufacturer and the consumer that is realized upon selling a product that addresses the performance and quality of the product—any failure within the scope of the warranty is applicable for correction or compensation. The warranty cost of a new product is not difficult to calculate if a quality control system is well-established. However, in EOL products, warranty costs are adjustable based on the remaining life, maintenance record, and usage.

Consumers become more aware and demand more for warranty as it is associated with the quality of the product and environmental awareness, especially when it comes to remanufactured products where the concern of product failure increases with the cost of repair and replacement. This led manufacturers to implement best practices in managing warranties' cost structure for remanufacturing as failure to do so may result in monetary loss rather than profit.

However, offering a similar warranty policy for remanufactured products as with new products may not be economically reasonable for manufacturers. Thus, further testing and comparison of warranty policies are mandatory to estimate the most appropriate warranty costs of remanufactured products. Murthy and Blischke (2006) have presented other issues regarding the service strategies in spare parts through repair and replacement in warranty duration.

This chapter represents the two-stage repair-or-full-refund warranty (RFRW) strategy within the duration as follows:

A consumer buys a remanufactured product that is offered with a fixed warranty period for a price that includes the product's upgrade before its sale. We assume that a specific upper limit of repair time has been set for minimal repair. Moreover, the term "length of the maintenance period" is used instead of the product maintenance cycle. For any product's failure within the warranty duration, the product either gets repaired within the minimum time which the manufacturer will counter minimal costs, or the buyer receives a full refund, and the maintenance cycle ends. Nonetheless, the buyer could receive no refund during the warranty period based on the warranty policy but unless the warranty expires, the maintenance cycle is considered valid.

5 NOTATIONS

This section starts with the model assumptions, then the notation of all the parameters used in this chapter follows.

5.1 ASSUMPTIONS

The following assumptions have been considered to simplify the analysis:

i.   The failures are statistically independent.
ii. Every item failure under the warranty period results in a claim.
iii. All claims are valid.

iv. The failure of a remanufactured item is only a function of its age.
v. The time to carry out the replacement/repair action is relatively small compared to the mean time between failures.
vi. The cost to service warranty claim (for repair/replacement of failed component) is a random variable.

5.2 Notations

pdf	probability density function
cdf	cumulative distribution function
ECR	expected cost rate
NHPP	non-homogeneous Poisson process
RFRW	repair-full refund warranty
IFR	increasing failure rate
T, Y	failure time and repair time, respectively
$f_0(\ )$	pdf of an age of new product
$F_0(\ )$	cdf of an age of new product
$v_0(\ )$	intensity function of an age of new product
$f(\ )$	pdf of an age of the upgraded product
$F(\ )$	cdf of an age of the upgraded product
$v(\ )$	intensity function of an age of the upgraded product
x	virtual age after the upgrade (i.e. age corresponding to the reduced failure rate after upgrade when purchased)
ux	reduced age to make the virtual age x
$H(a, b)$	expected number of failures within an interval (a, b) (i.e. $H(a, b) = \int v(u)\mathrm{d}u$)
$g(y)$	pdf of the repair time, respectively, of y
$G(y)$	cdf of the repair time, respectively, of y
r_0	repair time threshold
$N_{r0}\ (a,\ b)$	number of minimal repairs conducted in the interval (a,b) when the repair time threshold equals r_0
x	upgraded age of the remanufactured product
ω	length of warranty period
t_f	time when the full refund is made within the warranty period
$q(\cdot)$	pdf of t_f
$Q(\cdot)$	cdf of t_f
$\underset{\sim}{Q}(\cdot)$	survival function of t_f
C_m	minimal repair cost
C_f	full refund cost
c_o	purchasing price for the remanufactured product
c_u	upgrade cost prior to its sale for the remanufactured product
c_m	unit minimal repair cost for the remanufactured product
c_f	unit full refund cost for the remanufactured product
β	shape parameter for the power-law process
θ	scale parameter for the power-law process
κ	shape parameter for the Weibull distribution
λ	scale parameter for the Weibull distribution

6   COST MODEL FOR A SENSOR-EMBEDDED REMANUFACTURED PRODUCTS UNDER RFRW

To estimate the expected cost for a product under RFRW during the maintenance cycle, mathematical formulas were derived for both the expected length of the maintenance cycle and the expected total cost. The dealer, based on a specific cost structure, would handle any warranty cost associated with the warranty period

As mentioned earlier, failed products under the RFRW will either get a minimal repair or full refund and then the maintenance cycle ends. In some cases, no refund is issued but the maintenance cycle ends after the expiration of the warranty period. The failure of the product is following a non-homogeneous Poisson process with the minimal repair cost. Moreover, the total cost for the dealer includes the cost of purchasing the product, the cost of upgrade, the minimal repair cost, and a full refund (if it happens). Each of these costs will be described in the next subsections.

6.1   INITIAL REMANUFACTURED PRODUCT PURCHASE COST

For the cost of purchasing a manufactured product, we assume that the dealer purchases a new or secondhand item. Let Co denote the dealer's initial purchase cost that can be estimated by considering the depreciation rate of the product. Chattopadhyay and Murthy (2000) used the following formula to model the initial purchasing cost by considering the initial purchasing age ox adapted in this chapter

$$c_0 = k_0 P_0 \left(1 - \frac{o_x}{L} \right) \qquad (12.1)$$

where $P0$ is the price, L is the length of the expected life of a new product, and $0 < k_0 < 1$ is an index of the immediate loss of resale value of new product sold immediately after the purchase.

6.2   PRODUCT UPGRADE COST

Upgrading the remanufactured product before selling them is a typical practice by the dealer for promotion. We assume that the initial product's age is o_x at the dealer's purchasing time. The remaining life of SEP $x < o_x$ when consumers buy the product from the dealer. As mentioned before, the dealer incurs the upgrade cost that may depend on the product's age and the degree of improvement. The following formula adapted for upgrade cost has been represented by Smith and Bain (1975):

$$c_u = \frac{\alpha \cdot u_x}{1 - exp\left(-\varphi\left(o_x - u_x\right)\right)} \qquad (12.2)$$

where o_x is the age of the remanufactured product, u_x the decrease in the age due to upgrade, and α and φ are positive constants.

Note: when $u_x = 0$, there is no upgrade and, therefore, $c_u = 0$. As $u_x \to o_x$, c_u goes to ∞, which means that it's impossible to upgrade the secondhand product to as good as new. Let x denote the upgraded remaining life of the remanufactured product; thus, $x = o_x - u_x$.

6.3 WARRANTY COST UNDER RFRW

The dealer's minimal repair costs as well as the potential full refund cost were derived during the maintenance cycle within RFRW. Nonetheless, if all minimal repairs are performed within the specified time period, there will be no full refund and the maintenance cycle ends upon warranty expiry. Let C_m represent the minimal repair cost, the C_f represents the full refund cost, then the total warranty cost during the maintenance cycle can be estimated as the sum of both costs $C_m + C_f$. To evaluate these costs, it's essential to define a random variable that represents the length of the maintenance cycle. This maintenance cycle begins right after the consumer buys a secondhand product of upgraded age x and ends when either the warranty expires or a full refund takes place—whichever comes first. Let T_f represent the age when the full refund occurs within the warranty period, $x + u_x < T_f < x + u_x + \omega$ and let tf be its realization. If we let $T_0(\omega)$ represent the maintenance cycle length for the product under the RFRW, then by definition, $T_0(\omega)$ can be expressed as

$$T_o(\omega) = \begin{cases} T_f - (x + u_x), & \text{if full refund service occurs} \\ \omega, & \text{otherwise} \end{cases} \quad (12.3)$$

Note that the virtual age of x corresponds to the actual age of $x + u_x$ prior to the upgrade occurred by the dealer. Let $v_0(\ )$ and $v(\ )$ represent the intensity (failure rate function), respectively, of the new and upgraded products ages, then $v_0(x)$ is equal to $v(x + u_x)$. Likewise, $f_0(\ )$ and $f(\ )$ represent pdfs, respectively, of the new and upgraded products ages; $f_0(x)$ is equal to $f(x + u_x)$ and $F_0(x)$ is equal to $F(x + u_x)$.

Let $e(\omega)$ represent the expected time length until the full refund event occurs before the warranty expires. Then, $e(\omega)$ can be expressed as

To examine the expected costs during the maintenance cycle, we drive the expected quantity of product failures. Denote the number of failures during an interval $[a, b]$ by $N_{r0}(a, b)$ when the repair time threshold is set to r_0. Then, the total warranty cost for the dealer during the product maintenance cycle, denoted by $C(\omega)$, can be evaluated as

$$C(\omega) = \begin{cases} c_m \cdot N_{r0}(x + u_x, \ t_f) + c_f, & \text{if full refund service occurs} \\ c_m \cdot N_{r0}(x + u_x, \ x + u_x + \omega), & \text{otherwise} \end{cases} \quad (12.5)$$

Here, c_f represent the unit's cost of full refund. We assume that $c_f = \kappa_0 P_0$, regardless of the value of o_x for simplicity. Let $H(a, b) = \int v(u)du$ represent the expected quantity of failures within an interval (a, b), where $v(\ )$ is the intensity function for the failure process of the upgraded product. Then, due to all the failures in the interval, (a, b) are only minimally repaired before the full refund—no repair times exceed the repair time threshold r_0. Thus, we obtain

$$E[N_{r0}(a, b)] = \sum_{k=1}^{\infty} k \cdot P(N_{r0}(a, b) = k) = \sum_{k=1}^{\infty} k \cdot \frac{H(a, b)^k e^{-H(a, b)}}{k!} [G(r_o)]^k$$
$$= H(a, b) \cdot G(r_o) \cdot [e^{-H(a,b)}]^{\bar{G}(r_0)} \tag{12.6}$$

By applying Eq. (12.6), we have

$$E[N_{r0}(x + u_x, x + u_x + \omega)] = \int_{x+u_x}^{x+u_x+\omega} v(u)du \cdot G(r_o) \cdot \left[e^{-H(x+u_x, x+u_x+\omega)}\right]^{\bar{G}(r_0)}$$
$$\tag{12.7}$$

Likewise, the expected quantity of minimal repairs prior the full refund is occurred can be estimated by

$$E[N_{r0}(x + u_x, t_f)] \approx \int_{x+u_x}^{e(\omega)} v(u)du \cdot G(r_o) \cdot \left[e^{-H(x+u_x, e(\omega))}\right]^{\bar{G}(r_0)} \tag{12.8}$$

ARENA 14.7 is performed to generate the remaining life of the remanufactured product at failure (t_i), implementing a bivariate random number generator and replacement time history under warranty and recurrence sales through the simulation time interval. The ARENA simulation outcomes show the remaining life of failures under warranty, the remaining life after PM events, the number of replacements within the warranty period for each purchase, and the time between recurrences purchases.

7 RESULTS

In this part, the results to compute the expected number of failures and expected costs to the remanufacturer were obtained using the ARENA 14.7 program.

Table 12.3 presents the expected number of failures and costs for remanufactured DW and components for the Two-Stage Repair–Full Refund warranty policy. In Table 12.3, the expected number of failures represents the expected number of failed items per unit of sale. In other words, it is the average number of free replacements that the remanufacturer would have to provide during the warranty period per unit sold. The expected cost to the remanufacturer includes the cost of supplying the original item, C_s. Thus, the expected cost of warranty is calculated by subtracting C_s from the expected cost to the remanufacturer. For example, from Table 12.3, for $W = 0.5$ and $RL = 1$, the warranty cost of remanufactured for DW is $\$55.69 - C_s = |\$56.76 - \$55.00| = \$1.76$, which is $([\$11.76/\$ 55.00] \times 100) = 3.20\%$ of the cost of supplying the item, C_s, which is significantly less than that $\$55.00$, C_s. This saving might be acceptable, but the corresponding values for longer warranties are lower. For example, if $W = 2$ years and $RL = 1$, the corresponding percentage is $([|\$74.29 - \$55.00|/\$ 55.00] \times 100) = 35.07\%$.

TABLE 12.3

Expected Number of Failures and Cost for Remanufactured DW's Components for Two-Stage Repair-Full Refund Warranty

Components	W	Two-Stage Repair-Full Refund Warranty (RFRW)					
		Expected Probability of Failures			Expected Cost		
		RL = 1	RL = 2	RL = 3	RL = 1	RL = 2	RL = 3
Metal cover	**0.5**	0.7139	0.0047	0.0009	$10.05	$11.52	$9.21
	1	0.1432	0.0187	0.0086	$11.15	$12.62	$9.37
	2	0.2143	0.0415	0.0287	$16.73	$16.63	$9.64
Door latch	**0.5**	0.7055	0.0044	0.0055	$9.86	$11.44	$9.19
	1	0.1518	0.0181	0.0457	$11.63	$12.27	$9.32
	2	0.2057	0.0411	0.1540	$16.55	$16.33	$9.59
Spray arm	**0.5**	0.6969	0.0043	0.0297	$4.94	$4.62	$4.73
	1	0.1346	0.0188	0.2448	$6.50	$8.19	$4.83
	2	0.1973	0.0418	0.8232	$8.97	$10.07	$5.00
Float switch	**0.5**	0.6969	0.0019	0.1585	$2.71	$2.66	$2.26
	1	0.1003	0.0189	1.3083	$3.92	$3.63	$2.50
	2	0.1800	0.0361	0.5361	$5.21	$5.16	$2.66
Pump	**0.5**	0.6763	0.0044	0.8478	$10.29	$9.86	$9.64
	1	0.1467	0.0181	1.0998	$11.31	$10.56	$9.80
	2	0.2067	0.0418	0.0230	$15.90	$13.64	$9.91
Drain hoses	**0.5**	0.7045	0.0047	1.0540	$3.22	$2.74	$2.66
	1	0.1371	0.0185	1.1089	$4.67	$3.92	$2.93
	2	0.2160	0.0418	0.1225	$5.40	$4.54	$3.09
Gasket	**0.5**	0.7217	0.0043	1.0998	$6.15	$5.21	$5.00
	1	0.1577	0.0183	0.2236	$8.51	$6.07	$5.13
	2	0.2074	0.0415	0.6547	$10.58	$8.46	$5.40
Filtration system	**0.5**	0.7286	0.0047	0.7835	$1.64	$1.29	$0.91
	1	0.1346	0.0184	1.1952	$2.52	$2.04	$1.10
	2	0.2048	0.0418	1.1042	$4.43	$2.95	$1.18
Compressor	**0.5**	0.7037	0.0047	0.0232	$7.14	$6.69	$6.39
	1	0.1424	0.0185	0.1225	$9.21	$8.65	$6.93
	2	0.2067	0.0416	1.1089	$12.52	$11.31	$7.17
MO	**0.5**	1.019	0.012	0.000	$56.76	$64.25	$63.70
	1	0.264	0.036	0.000	$69.44	$70.12	$66.45
	2	28.746	28.351	24.491	$74.29	$73.90	$82.12

To assess the impact of RFRW on total cost, Table 12.4 presents the average values of all the cost by offering RFRW for the conventional product model and SEPs product model. According to this table, offering warranty achieves statistically significant savings in holding, backorder, disassembly, disposal, remanufacturing, transportation, warranty, and number of warranty claims. In

TABLE 12.4

Results of Warranty Models on Conventional Model and SEPs

Performance Measure	Mean Value for Warranty Model		95% Confidence Interval		
	Conventional Model ($\mu1$)	SEPs with RFRW Policy ($\mu2$)			
Holding cost	$252,347.75	$135,326.95	115,576.10	$<\mu1–\mu2<$	$131,467.81
Backorder cost	$46,810.12	$27,220.06	19,348.21	$<\mu1–\mu2<$	$22,008.59
Disassembly cost	$544,894.50	$288,603.69	253,126.73	$<\mu1–\mu2<$	$287,931.65
Disposal cost	$88,270.00	$54,646.39	33,208.51	$<\mu1–\mu2<$	$37,774.68
Testing cost	$1,873,350.02	$0.00	1,850,222.25	$<\mu1–\mu2<$	$2,104,627.80
Remanufacturing cost	$47,269.58	$806,865.55	-750,218.25	$<\mu1–\mu2<$	-$853,373.25
Transportation cost	$42,075.12	$28,368.28	13,537.62	$<\mu1–\mu2<$	$15,399.05
Warranty cost	$29,100.00	$16,338.87	12,603.58	$<\mu1–\mu2<$	$14,336.57
Number of claims	32062	10754	21,045.57	$<\mu1–\mu2<$	$23,939.34
PM cost	$96,186.66	$32,260.74	63,136.72	$<\mu1–\mu2<$	$71,818.01
Total cost	$3,052,365.98	$1,400,384.11	1,631,587.03	$<\mu1–\mu2<$	$1,855,930.25
Total revenue	$3,978,251.86	$4,523,606.61	-538,621.98	$<\mu1–\mu2<$	-$612,682.50
Profit	$925,885.88	$3,123,222.50	-2,170,209.01	$<\mu1–\mu2<$	-$2,468,612.74

addition, warranty with SEPs provide statistically significant improvements in total cost and profit by saving 54.12% in total cost and increase 237.32% in total profit for SEPs.

8 CONCLUSIONS

Sensors are implanted into sensor-embedded products during the initial production process. The value of sensors is realized through their ability to determine the best warranty policy and warranty period to present to consumers when selling remanufactured components and products. The remaining life and condition of components and products may be estimated prior to presenting a warranty, based upon the data collected through the sensors. Such information allows the remanufacturer to avoid unnecessary costs by enabling the re-manufacturer to control the number of claims during warranty periods. Herein, the costs of the Two-Stage Repair–Full Refund warranty policy were explored through the offering of PM for different periods. The impact of offering the consumer Two-Stage Repair–Full Refund policy to each disassembled compo-nent and SEP was also analyzed to identify the impact of SEPs on warranty costs. To further examine the issue, a case study was constructed in addition to a number of simulation scenarios to illustrate the prospective value of the model proposed herein.

REFERENCES

Agrawal, V., Ferguson, M., Thomas, V., & Toktay, L. B. (2017). Is Leasing Green? The Environmental Impact of Product Recovery, Remarketing and Disposal under Leasing and Selling. Management Science; working paper; Georgia Institute of Technology: Atlanta, GA, USA.

Alqahtani, A. Y., & Gupta, S. M. (2017a). Warranty as a marketing strategy for remanufactured products. *Journal of Cleaner Production*, 161, 1294–1307.

Alqahtani, A. Y., & Gupta, S. M. (2017b). One-dimensional renewable warranty management within sustainable supply chain. *Resources*, 6(2), 16.

Alqahtani, A. Y., & Gupta, S. M. (2017c). Evaluating two-dimensional warranty policies for remanufactured products. *Journal of Remanufacturing*, 7(1), 19–47.

Alqahtani, A. Y., & Gupta, S. M. (2017d). Warranty and maintainability analysis for sensor embedded remanufactured products in reverse supply chain environment. *International Journal of Supply Chain Management*, 6(4), 22–42.

Alqahtani, A. Y., & Gupta, S. M. (2018a). *Warranty and Preventive Maintenance for Remanufactured Products: Modeling and Analysis.* CRC Press, Boca Raton, Florida.

Alqahtani, A. Y., & Gupta, S. M. (2018b). Money-back guarantee warranty policy with preventive maintenance strategy for sensor-embedded remanufactured products. *Journal of Industrial Engineering International*, 14(4), 767–782.

Alqahtani, A. Y., & Gupta, S. M. (2019a). Warranty cost sharing policies with preventive maintenance strategy for sensor embedded remanufactured products. *International Journal of Industrial and Systems Engineering*, 31(3), 360–394.

Alqahtani, A. Y., Gupta, S. M., & Nakashima, K. (2017a). One-dimensional warranty policies analysis for remanufactured products in reverse supply chain. *Innovation and Supply Chain Management*, 11(2), 13–22.

Alqahtani, A. Y., Gupta, S. M., & Yamada, T. (2017b). Combined two-dimensional non-renewable warranty policy analysis for remanufactured products. *Procedia CIRP*, 61, 189–194.

Alqahtani, A. Y., Kongar, E., Pochampally, K. K., & Gupta, S. M. (Eds.). (2019a). *Responsible Manufacturing: Issues Pertaining to Sustainability.* CRC Press, Boca Raton, Florida.

Alqahtani, A. Y., Gupta, S. M., & Nakashima, K. (2019b). Applicability of using the Internet of Things in warranty analysis for product recovery. *Responsible Manufacturing* (pp. 291–322). CRC Press, Boca Raton, Florida.

Alqahtani, A. Y., Gupta, S. M., & Nakashima, K. (2019c). Warranty and maintenance analysis of sensor embedded products using internet of things in industry 4.0. *International Journal of Production Economics*, 208, 483–499.

Blischke, W. (1995). *Product Warranty Handbook.* CRC Press, Boca Raton, Florida.

Blischke, W. R., & Murthy, D. N. P. (1993). *Warranty Cost Analysis.* Marcel Dekker, Inc., New York.

Cai, L., Shi, X., & Zhu, J. (2019). Quality recovery or low-end recovery? Profitability and environmental impact of durable product recovery. *Sustainability*, 11(6), 1726.

Chattopadhyay, G. N., & Murthy, D. P. (2000). Warranty cost analysis for second-hand products. *Mathematical and Computer Modelling: An International Journal*, 31(10–12), 81–88.

Feng, Y., Gao, Y., Tian, G., Li, Z., Hu, H., & Zheng, H. (2018). Flexible process planning and end-of-life decision-making for product recovery optimization based on hybrid disassembly. *IEEE Transactions on Automation Science and Engineering*, 16(1), 311–326.

Gal-Or, E. (1989). Warranties as a signal of quality. *Canadian Journal of Economics*, 22(1), 50–61.

Govindan, K., Garg, K., Gupta, S., & Jha, P. C. (2016). Effect of product recovery and sustainability enhancing indicators on the location selection of manufacturing facility. *Ecological Indicators*, 67, 517–532.

Govindan, K., Jha, P. C., & Garg, K. (2016). Product recovery optimization in closed-loop supply chain to improve sustainability in manufacturing. *International Journal of Production Research*, 54(5), 1463–1486.

Heal, G. (1977). Guarantees and risk-sharing. *The Review of Economic Studies*, 44(3), 549–560.

Ilgin, M. A., & Gupta, S. M. (2010). Environmentally conscious manufacturing and product recovery (ECMPRO): A review of the state of the art. *Journal of environmental management*, 91(3), 563–591.

Ilgin, M. A., Gupta, S. M., & Battaïa, O. (2015). Use of MCDM techniques in environmentally conscious manufacturing and product recovery: State of the art. *Journal of Manufacturing Systems*, 37, 746–758.

Jin, M., & Zhou, Y. (2020). Does the remanufactured product deserve the same warranty as the new one in a closed-loop supply chain?. *Journal of Cleaner Production*, 262, 1–11.

Liao, B. F., Li, B. Y., & Cheng, J. S. (2015). A warranty model for remanufactured products. *Journal of Industrial and Production Engineering*, 32(8), 551–558.

Liu, J., & Wang, S. (2017). Balancing disassembly line in product recovery to promote the coordinated development of economy and environment. *Sustainability*, 9(2), 309.

Lutz, N. A., & Padmanabhan, V. (1995). Why do we observe minimal warranties? *Marketing Science*, 14(4), 417–441.

Meng, K., Lou, P., Peng, X., & Prybutok, V. (2017). Multi-objective optimization decision-making of quality dependent product recovery for sustainability. *International Journal of Production Economics*, 188, 72–85.

Murthy, D. P., & Blischke, W. R. (2006). *Warranty Management and Product Manufacture*. Springer, London.

Park, M., Jung, K. M., & Park, D. H. (2020). Warranty cost analysis for second-hand products under a two-stage repair-or-full refund policy. *Reliability Engineering & System Safety*, 193, 106596.

Pazoki, M., & Zaccour, G. (2019). A mechanism to promote product recovery and environmental performance. *European Journal of Operational Research*, 274(2), 601–614.

Sharma, S. K., Mahapatra, S. S., & Parappagoudar, M. B. (2016). Benchmarking of product recovery alternatives in reverse logistics. *Benchmarking: An International Journal*, 23(2), 406–424.

Smith, R. M., & Bain, L. J. (1975). An exponential power life-testing distribution. *Communications in Statistics-Theory and Methods*, 4(5), 469–481.

Soberman, D. A. (2003). Simultaneous signaling and screening with warranties. *Journal of Marketing Research*, 40(2), 176–192.

Spence, M. (1977). Consumer misperceptions, product failure and producer liability. *The Review of Economic Studies*, 44(3), 561–572.

Szabo, S., & Webster, J. (2020). Perceived greenwashing: The effects of green marketing on environmental and product perceptions. *Journal of Business Ethics*, 1–21. doi.org/10.1007/s10551-020-04461-0

Tang, J., Li, B. Y., Li, K. W., Liu, Z., & Huang, J. (2020). Pricing and warranty decisions in a two-period closed-loop supply chain. *International Journal of Production Research*, 58(6), 1688–1704.

Yazdian, S. A., Shahanaghi, K., & Makui, A. (2016). Joint optimisation of price, warranty and recovery planning in remanufacturing of used products under linear and non-linear demand, return and cost functions. *International Journal of Systems Science*, 47(5), 1155–1175.

Zong, S., Li, S., & Shang, Y. (2020). Optimal quality-based recycling and reselling prices of returned SEPs in IoT environment. *International Conference on Intelligent Decision Technologies* (pp. 91–103). Springer, Singapore.

13 An Application of Warehouse Layout and Order Picking in a Distribution Center

Saadettin Erhan Kesen and Ece Yağmur
Industrial Engineering Department, Konya Technical University, Konya, Turkey

1   INTRODUCTION

Companies must deliver their products or services to the right place at the right time with the lowest cost to maintain their position in the global market and compete with their rivals who are constantly changing and developing in today's conditions. Therefore, supply chain management (SCM) has become a particularly important concept for businesses in recent years. Warehouse management is one of the most critical issues addressed in the supply chain to effectively manage product flow. In literature, warehouses are examined in two groups—classic warehouses and distribution centers—according to their characteristics (Dawe,1995; Higginson and Bookbinder, 2005). In this study, we examine a distribution center where Fast Moving Consumer Goods (FMCG) are stored within a logistics company. In distribution centers, many different products are delivered to customers quickly by consolidating related orders in mixed pallets. So, a good warehouse layout and an efficient order picking strategy are needed to sort these products from related locations quickly. This concept, mentioned in literature as an order picking process, is a laborious activity that constitutes 65% of the total operating cost in warehouses (Žulj et al., 2018). Despite advancing technology and automation systems, it is estimated that 80% of warehousing and order picking operations are carried out manually in Western Europe today (De Koster et al., 2007). Order picking operations arbitrarily made without planning causes extra time for vehicles and operators, resulting in inefficient use of resources.

In this study, we first emphasized the warehouse layout problem that directly affects the order picking process. The placement of the products in the distribution center was carried out in two stages. In the first stage, we determined which product would be placed in which corridor considering the stock turnover rates (demands) of the orders. Following the first stage, the assignment of the location in each corridor is made for the products by considering their weights. After the warehouse layout process is completed, a new mixed-integer linear programming (MILP) model is proposed to determine the order picking visitation sequence performed by the operator.

Order picking problem is often considered as a special case of the Traveling Salesman Problem (TSP) in literature. Ratliff and Rosenthal (1983) used the TSP algorithm to minimize order picking time in a rectangular warehouse with a linear corridor. This approach has been extended by De Koster and Van Der Poort (1998) for modern warehouses where there is no central warehouse concept and returns can be made within the corridors. The researchers compared the dynamic programming approach proposed for this system with the S-shaped picking strategy. Roodbergen and Koster (2001) examined the order picking process in a warehouse with more than two multiple cross corridors and proposed a new heuristic method for the routing problems. While the largest gap heuristic proposed by Hall (1993) yields the best results in two cross-corridor warehouses, the heuristic approach proposed by researchers yields the best results in more than two multiple corridors warehouse systems. Goetschalckx and Donald Ratliff (1988) investigated the order picking problem in large corridors and proposed the S-shaped picking heuristics, which is often used in practice.

Petersen (1997) compared the performances of six different routing methods that are frequently used in literature for order picking problems. For further information on the classification and related studies the order picking problem, refers to the survey study of De Koster et al. (2007).

There are very few studies that consider priority relationships in the order picking process. Dekker et al. (2004) examined the situation in which a fragile product should be collected as the last order in a wholesaler store with garden supplies. Chabot et al. (2017) examined the routing problem under product weight, fragility, and product category constraints during the order picking process.

There are also a few studies in literature where different problems such as location assignment and routing in order picking processes are examined simultaneously. Van Gils et al. (2018) made a classification of the studies where different problems were examined concurrently during the order picking process. These combinations were examined under four groups: "location assignment and routing", "location assignment and order grouping", "order grouping and routing", and other combinations. In literature, many articles have reported that there is a statistically significant interaction between location assignment and routing problems, both in single block (Manzini et al., 2007; Petersen and Schmenner, 1999) and multi-block warehouses (Shqair et al., 2014; Theys et al. 2010). In these articles, stock turnover rates of the products were also considered to determine the order picking route.

The contribution of the study can be expressed in the following: first, this study is important for warehouses and distribution centers to easily apply and provide efficient solutions. Second, significant improvements have been achieved in the distribution center, where both the new layout and method of order picking have been implemented. Finally, a new model—which is not included in the literature—has been developed, in which weights of the products are considered to prevent deformation during the order picking process.

The remaining parts of the study are as follows: in Section 2, the general structure of the examined distribution center and current state are explained. In Sections 3 and 4, the proposed layout plan and a new mathematical model for the order picking process are defined, respectively. Section 5 examines the mixed pallet generation process for a real case and compares current and proposed situations. In Section 6, comparative results are given on 50 randomly selected mixed pallet samples taken from the distribution center. And finally, the general results and the future directions are mentioned in Section 7.

2 CURRENT LAYOUT

In this section, technical specifications and the current layout plan of the distribution center are given. As shown in Figure 13.1, the storage area consists of eight corridors and 442 locations. Corridors in the warehouse are numbered from 1 to 6, respectively. Unlike other corridors, corridor 3 and corridor 4 consist of two sections on the right and left side and there is a collection area outside the storage zone. To prepare the mixed pallet, the operator starts its tour from this collection area, and the tour ends in this area by leaving the mixed pallet prepared.

427 out of the 442 locations are actively used in the storage area. Other locations are reserved for products with periodic demands. In the current situation, the usage status of the corridors in the storage area is given in Table 13.1.

3 PROPOSED LAYOUT

In this section, we propose a new layout plan for the distribution center. When the current layout plan is examined, it is seen that the products are randomly placed in the corridors without any rules. Therefore, for the order picking process, a new layout plan for the storage area was proposed before the determination of the route for picking. Therefore, the distance of each pair of locations was calculated. Afterward, two criteria (product weights and demands) were considered in deciding the location where the products would be placed. As mentioned before, the placement process was carried out in two stages of corridor assignment and location assignment described in the subsequent subsections.

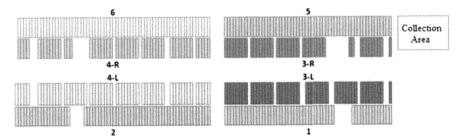

FIGURE 13.1   Layout Plan of Storage Area.

TABLE 13.1

The Usage Status of The Corridors

Corridor	Current Capacity	Active Used Capacity
1	52	50
2	63	62
3R	43	40
3L	49	46
4R	52	50
4L	58	56
5	58	57
6	67	66

3.1 CORRIDOR ASSIGNMENT

When determining which corridors will be assigned to which products, demands (stock turnover rates) were taken into consideration. The 427 types of products in the warehouse are sorted in ascending order in terms of stock turnover rates over a period of six months. Accordingly, the products with the highest demands were placed closer to the collection area, as shown in Table 13.2.

During this placement process, the capacities of the corridors and the distances between each pair of locations—including the starting point (collection area)—are taken into consideration. We also paid attention to the empty space in each corridor for the products with periodic demands.

3.2 LOCATION ASSIGNMENT

After making the assignment of corridors, all products in each corridor are sorted according to their weights for the location assignments process. The products

TABLE 13.2

Proposed Corridor Assignment

Corridor	Product ID	Proposed Usage	Empty Space
1	144–193	50	2
2	366–427	62	1
3R	58–97	40	3
3L	98–143	46	3
4R	250–299	50	2
4L	194–249	56	2
5	1–57	57	1
6	300–365	66	1

located in the same corridor are placed with the heaviest closest to the collection area, and so forth. The reason for this approach is to give priority to heavy products during the collection process to prevent light products from being damaged under heavy ones.

4 PROPOSED MATHEMATICAL MODEL FOR ORDER PICKING

To minimize the total distance traveled by the operator during the preparation of the mixed pallet, the distances between each pair of locations in the storage area must be calculated. For this purpose, a distance matrix of 427×427 is generated. The objective of the mixed pallet generation process is to minimize the total distance traveled by the operator. In the distribution center, orders are assigned by considering the capacity of the pallet. Therefore, the pallet capacity is not exceeded in any case. In this respect, the problem is similar to the TSP or Vehicle Routing Problem (VRP) with infinite capacity. However, in the proposed method, the product weights will be taken into consideration in the picking process. So, we develop a new mathematical formulation. N denotes the set of locations; the proposed mathematical model is given below:

4.1 NOTATIONS

i,j: location index ($i,j \in N$)

Variables
X_{ij}: 1 If the forklift goes directly from location i to location j,0 otherwise ($\forall i, j \in N: i \neq j$)
U_i: the load of forklift after picking from location i ($\forall i \in N$)

Parameters
t_{ij}: the distance between locations i and j ($\forall i,j \in N: i \neq j$)
Q: capacity of pallet
d_i: weight of the product in location i
a: threshold value (maximum rate of weights of orders which are consecutively picking)
M: Sufficiently large number

4.2 MATHEMATICAL MODEL
Objective function

$$Minimize \sum_{i \in N} \sum_{j \in N} t_{ij} X_{ij} \qquad (13.1)$$

Constraints

$$\sum_{i \in N} X_{ij} = 1 \quad \forall j \in N \tag{13.2}$$

$$\sum_{j \in N} X_{ij} = 1 \quad \forall i \in N \tag{13.3}$$

$$U_i - U_j + QX_{ij} \leq Q - d_j \quad \forall i, j \in N; i \neq j \tag{13.4}$$

$$d_i \leq U_i \quad \forall i \in N \tag{13.5}$$

$$U_i \leq Q \quad \forall i \in N \tag{13.6}$$

$$U_j - U_i - (d_i * \alpha) - (M * (1 - X_{ij})) \leq 0 \quad \forall i, j \in N \tag{13.7}$$

$$X_{ij} \in \{0, 1\} \quad \forall i, j \in N \tag{13.8}$$

$$U_i \geq 0 \quad \forall i \in N \tag{13.9}$$

The objective in Eq. (13.1) is to minimize the total distance traveled by the operator in the order picking process. While constraint set (2) ensures that each location must be visited exactly once, constraint set (3) guarantees that entering and leaving arcs to each location are equal. Constraint sets (4)–(6) are capacity and sub-tour elimination constraints. Constraint set (4) indicates that the total load of vehicles on any location must not exceed the pallet capacity (sufficiently large). Constraint sets (5) and (6) are bounding constraints for U_i variables. Constraint set (7) states that the rate of the weights of the products j and i, which are taken to the mixed pallet consecutively by the forklift, must not exceed the threshold value of α. Constraints sets (8) and (9) represent the integrality and non-negativity restrictions on the variables, respectively.

5 CASE STUDY

In this section, we examine the case with a mixed pallet sample with ten orders prepared by the operator to demonstrate the improvement achieved as a result of the new layout plan and order picking approach. The current locations of the orders in the related mixed pallet are given in Figure 13.2.

According to Figure 13.2, the distance of the locations to each other and the weights of the related orders in the examined mixed pallet are given in Table 13.3.

Here, the orders are collected in the following order: 1–2–6–5–7–10–3–9–4–8. We can see that in this case, the total distance traveled by the operator is calculated as $140 + 12 + 71 + 68 + 13 + 41 + 89 + 23 + 36 + 26 + 94 = 613$ units via the matrix in Table 13.3. After the proposed layout plan, new locations of the products in the mixed pallet are given in Figure 13.3.

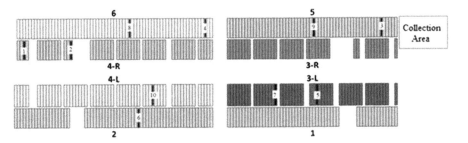

FIGURE 13.2 Order Locations in Current Layout.

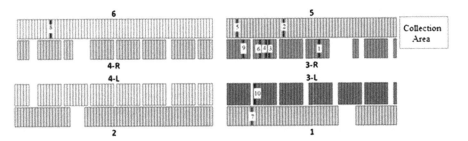

FIGURE 13.3 Order Locations in Proposed Layout.

TABLE 13.3
Current Distance Matrix and Product Weights

t_{ij}	0	1	2	3	4	5	6	7	8	9	10	d_i
0	0	140	124	8	68	46	110	59	94	31	96	—
1	140	0	12	130	75	101	55	88	51	107	45	13539
2	124	12	0	115	60	85	71	72	66	92	30	10195
3	8	130	115	0	59	44	103	57	85	23	89	4025
4	68	75	60	59	0	47	47	34	26	36	34	1569
5	46	101	85	44	47	0	68	13	73	67	54	1547
6	110	55	71	103	47	68	0	55	73	80	54	1253
7	59	88	72	57	34	13	55	0	60	57	41	1130
8	94	51	66	85	26	73	73	60	0	62	60	1114
9	31	107	92	23	36	67	80	57	62	0	66	475
10	96	45	30	89	34	54	54	41	60	66	0	1241

TABLE 13.4
Proposed Distance Matrix

t_{ij}	0	1	2	3	4	5	6	7	8	9	10
0	0	37	41	54	55	57	57	66	72	120	63
1	37	0	61	17	18	45	20	24	51	99	26
2	41	61	0	43	42	16	40	40	47	78	34
3	54	17	43	0	1	27	3	7	35	82	9
4	55	18	42	1	0	26	2	6	34	81	8
5	57	45	16	27	26	0	24	24	31	62	18
6	57	20	40	3	2	24	0	4	32	79	4
7	66	24	40	7	6	24	4	0	26	79	6
8	72	51	47	35	34	31	32	26	0	87	26
9	120	99	78	82	81	62	79	79	87	0	73
10	63	26	34	9	8	18	4	6	26	73	0

According to Figure 13.3, with the proposed layout plan the distance of each pair of the locations in the examined mixed pallet is given in Table 13.4.

To demonstrate the effect of product weights on the order picking process, we determined the picking routes according to different threshold values (α). As a result of the proposed mathematical model, orders are collected in (1–2–3–4–5–6–10–7–8–9) when α is set to 1.1. In this case, the total distance traveled by the operator is calculated as 435 units ($37 + 61 + 43 + 1 + 26 + 24 + 4 + 6 + 26 + 87 + 120$) via the matrix in Table 13.4. While α is set to 1.3, orders are collected in the sequence of (1–2–3–4–6–7–8–10–5–9); when α is set as 1.5, orders are collected in the sequence of (2-1-3-4-6-7-10-8-5-9). In these cases, the total distances traveled by the operator are calculated as 400 ($37 + 61 + 43 + 1 + 2 + 4 + 26 + 26 + 18 + 62 + 120$) and 371 ($41 + 61 + 17 + 1 + 2 + 4 + 6 + 26 + 31 + 62 + 120$), respectively. As can be seen, the α parameter has a direct effect on the problem and the length of the route decreases as the ratio of the weights of consecutive products increases.

6 COMPARATIVE RESULTS

In this section, randomly selected pallet samples are examined to test the effect of the proposed layout plan and order picking process in the distribution center. Accordingly, a total of 50 different pallet samples ranging from two pallets to 37 pallets were examined, as can be seen in Table 13.5.

For each pallet sample, three cases—current situation, current layout and proposed route, and proposed layout and proposed route—are experimented. The current situation represents the current layout and current route. The mathematical model developed for the order picking process was solved by CPLEX. Optimal solutions were reached in less than a minute regardless of the instance size.

TABLE 13.5
Randomly Selected Instances

Pallet	No. of Products	Pallet	No. of Products	Pallet	No. of Products	Pallet	No. of Products
1	2	16	6	31	14	46	33
2	2	17	6	32	14	47	34
3	2	18	6	33	14	48	35
4	2	19	6	34	14	49	36
5	2	20	6	35	19	50	37
6	3	21	10	36	20		
7	3	22	10	37	21		
8	3	23	10	38	22		
9	3	24	10	39	23		
10	3	25	10	40	25		
11	4	26	11	41	25		
12	4	27	14	42	26		
13	4	28	14	43	26		
14	4	29	14	44	32		
15	4	30	14	45	32		

In Table 13.6, the first column represents the pallet instance. In the second column, the distance taken by the operator is given according to the existing layout plan and the current order picking process; the product weights are not considered. The distance taken by the operator is given according to the current layout and proposed route of different α levels in the third column; in the fourth column, the distance taken by the operator is given according to the proposed layout and picking process. Finally, the last column shows the improvement between the proposed and current layout. Bold values represent the improvement in the proposed layout and route against the current one.

7 CONCLUSIONS AND FUTURE DIRECTIONS

In this work, warehouse layout and order picking applications were studied in the distribution center of a logistics company. The products that are placed in the locations without being subject to a certain rule have been replaced according to the stock turnover rate and weight. After the layout phase is complete, a new mathematical model has been proposed for the order picking process, considering the product weights during the preparation of the mixed pallet. When the results were investigated, about 28% improvement was achieved by the proposed layout and routing process. The proposed model is capable of finding optimal solutions in seconds even for the largest pallet of 37 products. Therefore, the proposed approach has been found to be useful in terms of relevance in warehouses and distribution

TABLE 13.6

The Comparison of Current and Proposed Situation

Pallet	Current Situation	Current Layout, Proposed Route					Proposed Layout, Proposed Route					Improvement %				
		$\alpha=1.1$	$\alpha=1.3$	$\alpha=1.5$	$\alpha=1.7$	$\alpha=1.9$	$\alpha=1.1$	$\alpha=1.3$	$\alpha=1.5$	$\alpha=1.7$	$\alpha=1.9$	$\alpha=1.1$	$\alpha=1.3$	$\alpha=1.5$	$\alpha=1.7$	$\alpha=1.9$
1	107	107	107	107	107	107	121	121	121	121	121	–	–	–	–	–
2	302	302	302	302	302	302	159	159	159	159	159	47.35	47.35	47.35	47.35	47.35
3	242	242	242	242	242	242	103	103	103	103	103	57.44	57.44	57.44	57.44	57.44
4	241	241	241	241	241	241	156	156	156	156	156	35.27	35.27	35.27	35.27	35.27
5	223	223	223	223	223	223	58	58	58	58	58	73.99	73.99	73.99	73.99	73.99
6	261	261	261	261	261	261	205	193	193	193	193	21.46	26.05	26.05	26.05	26.05
7	316	310	310	298	298	298	113	113	113	113	113	63.55	63.55	62.08	62.08	62.08
8	301	302	301	301	301	301	157	157	157	157	157	48.01	47.84	47.84	47.84	47.84
9	250	317	317	250	250	250	335	335	238	238	238	–	–	4.80	4.80	4.80
10	297	297	297	297	297	297	237	237	237	237	237	20.20	20.20	20.20	20.20	20.20
11	209	175	175	175	175	175	204	204	204	204	204	–	–	–	–	–
12	269	262	262	262	262	262	218	218	218	218	218	16.79	16.79	16.79	16.79	16.79
13	333	343	260	260	259	259	213	213	213	209	209	37.90	18.08	18.08	19.31	19.31
14	332	320	320	320	295	295	311	311	311	286	286	2.81	2.81	2.81	3.05	3.05
15	246	246	246	178	178	178	278	278	278	278	278	–	–	–	–	–
16	388	345	345	304	304	304	239	239	195	195	195	30.72	30.72	35.86	35.86	35.86
17	460	439	366	366	342	329	330	237	237	237	237	24.83	35.25	35.25	30.70	27.96
18	349	289	289	289	289	289	392	392	392	392	392	–	–	–	–	–
19	417	466	390	375	375	316	263	263	263	263	263	43.56	32.56	29.87	29.87	16.77
20	248	291	291	230	228	228	295	289	273	270	270	–	0.69	–	–	–
21	451	497	400	327	297	297	417	417	281	266	265	16.10	–	14.07	10.44	10.77

(Continued)

TABLE 13.6 (Continued)

Pallet	Current Situation	Current Layout, Proposed Route					Proposed Layout, Proposed Route					Improvement %				
		α=1.1	α=1.3	α=1.5	α=1.7	α=1.9	α=1.1	α=1.3	α=1.5	α=1.7	α=1.9	α=1.1	α=1.3	α=1.5	α=1.7	α=1.9
22	565	427	427	371	371	347	328	321	321	321	321	23.19	24.82	13.48	13.48	7.49
23	718	706	462	462	398	396	499	482	436	380	330	29.32	–	5.63	4.52	16.67
24	484	561	514	434	404	404	393	393	316	316	316	29.95	23.54	27.19	21.78	21.78
25	592	375	331	331	331	331	360	316	316	316	316	4.00	4.53	4.53	4.53	4.53

Pallet	Current Situation	Current Layout, Proposed Route					Proposed Layout, Proposed Route					Improvement %				
		α=1.1	α=1.3	α=1.5	α=1.7	α=1.9	α=1.1	α=1.3	α=1.5	α=1.7	α=1.9	α=1.1	α=1.3	α=1.5	α=1.7	α=1.9
26	438	600	462	357	357	326	459	459	455	408	390	23.50	0.65	–	–	–
27	918	713	649	584	584	521	608	548	531	531	486	14.73	15.56	9.08	9.08	6.72
28	620	709	525	390	382	382	610	376	376	376	376	13.96	28.38	3.59	1.57	1.57
29	477	565	439	402	361	350	440	349	270	246	246	22.12	20.50	32.84	31.86	29.71
30	770	609	523	477	445	434	494	444	444	380	380	18.88	15.11	6.92	14.61	12.44
31	867	640	579	459	414	414	546	519	385	367	349	14.69	10.36	16.12	11.35	15.70
32	784	900	675	580	517	479	721	504	437	396	377	19.89	25.33	24.66	23.40	21.29
33	583	568	371	361	361	361	344	344	344	344	344	39.44	7.28	4.71	4.71	4.71
34	613	614	518	481	432	410	529	452	452	406	406	13.84	12.74	6.03	6.02	0.98
35	910	1002	810	662	658	658	771	747	640	640	640	23.05	7.78	3.32	2.74	2.74
36	748	765	528	442	442	428	604	502	463	463	443	21.05	4.92	–	–	–
37	1117	1160	857	781	746	742	901	719	650	584	503	22.33	16.10	16.77	21.72	32.21
38	859	871	831	650	608	582	851	652	563	532	532	2.30	21.54	13.38	12.50	8.59
39	1016	1054	812	670	492	462	739	540	476	424	383	29.89	33.50	28.96	13.82	17.10
40	1229	831	747	494	493	392	536	479	319	319	315	35.50	35.88	35.43	35.29	19.64
41	766	1023	771	717	626	626	755	676	676	607	604	26.20	12.32	5.72	3.04	3.51

(Continued)

TABLE 13.6 (Continued)

Pallet	Current Situation	Current Layout, Proposed Route					Proposed Layout, Proposed Route					Improvement %				
		$\alpha=1.1$	$\alpha=1.3$	$\alpha=1.5$	$\alpha=1.7$	$\alpha=1.9$	$\alpha=1.1$	$\alpha=1.3$	$\alpha=1.5$	$\alpha=1.7$	$\alpha=1.9$	$\alpha=1.1$	$\alpha=1.3$	$\alpha=1.5$	$\alpha=1.7$	$\alpha=1.9$
42	1701	1013	865	779	779	700	945	725	655	655	655	6.71	16.18	15.92	15.92	6.43
43	1204	965	886	662	574	516	752	702	566	552	507	22.07	20.77	14.50	3.83	1.74
44	1692	1570	1158	943	880	880	1131	803	735	735	726	27.96	30.66	22.06	16.48	17.50
45	1375	1238	1128	988	799	636	561	488	384	384	348	54.68	56.74	61.13	51.94	45.28
46	1520	1450	1070	764	694	658	1203	842	722	722	711	17.03	21.31	5.50	-	-
47	1660	1500	1319	956	784	686	944	878	698	591	532	37.07	33.43	26.99	24.62	22.45
48	1549	1236	996	850	850	803	976	744	697	660	638	21.04	25.30	18.00	22.35	20.55
49	1844	1508	1176	1080	1022	924	982	798	734	649	600	34.88	32.14	32.04	36.50	35.06
50	1924	1683	1306	982	916	840	1300	1009	811	723	719	22.76	22.74	17.41	21.07	14.40

centers and significant improvements in a noticeably brief time. The availability of using different α levels according to the degree of weight sensitivity of the products provides decision-makers with the flexibility for distribution centers where a wide range of products are stored. As a result, the processes carried out during the storage phase are very important for the timely flow of the product in the supply chain. Therefore, the proposed approach minimizes idle times and unnecessary resource usage and reduces activities that do not add value to the chain.

REFERENCES

Chabot, T., Lahyani, R., Coelho, L. C., & Renaud, J. (2017). Order picking problems under weight, fragility and category constraints. *International Journal of Production Research*, 55(21), 6361–6379.

Dawe, R. L. (1995). Reengineer warehousing. *Transportation and Distribution*, 36(1), 98–102.

De Koster, R., & Van Der Poort, E. (1998). Routing orderpickers in a warehouse: a comparison between optimal and heuristic solutions. *IIE Transactions*, 30(5), 469–480.

De Koster, R., Le-Duc, T., & Roodbergen, K. J. (2007). Design and control of warehouse order picking: A literature review. *European Journal of Operational Research*, 182(2), 481–501.

Dekker, R., De Koster, M., Roodbergen, K. J., & Van Kalleveen, H. (2004). Improving order-picking response time at Ankor's warehouse. *Interfaces*, 34(4), 303–313.

Goetschalckx, M., & Donald Ratliff, H. (1988). Order picking in an aisle. *IIE Transactions*, 20(1), 53–62.

Hall, R. W. (1993). Distance approximations for routing manual pickers in a warehouse. *IIE Transactions*, 25(4), 76–87.

Higginson, J. K., & Bookbinder, J. H. (2005). Distribution centres in supply chain operations. In *Logistics Systems: Design and Optimization* (pp. 67–91). Springer, Boston, MA.

Manzini, R., Gamberi, M., Persona, A., & Regattieri, A. (2007). Design of a class-based storage picker to product order picking system. *The International Journal of Advanced Manufacturing Technology*, 32(7-8), 811–821.

Petersen, C. G. (1997). An evaluation of order picking routeing policies. *International Journal of Operations & Production Management*, 17(11), 1098–1111.

Petersen, C. G., & Schmenner, R. W. (1999). An evaluation of routing and volume-based storage policies in an order picking operation. *Decision Sciences*, 30(2), 481–501.

Ratliff, H. D., & Rosenthal, A. S. (1983). Order-picking in a rectangular warehouse: a solvable case of the traveling salesman problem. Operations research, 31(3), 507–521.

Roodbergen, K. J., & Koster, R. (2001). Routing methods for warehouses with multiple cross aisles. *International Journal of Production Research*, 39(9), 1865–1883.

Shqair, M., Altarazi, S., & Al-Shihabi, S. (2014). A statistical study employing agent-based modeling to estimate the effects of different warehouse parameters on the distance traveled in warehouses. *Simulation Modelling Practice and Theory*, 49, 122–135.

Theys, C., Bräysy, O., Dullaert, W., & Raa, B. (2010). Using a TSP heuristic for routing order pickers in warehouses. *European Journal of Operational Research*, 200(3), 755–763.

Van Gils, T., Ramaekers, K., Caris, A., & de Koster, R. B. (2018). Designing efficient order picking systems by combining planning problems: State-of-the-art classification and review. *European Journal of Operational Research*, 267(1), 1–15.

Žulj, I., Glock, C. H., Grosse, E. H., & Schneider, M. (2018). Picker routing and storage-assignment strategies for precedence-constrained order picking. *Computers & Industrial Engineering*, 123, 338–347.

14 Application of Analytical Network Process-Based BOCR Model on Green Supplier Selection Problem

A Case of ISO 14001 Certified Hotel

Cihan Çetinkaya[1], Eren Özceylan[2], and Buşra Baytur[2]

[1]Management Information Systems Department, Adana Alparslan Türkeş Science and Technology University, Adana, 01250, Turkey

[2]Industrial Engineering Department, Gaziantep University, Gaziantep, 27010, Turkey

1 INTRODUCTION

In addition to ensuring customer satisfaction in their products/services to survive and achieve sustainable profit, businesses should manage their costs and capacities effectively. It is particularly important to understand the demands and expectations of customers and meet them in the most convenient and fastest way through service enterprises directly interacting with customers. In this respect, firms should have a clear understanding of supply chain management. Customer satisfaction is affected by many factors especially in the service sector, which is the third biggest sector of the economy. The enterprises in this sector purchase by cooperating with their suppliers to meet their needs. To reduce operating costs, increase profits, improve service quality, and increase customer satisfaction, they should develop a procurement process that meets their goals (Ragatz et al. 1997).

Consumers' sensitivity to the environment will encourage producers to move to more sustainable production, thus, natural resources will be preserved and passed on to the next generations (Tozanli et al. 2019). As environmental awareness increases, companies also become aware of their environmental obligations. To maintain managerial activities environmentally-conscious is important to leave a sustainable world for future generations. In recent years, the environmental preferences of the companies have increased (Robinson, 2019).

The green management approach means that the enterprises operate with an environmental sense of responsibility. Green management concept, in short, is based on the principle of conserving the environment and considering the concepts of economic growth in the long-term. The starting point of green management is sustainable development. Sustainability means protecting natural resources and improving the quality of life. In this context, the evaluation of social, economic, and ecological objectives—and the organization of activities in the interest of each—will help in sustainable development (Karabulut, 2003).

The aim of the green management approach is to regulate the activities of the enterprises—their aims, functions, organizational structure, and production processes—to prioritize the environment. In addition, other objectives of green management are strengthening the enterprises economically, ensuring more effective use of natural resources, and reducing the volume of emissions, contributing to the development of technologies that will ensure environmental protection and cleaner production (Karabulut, 2003).

Both the regulations and the sensitivity of the enterprises to the environment resulted in a more responsible supplier selection. This perspective is expressed as Green Purchase (GP).

The Green Purchase is defined as, "among the goods, services, and works that have the same basic function, they have less environmental impact than their counterparts throughout their life cycle" in The European Commission's statement on Public Procurement for a Better Environment. If enterprises use their purchasing power by preferring the goods, services, and works that have less negative environmental impact, they can contribute to local, regional, national, and international sustainability goals. GP is the selection of materials to be purchased from recyclable, reusable, or recycled materials (Sarkis, 2003).

Hotels are also adopting environmental-friendly operations and GP applications to reduce their harm to the environment. These types of hotels are called green/sustainable/eco-friendly hotels (Verma and Chandra, 2017). Applications by green hotels include recycling waste, towel and linen reuse programs, low-flow taps and showerheads, water discharge urinals, refillable bath amenities, automatic climate control and light sensors, and natural ventilation (Rahmana et al. 2012). The facilities that are awarded "Green Star Certificate" (Environmentally Friendly Accommodation Facility) from the Culture and Tourism Ministry in Turkey apply saving water, increasing energy efficiency, reducing the consumption of harmful substances, reducing waste, encouraging the use of renewable energy sources, planning the accommodation establishments as environmentally sensitive from the investment stage, raising awareness on environment, and cooperating with related organizations. Facilities that received Green Star Certificate can contribute to the protection of the environment, make a difference in their promotion and marketing, and can contribute to their businesses and the country's economy without sacrificing service quality. They can be instrumental in the protection of the environment and be exemplary facilities (KTB, 2007).

Within the scope of this study, a five-star hotel is investigated for its Green Star acquisition procedures. The green supplier selection problem of the hotel—that has an ISO 14001 certification—is solved by using Analytical Network Process (ANP) and Benefits, Opportunities, Costs, and Risks (BOCR) methodology. The analytic hierarchy process (AHP) methodology is also used while determining the criterion weights.

The organization of the study is as follows: in the next section, a literature review related to supplier selection is examined. Later the AHP and ANP methods are described in Section 3. In Section 4, BOCR criteria and sub-criteria are defined and the analyses obtained by using the Super Decisions package program are interpreted. In Section 5, the results and recommendations are given.

2 LITERATURE REVIEW

This section is devoted to a detailed literature review.

2.1 MULTI-CRITERIA DECISION MAKING

Multi-Criteria Decision Making (MCDM) is a set of methods that forms a branch of Decision Science and incorporates different approaches. MCDM is based on the process of modeling the decision process according to the criteria and analyzing it to maximize the benefit that the decision-maker will achieve at the end of the process.

MCDM both represents an approach and describes a top concept that includes techniques or methods designed to assist people who face problems with multiple and conflicting criteria (Bogetoft and Pruzan, 1997).

The MCDM technique is used both for evaluation and design problems, the major distinction being whether the solutions/alternatives are explicitly (evaluation problems) or implicitly (design problems) (Ilgin et al. 2015).

2.2 BOCR ANALYSIS

The complex decision-making process usually requires an analysis of a decision against benefits, opportunities as a potentially good chance that may result in the future, costs that may arise, and risks in the long-term (Saaty and Vargas, 2001).

A full BOCR analysis is similar to that of SWOT, although both of them are usually applied on the micro-level to assess internal and external factors that affect the performance of a company. However, these analyses may be perfectly adapted for evaluation at the macro-level (Šimelytė et al. 2014).

2.3 LITERATURE REVIEW ON MULTI-CRITERIA DECISION MAKING IN SUPPLIER SELECTION

MCDM techniques consist of approaches and methods that try to reach the best possible solution that meets multiple conflicting criteria. Therefore, many researchers have used MCDM methods to select the best supplier. Some of these are discussed below.

In Dickson's (1966) supplier selection study, 23 criteria were used, and the most important criteria were determined as quality, cost, and performance with the survey. Weber et al. (1991) examined 74 articles dealing with vendor selection criteria in manufacturing and retail environments and classified them according to 23 criteria identified by Dickson (1966) in his study. He presented a comprehensive review of the literature that provides the most important criteria for supplier selection.

Handfield et al. (2002) proposed an Analytical Hierarchy Process (AHP)-based model to evaluate the relative importance of various environmental factors. They evaluated suppliers based on these factors. They presented various case studies to illustrate the strengths and weaknesses of the approach. This article also explored that AHP can be part of a comprehensive Environmentally Friendly Purchasing (ECP) system.

Talluri and Baker (2002) used Data Envelopment Analysis (DEA) to perform a logistic network design. When evaluating suppliers, producers, distributors, and potential stakeholders, they used six evaluation factors that were proposed by other researchers for supplier evaluation that included two inputs and four outputs.

Bhutta and Khurrum (2003) reviewed 154 articles on the selection of suppliers published in peer-reviewed journals. He classified these articles as conceptual articles, case studies, criteria papers, literature review, mathematical modeling, and methodology.

Chan and Chung (2004) developed multi-criteria genetic optimization to solve distribution network problems in supply chain management. In this study, they combined AHP with the Genetic Algorithm (GA) to achieve MCDM ability and reduce computational time.

Vaidya and Kumar (2004) presented a literature review of the applications of the AHP in various areas where AHP is used as a multi-criteria decision-making tool. Haq and Govindan (2006) presented a structured model to evaluate vendor selection of the firm in the rubber industry in Southern India using AHP and the model was confirmed by fuzzy AHP. In this study, they identified seven main criteria and 32 sub-criteria.

Lu et al. (2007) considered the electronics industry and they defined their criteria as environmental criteria material, energy, solid residue, liquid residue, gas residue, and technology and applied the AHP method.

Govindan et al. (2008) analyzed the interaction of criteria used to select the green suppliers who address environmental performance using Interpretive Structural Modeling (ISM) and AHP that sample an automobile company in the southern part of India. Some of the major criteria have been highlighted and put into an ISM model to analyze the interaction between them.

Yan (2009) conducted a study to reduce asymmetric information about the green supply chain and to increase the efficiency of green suppliers. In this study, AHP was applied to dynamically adjust the green supplier evaluation index weights. The author presented an integrated model with AHP and GA.

Kuo et al. (2010) presented a green supplier selection model combining Artificial Neural Networks (ANN) with DEA and ANP. The developed model exceeded the data accuracy limitations and the decision-making units (DMUs) quantity constraints encountered in traditional DEA.

Sun et al. (2010) considered 3rd Party Logistics firm selection as a multi-criteria decision problem with dependency and feedback and examined the resolution of the ANP approach with BOCR. In this study, they identified 12 sub-criteria: just in time delivery, accurate delivery rate, the efficiency of order process under the benefit criteria; service costs under the criteria of costs, transaction costs, variable transport costs; service activity, business volume, operating experience under the criteria of opportunities; under the risk criterion, they examined data security, information accuracy, and employee level criteria.

Ho et al. (2010) reviewed the literature of multi-criteria decision-making approaches for supplier evaluation and selection from 2000 to 2008. This study provides evidence that multi-criteria decision-making approaches are better than traditional cost-based approaches.

Kannan et al. (2013) developed an order allocation problem to improve GSCM initiatives. A fuzzy TOPSIS (Technique for Order Preference by Similarity to Ideal Solution) method was used to evaluate selected suppliers, with the relative importance weights determined through AHP.

Kumar et al. (2014) proposed a unified Green DEA approach to model suppliers' carbon footprints and supplier selection problem as a necessary dual role factor in a well-known automobile spare parts manufacturer in India.

Freeman and Chen (2015) explained how to design a systematic and comprehensive methodology to assist in the selection of the most appropriate green supplier through a two-stage survey approach and the AHP-Entropy/TOPSIS methodology. Senior managers were found to rank traditional criteria higher than environmental alternatives.

Govindan and Sivakumar (2015) used the fuzzy TOPSIS method to select green suppliers to reduce carbon emissions in the paper industry. In this study, an integrated structure was built by integrating multi-objective linear programming.

Uygun and Dede (2016) proposed a practical hybrid fuzzy multi-criteria decision-making approach consisting of fuzzy DEMATEL (The Decision-Making Trial and Evaluation Laboratory), fuzzy ANP, and fuzzy TOPSIS methods to evaluate the overall green performance of four companies.

Awasthi and Govindan (2016) addressed the problem of evaluating green supplier development programs and proposed a solution approach based on a fuzzy NGT (Nominal Group Technique)—VIKOR (Vise Kriterijumska Optimizacija I Kompromisno Resenje). NGT was used to determine the criteria. Sensitivity analysis was performed to determine the influence of criteria weights on results.

Wan Xu and Dong (2017) investigated a type of supplier selection problem with two-level criteria and proposed a hybrid method by combining ANP with ELECTRE II (ELimination Et Choix Traduisant la REalité II) in an interval 2-tuple linguistic environment to compare suppliers. This method was in the form of a two-stage range that could appropriately model the quantitative and qualitative criteria involved in supplier selection.

Abdel-Basset Mohamed and Smarandache (2018) suggested a framework consisting of four phases by integrating ANP with TOPSIS using the interval-valued neutrosophic numbers. ANP was used to calculate the weights of the selected criteria, considering their dependence. A case study of a dairy and foodstuff corporation had been solved employing the proposed framework.

Abdullah Chan and Afshari (2019) used the PROMETHEE (Preference Ranking Organization Method for Enrichment of Evaluations) method for green supplier selection. Using the five decision-makers for a group of senior managers at an organic farm in Malaysia and the five-point Likert scale, seven determined economic and environmental criteria were evaluated, and the most appropriate green supplier was selected.

Rouyendegh et al. (2020) used the IFTOPSIS (Intuitionistic Fuzzy TOPSIS) method to select the optimum green supplier for the firm located in Ankara. Fuzzy logic is integrated into the TOPSIS method. The criteria are determined in order of importance and a Fuzzy Set—effective in eliminating uncertainty—chooses the most suitable alternative. The combined multi-criteria decision approaches and their applications are summarized in Table 14.1.

3 PROBLEM STATEMENT

Qualitative techniques are by far the most frequently used MCDM method—AHP, ANP, and TOPSIS are the most popular ones (Gupta and Ilgin, 2018).

The Analytical Network Process (ANP) is a generalized form of the Analytic Hierarchy Process (AHP) method developed by Thomas L. Saaty in 1971 (Wind and Saaty, 1980). Both methods are based on pairwise comparison. When making pairwise comparisons between the criteria, the basic scale with values from 1 to 9 is used, as recommended by Saaty (Felek et al. 2007). The basic scale is shown in Table 14.2. Each criterion is compared to the other and "The Pairwise Comparisons Matrix" is created as shown in Table 14.3.

Aside from being valid in both AHP and ANP, the problems in AHP are handled in a hierarchical structure, while the relations between the criteria and the aspects of these relations are defined in the ANP to form a network structure. The most significant difference of ANP according to AHP is that it uses an interactive hierarchical structure instead of a hierarchical structure from top to bottom (Timor, 2011).

The ANP method includes feedback and dependence between criteria. In ANP, all components of the problem and possible relationships are defined, and it is decided whether these are one-way or two-way. Subsequently, pairwise comparisons of all other components acting on one component are analyzed to determine their superiority in effect. Since two-way interactions are considered, the interaction of non-visible elements directly connected to each other within the entire system is considered. Therefore, the method allows feedback. The dependence between criterion sets is called external dependency and the dependence on the criterion's own set is called internal dependency (Niemira and Saaty, 2004). Figure 14.1 shows the difference between the structure of AHP (a) and the structure of the ANP (b).

TABLE 14.1
Summary of Previous Research

Author(s)	Approach (es)	Additional Features of Decision Approach	Application Domain
Handfield et al. (2002)	AHP	AHP that can be part of a comprehensive Environmentally Friendly Purchasing (ECP) system	Automotive, paper, and apparel manufacturer
Talluri and Baker (2002)	DEA	A multiphase mathematical programming based on game theory, linear and integer programming	Logistics distribution network design
Chan and Chung (2004)	AHP, GA	Multi-criteria genetic optimization	Distribution network problems
Haq and Govindan (2006)	Fuzzy AHP	AHP and Fuzzy AHP has been applied	Rubber industry
Lu et al. (2007)	AHP	Environmentally friendly design indexed AHP	Surveyed 500 companies in Taiwan
Govindan et al. (2008)	AHP, ISM	ISM based ANP analysis with environmental criteria	An automobile company
Yan (2009)	AHP, GA	Multi-criteria genetic optimization	Green supplier evaluation
Kuo et al. (2010)	DEA, ANN, ANP	Hybrid method	Digital camera manufacturing
Sun et al. (2010)	ANP, BOCR	ANP by using rated BOCR	Third-party logistics service
Kannan et al. (2013)	Fuzzy AHP, Fuzzy TOPSIS, MOLP	Multi-objective linear programming model using weighted max-min method	Automobile manufacturing company
Kumar et al. (2014)	Green DEA	Comparisons of AHP-TOPSIS, GDEA, and Saen's model for carbon footprint measurement	Automobile spare parts manufacturer
Freeman and Chen (2015)	AHP-Entropy/ TOPSIS	Developing mathematical model	Electronic machinery manufacturer
Awasthi and Govindan (2016)	Fuzzy NGT, VIKOR	Hybrid method	A leading manufacturer of automobiles
Wan Xu and Dong (2017)	ANP, ELECTRE II	Hybrid method	An auto manufacturer company
Abdel-Basset Mohamed and Smarandache (2018)	ANP, TOPSIS	Integrated model	A dairy and foodstuff corporation

(Continued)

TABLE 14.1 (Continued)

Author(s)	Approach (es)	Additional Features of Decision Approach	Application Domain
Abdullah Chan and Afshari (2019)	PROMET-HEE	Different types of preference functions	Organic products and services farm
Rouyendegh et al. (2020)	IFTOPSIS	Hybrid method	A firm in Ankara

TABLE 14.2
1–9 Scale Used in Pairwise Comparison Method (Saaty, 2008)

Degrees	Definition
1	Equal important
3	Slight importance of one over another
5	Moderate Importance of one over another
7	Very strong importance
9	Extreme importance of one over another
2,4,6,8	Intermediate values between two adjacent values

TABLE 14.3
Establishment of Pairwise Comparisons Matrix for Criteria (Vargas, 1990)

	Criteria 1	Criteria 2	...	Criteria n
Criteria 1	W_1/W_1	W_1/W_2	...	W_1/W_n
Criteria 2	W_2/W_1	W_2/W_2	...	W_2/W_n
...	...	...	...	...
Criteria n	W_n/W_1	W_n/W_2	...	W_n/W_n

The structure in which the benefits, opportunities, costs, and risks that each option can consider is called the BOCR model. Here, BOCR clusters have different weights and the criteria in each cluster are correctly determined (Saaty, 2001).

In complex decision systems, a decision can be structurally divided into three parts—a real measure of success, our own value system, our ethical and moral assessments of decision, and the network structure for hierarchies or influences, along with objective realities that make one decision preferable than others. Generally, the group criteria (clusters) used for assigning weight or grading the BOCR at the first level can be a structure formed by an analytical hierarchy method or a simple structure that can be graded directly. Here, our ethical and moral assessments of the decision are handled with value judgments that can be defined based on BOCR nodes called control criteria. Finally, by combining these two parts with the network structure and establishing relations, a complex decision system structure will be established (Saaty, 2001).

The ANP and BOCR approach, which allows comparison of qualities that cannot be expressed quantitatively, takes into account positive (benefit) and negative (cost) factors when choosing among alternatives. It provides results for the selection of those with a benefit/cost (B/C) ratio above one.

The BOCR model is characterized by a controlled hierarchical structure in an ANP framework. Accordingly, the main objective is at the first level, followed by criteria and sub-criteria that are at the second level of the hierarchy, while alternatives are located at the third level in the model (Saaty, 2005; 2008; 2009).

The critical point in the ANP structure is the accurate determination and pairwise comparison of the criteria and sub-criteria. Positive and negative factors are taken into consideration when selecting ANP models by the BOCR approach. Benefits, opportunities, costs, and risks may not be of equal importance.

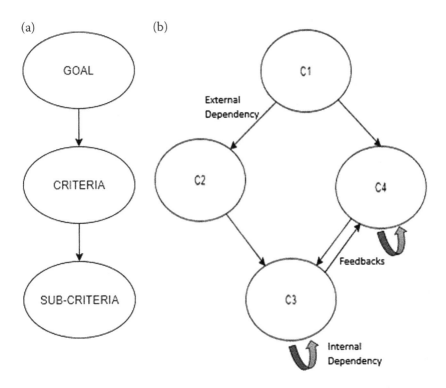

FIGURE 14.1 Analytic Hierarchy Process (a) and Analytical Network Process (b) Structure (Saaty, 2004).

Opportunities in BOCR analysis usually catch expectations about positive spin-off, future profits, and revenue of future positive developments, where benefits represent current revenue or those from positive developments one is relatively certain of. Likewise, a firm's weak (W) points may not tell the whole story of negative aspects in SWOT analysis; external threats (T) concerning competition or unfavorable developments in society must be dealt with as well. Risks in BOCR analysis are supposed to catch expected consequences of future negative developments, whereas costs represent (current) losses and efforts and consequences of negative developments one is relatively certain of (Wijnmalen, 2007).

4 PROPOSED APPROACH

This selection model is designed in four stages. In the first stage, expert opinions are taken, relevant literature is searched, and the main criteria and sub-criteria are determined. In the second stage, inter-elemental interactions are determined, and a decision network structure is formed. In the third stage, pairwise comparisons are made based on expert opinions, and consistencies are checked. Candidate suppliers are examined for alternatives. Since two of the five suppliers did not comply with the basic working conditions of the firm, they were eliminated without being evaluated. Three suppliers considered worthy of study are included in the evaluation. In the fourth stage, the Super Decisions package program is used to solve the problem. The proposed model is implemented in the following steps, with the algorithm in the flow diagram presented in Figure 14.2.

4.1 General Outline of Case

The hotel is in Mardin City and has a capacity of 450 beds. It is the largest congress center of the Eastern and South-Eastern Anatolia Region and the first and only five-star hotel of Mardin.

To increase its brand image, make its operations efficient, and not to harm the environment, it has established an Integrated Management System (IMS). IMS consists of 9,001 Quality Management System, 14,001 Environmental Management System, 22,000 Food Safety Management System, and 10,002 Customer Satisfaction Management System.

The company has established the 14,001 Environmental Management System based on risk analysis, aiming to reduce the use of natural resources and minimize the damage to soil, water, and air. Their objective is to receive the Green Star symbol that promotes positive contributions to the environment in the context of sustainable tourism.

In this study, environmental activities in the purchasing processes of this hotel are considered and environmental criteria are taken into consideration in supplier selection. The criterion weights are obtained by using the AHP method used effectively in decision-making problems.

To select the most suitable supplier selling cleaning materials, BOCR criteria and sub-criteria are determined by searching relevant literature and by taking expert

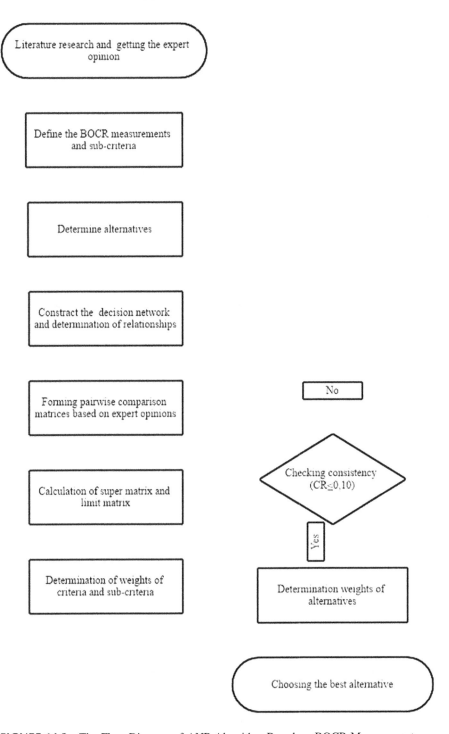

FIGURE 14.2   The Flow Diagram of ANP Algorithm Based on BOCR Measurements.

opinions. Pairwise comparison matrixes are formed by acquiring professional judgments and results are obtained by using the Super Decisions package program. Consistency ratios are calculated and the ANP method is applied by using criterion weights to select the best alternative among the suppliers.

4.2 Determining Selection Criteria

The sub-criteria that the company should pay attention to in the selection of suppliers are determined by the BOCR criteria given in Figure 14.3. The BOCR criteria were determined by considering the literature review and expert opinions.

The criteria are based on literature review, operating procedures of hotel management, and the experience of the purchasing specialist. Face-to-face interviews

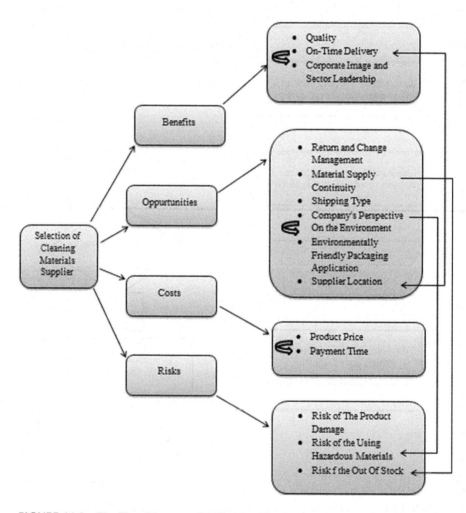

FIGURE 14.3 The Flow Diagram of ANP Algorithm Based on BOCR Measurements.

were conducted with the hotel's purchasing specialist who has been working for more than five years. When performing pairwise comparisons, the past performances of the suppliers are examined (number of delivery days, damage-free, product supply, confidence, etc.). The performance data and the opinions of the purchasing specialist are also taken into consideration.

4.2.1 Identification of BOCR Measurements and Sub-criteria

4.2.1.1 Benefits

The sub-criteria quality, on-time delivery, corporate image, and sector leadership are determined under the benefits measurement. Brief descriptions are as follows:

- Quality: Buyers must purchase goods and services from suppliers that can produce these products at the lowest costs, highest quality, within the shortest lead time, and who are environmentally responsible in managing their associated processes (Handfield et al. 2002). Here, quality criteria are the ability to meet the expectations and requirements of the requested product. These are features such as stain removal, cleaning without damage to the fabric and surface, non-irritation of human tissue, and ease of use.
- On-Time Delivery: It refers to the delivery of products at the deadline. Lead Time and On-Time Delivery Time between procurement and delivery of an order lags in delivery schedule (Mani et al. 2014).
- Corporate Image and Sector Leadership: It shows the percentage and recognition of the companies that provide cleaning supplies in the market they are currently competing with.

4.2.1.2 Opportunities

The sub-criteria return and change management, material supply continuity, shipping type, company's perspective on the environment, environmental-friendly packaging application, and supplier location are determined under the opportunities measurement. Brief descriptions are as follows:

- Return and Change Management: It refers to the management of materials that require return and exchange for future cooperation.
- Material Supply Continuity: Ensuring the supply of the desired material at the desired time for future cooperation.
- Shipping Type: Dispatch of materials without partial shipment. It is an indicator of the importance given to the environment. Partial shipments cause an increase in exhaust gases released into the environment.
- Company's Perspective on the Environment: It shows the sensitivity of suppliers to environmental pollution and their observations in their activities.
- Environmental-Friendly Packaging Application: Grover et al. (2016) said that Green Product and Eco-Design is the use of environmental-friendly technology and materials; design capability for reduced consumption of material/energy, reuse, recycle of material; and design of products to avoid or reduce the use of

harmful materials, green packing. This criterion shows the provision and use of recyclable materials. By eliminating pollution (especially in the purchased goods and services), costs can be reduced and any associated legal problems avoided (or greatly reduced) (Handfield et al. 2002).

- Supplier Location: It shows the availability of suppliers to their customers.

4.2.1.3 Costs

The sub-criteria product price and payment time are determined under the cost measurement. Brief descriptions are as follows:

- Product Price: It is the criterion that demonstrates the suitability of cleaning materials price.
- Payment Time: The flexibility of suppliers in payment-time applications for future cooperation.

4.2.1.4 Risks

The sub-criteria risk of product damage, risk of using hazardous materials, risk of being out of stock are determined under the risk measurement. Brief descriptions are as follows:

- Risk of Product Damage: It is the sub-criterion showing the attitudes of the supplier to take responsibility for carryout/theft/damage in the products during transportation.
- Risk of Using Hazardous Materials: It is a sub-criterion that indicates the level of use of hazardous substances in chemical products.
- Risk of Being Out of Stock: It is a sub-criterion that indicates the risk of not being supplied with the product by the supplier.

4.2.2 Alternatives

The names of the companies selling cleaning materials are expressed as Supplier 1, Supplier 2, and Supplier 3 to prevent unfair competition and inconvenience for the firms. The suppliers' general information is as follows:

- Supplier 1: It is one of the pioneers in the sector and provides professional cleaning, hygiene, beverage products, and services with integrated solutions. It has a wide distribution network. It carries out environmental protection activities.
- Supplier 2: It carries out various production and support service activities in addition to hospital hygiene, food and beverage industry, industrial products, textile products, and pest control. There is no continuous distribution network.
- Supplier 3: It is a provider of cleaning, sanitation, and maintenance products, systems, and services that efficiently integrate chemicals, machines, and sustainability programs. It carries out environmental protection activities. Communication is difficult because the headquarters of the firm is abroad.

4.3 ANP MODEL APPLICATION AND SOLUTION

This study has been carried out to select the most suitable green supplier based on their purchasing processes, considering the environmental sensitivity of the hotel management.

The relationships between the criteria are defined and shown in Figure 14.3. Four main criteria and 14 sub-criteria are considered. The sub-criteria namely Company's Perspective on the Environment, Environmentally Friendly Packaging Application, and Risk of the Using Hazardous Materials reflect the environmental dimension. Pairwise comparison values are entered in the "Super Decisions" package program. In all pairwise comparisons, the consistency ratio is below 0.10. Some examples from the interface are shown in Figure 14.4, Figure 14.5, and Figure 14.6.

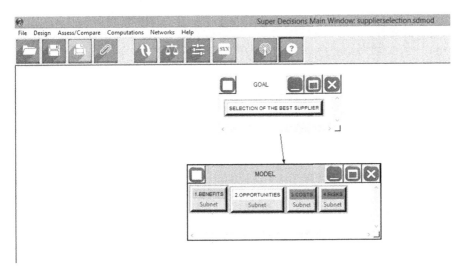

FIGURE 14.4 ANP Structure with BOCR Used in Selection of the Best Cleaning Material Supplier.

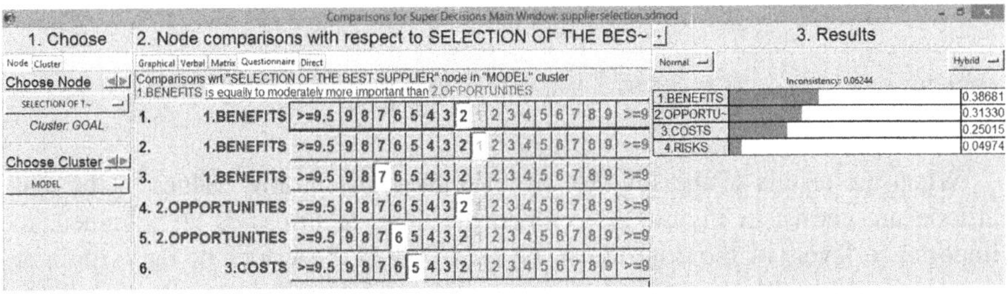

FIGURE 14.5 Pairwise Comparison Results of Main Criteria and Results of Consistency Analysis.

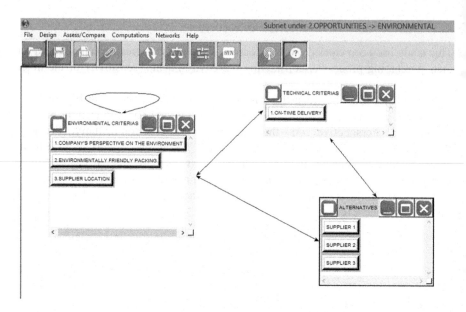

FIGURE 14.6   Environment Sub-criterion Structure of the Main Criterion of Opportunities.

Icon	Name	Normalized by Cluster	Limiting
No Icon	SELECTION OF THE BEST SUPPLIER	0.00000	0.000000
No Icon	1.BENEFITS	0.38681	0.386811
	2.OPPORTUNITIES	0.31330	0.313301
No Icon	3.COSTS	0.25015	0.250150
No Icon	4.RISKS	0.04974	0.049738

FIGURE 14.7   Table of Priorities According to ANP Structure.

When the results of the solution are examined, the priority values of the main criteria are shown in Figure 14.7. When the table of priorities is examined, the importance levels of the criteria can be seen. When the values of the criteria are examined in the table, benefits which the most important criterion in the selection of cleaning material has the most significance with a value of 38.68%. The main criterion that influences the least is the risks with a value of 4.97%. Alternatives are

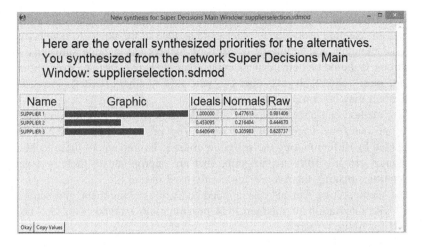

FIGURE 14.8   Ranking of Alternatives.

weighted by pairwise comparisons for each criterion and sorted by criteria weights. The results of the analysis are given in Figure 14.8. The results of the cleaning material suppliers to be selected by considering environmental sensitivities, ranked Supplier 1 first with a weight of 47.76%, Supplier 3 is ranked second with a weight of 30.59%, and Supplier 2 with a weight of 21.64% is ranked last. The most suitable supplier is Supplier 1.

5   CONCLUSIONS AND FUTURE RESEARCH

One of the factors that determine supply chain performance in our world is the consideration of environmental factors in resource use. Resource utilization decisions determine the responsiveness to the customer, financial efficiency, and degree of impact on nature. The most crucial resource utilization decision is supplier selection. Therefore, in this study, the supplier selection problem is discussed. Multi-criteria decision-making methods are used because the supplier selection problem has multiple and conflicting criteria. The ANP model based on the BOCR criteria is determined as a problem-solution methodology. According to the data obtained, the weights of the criteria are decided by the AHP method. The most important criterion is the benefits criterion, with a value of 38%.

Determined weights are used in the application of the ANP method, which is one of the multi-criteria decision-making methods. As a result of the study, Supplier 1 is selected as the best green supplier for the company, and Supplier 3 has been identified as the (second-best) alternative.

The study can be a guide for firm management to address environmental factors in selecting suppliers; it can also be adapted to selection problems in other matters. This method can also be used in other decision-making problems in literature and contributes by using BOCR criteria. The proposed approach can be implemented by differentiating the selection criteria. The study can be a guide for company management in addressing environmental factors in supplier selection; it can also be adapted to selection problems in other matters. One of the points to note is that the personal thoughts and intuitions of decision-makers can include subjective interpretations. It should be noted that accurate, clear, objective data and evaluations will provide more valid results. Also in cases where the number of criteria increases, the application may be complex for the firms.

This method can also be used in other decision-making problems in literature and contributes by using BOCR criteria. The proposed approach can be implemented by differentiating the selection criteria. To implement these techniques, companies can use software programs that are appropriate to their systems and expectations, making them more commonly used in real life.

The methodology can also be applied to various assessment and sequencing areas. The integration of mathematical programming methods can also be considered for future research.

REFERENCES

Abdel-Basset, M., Mohamed, M., & Smarandache, F. (2018). "A hybrid neutrosophic group ANP-TOPSIS framework for supplier selection problems", *Symmetry*, 10(6), 226.

Abdullah, L., Chan, W., & Afshari, A. (2019). "Application of PROMETHEE method for green supplier selection: A comparative result based on preference functions", *Journal of Industrial Engineering International*, 15, 271–285.

Awasthi, A., & Govindan, K. (2016). "Green supplier development program selection using NGT and VIKOR under fuzzy environment", *Computers & Industrial Engineering*, 91, 100–108.

Bhutta, M. & Khurrum S. (2003). "Supplier selection problem: methodology literature review", *Journal of International Information Management*, 12(2), 53–72.

Bogetoft, P. & Pruzan, P. (1997). Planning with Multiple Criteria: Investigation, Communication and Choice, Handelshojskolens Forlag, Copenhagen Business School Press.

Chan, F. T. S., & Chung, S. H. (2004). "Multi-criteria genetic optimization for distribution network problems", *The International Journal of Advanced Manufacturing Technology*, 24(7–8), 517–532.

Copacino, W. C. (1996). "Seven supply chain principles", *Trafic Management*, 35(1), 60.

Dickson, G. W. (1966). "An analysis of vendor selection and the buying process", *Journal of Purchasing*, 2(1), 5–17.

Erdoğmuş Ş., Kapanoglu M. & Koç E. (2005). "Evaluating high-tech alternatives by using analytic network process with BOCR and multiactors", *Evaluation and Program Planning*, 28, 391–399.

Erginel, N. & Şentürk, S. & Binici, Y. (2014). "The use of ANP method based on BOCR criteria for 3PL provider selection". *Journal of Science and Technology*-B-Theoretical *Science*, 3(1), 33–44.

Ergu, D., & Peng,Y. (2014). "A framework for SaaS software packages evaluation and selection with virtual team and BOCR of analytic network process", *The Journal of Supercomputing*, 67, 219–238.

Feglar, T., Levy, J. K., Feglar, T., & Feglar, T. (2006). "Advances in decision analysis and systems engineering for managing large-scale enterprises in a volatile world: integrating benefits, opportunities, costs and risks (BOCR) with the business motivation model (BMM)", *Journal of Systems Science and Systems Engineering*, 15(2), 141–153.

Felek, S., Yuluğkural Y. & Aladağ Z. (2007). "Comparing AHP and ANP results to estimate market share in mobile communication sector", *Makine Mühendisleri Odası Endüstri Mühendisliği Dergisi*, 18(1), 6–22.

Freeman, J., & Chen, T. (2015). "Green supplier selection using an AHP-entropy-TOPSIS framework", *Supply Chain Management: An International Journal*, 20(3), 327–340.

Govindan, K., & Sivakumar, R. (2015). "Green supplier selection and order allocation in a low-carbon paper industry: integrated multi-criteria heterogeneous decision-making and multi-objective linear programming approaches", *Annals of Operations Research*, 238(1–2), 243–276.

Govindan, K., Haq, A. N., Sasikumar, P., & Arunachalam, S. (2008). "Analysis and selection of green suppliers using interpretative structural modelling and analytic hierarchy process", *International Journal of Management and Decision Making*, 9(2), 163–182.

Grover, R., Grover, R., Rao, V. B. & Kejriwal, K. (2016). "Supplier selection using sustainable criteria in sustainable supply chain management", *International Journal of Economics and Management Engineering*, 10, 1775–1780.

Gupta, S. M. & Ilgin, M. A. (2018). *Multiple Criteria Decision Making Applications in Environmentally Conscious Manufacturing and Product Recovery*, CRC Press, Boca Raton, Florida, ISBN: 978-1498700658.

Handfield R, Walton S. V., Sroufe R. & Melnyk S. A. (2002). "Applying environmental criteria to supplier assessment: a study in the application of the analytical hierarchy process", *European Journal of Operational Research*, 141(1), 70–87.

Haq, A. N. & Govindan, K. (2006). "Fuzzy analytical hierarchy process for evaluating and selecting a vendor in a supply chain model", *The International Journal of Advanced Manufacturing Technology*, 29, 826–835.

Hernández C. T., Marins F. A. S. & Duran J. A. (2016). "Selection of reverse logistics activities using an ANP-BOCR model", *IEEE Latin America Transactions*, 14(8), 3886–3891.

Ho, W., Xu, X., & Dey, P. K. (2010). "Multi-criteria decision making approaches for supplier evaluation and selection: a literature review", *European Journal of Operational Research*, 202(1), 16–24.

Ilgin, M. A., Gupta, S. M. & Battaia, O. (2015). "Use of MCDM techniques in environmentally conscious manufacturing and product recovery: state of the art", *Journal of Manufacturing Systems*, 37(3), 746–758.

Kannan, D., Khodaverdi, R., Olfat, L., Jafarian, A., & Diabat, A. (2013). "Integrated fuzzy multi criteria decision making method and multi-objective programming approach for supplier selection and order allocation in a green supply chain", *Journal of Cleaner Production*, 47, 355–367.

Kara, M., Yurtsever, Ö. & Fırat, S. (2016). "Sustainable supplier evaluation and selection criteria". *Social and Economic Perspectives on Sustainability, Chapter: 11 Publisher: IJOPEC Publications Editors*: M. Mustafa Erdoğdu, 159–167.

Karabulut, E. (2003). Research on analyzing the impact of environmental awareness and green management practices in companies, Ph.D. Thesis, Business Administration, İstanbul University.

KTB. (2007). Türkiye Turizm Stratejisi (2023). Ankara: Kültür ve Turizm Bakanlığı. (In Turkish).

Kumar, A., Jain, V., & Kumar, S. (2014). "A comprehensive environment friendly approach for supplier selection", *Omega*, 42(1), 109–123.

Kuo R. J., Wang Y. C. & Tien F. C. (2010). "Integration of artificial network and MADA methods for green supplier selection", *International Journal of Cleaner Production*, 18, 1161–1170.

Lu, L. Y. Y., Wu, C. H., & Kuo, T.-C. (2007). "Environmental principles applicable to green supplier evaluation by using multi-objective decision analysis", *International Journal of Production Research*, 45(18–19), 4317–4331.

Mani, V., Agrawal, R., & Sharma, V. (2014). "Supplier selection using social sustainability: AHP based approach in India", *International Strategic Management Review*, 2(2), 98–112.

Mason, T. (1996). "Getting your suppliers on the team", *Logistics Focus*, 4(1), 10–12.

Niemira, M. P. & Saaty, T. L. (2004). "An analytic network process model for financial-crisis forecasting", *International Journal of Forecasting*, 20, 573–587.

Peker İ., Baki B., Tanyaş M., & Ar İ. M. (2016). "Logistics center site selection by ANP/ BOCR analysis: a case study of Turkey", *Journal of Intelligent & Fuzzy Systems*, 30, 2383–2396.

Ragatz, Gary L. & Handfield, R. B. & Scannell, T. V. (1997). "Success factors for integrating supplier into new product development", *The Journal of Product Innovation Management*, (May), 14(3), 190–203.

Rahmana, I., Reynolds, D. & Svarena, S. (2012). "How "green" are North American hotels? An exploration of low-cost adoption practices", *International Journal of Hospitality Management*, 31, 720–727.

Robinson, J., (2019). "Millennials worry about the environment, Should Your Company? https://www.gallup.com/workplace/257786/millennials-worry-environment-company. aspx Access date: 19.03.2020.

Rouyendegh, B. D., Yildizbasi, A., & Üstünyer, P. ((2020)). "Intuitionistic fuzzy TOPSIS method for green supplier selection problem", *Soft Computing*, 24, 2215–2228.

Saaty, T., & Özdemir, M. S. (2005). *The Encyclicon: A Dictionary of Applications of Decision Making with Dependence and Feedback Based on the Analytic Network Process*. PA: RWS Publications.

Saaty, T. L. (1990). *Decision Making for Leaders: The Analytical Hierarchy Process for Decisions in A Complex World*, Pittsburgh, PA: RWS Publications.

Saaty, T. L. (2001). *Decision-Making with Dependence and Feedback: The Analytic Network Process* (2nd edition), RWS Publications, Pittsburgh, PA, USA.

Saaty, T. L. (2004). "Decision making – The analytic hierarchy process and network processes (AHP.ANP)". *Journal of Systems Science and Systems Engineering*, 13(1), 1–34.

Saaty, T. L. (2005). *Theory and Applications of the Analytic Network Process: Decision Making with Benefits, Opportunities, Costs and risks*. PA: RWS Publications.

Saaty, T. L. (2008). "The analytic network process". *Iranian Journal of Operations Research*, 1(1), 1–27. Retrieved from: http://iors.ir.

Saaty T. L. (2008). "Relative measurement and its generalization in decision making why pairwise comparisons are central in mathematics for the measurement of intangible factors the analytic hierarchy/network process", *Review of the Royal Spanish Academy of Sciences Series a Mathematics (RACSAM)*, 102(2), 257.

Saaty, T. L. (2009). "Applications of analytic network process in entertainment", *Iranian Journal of Operations Research*, 1(2), 41–55.

Saaty, T. L. & Vargas, L. G. (2001). *Models, Methods, Concepts & Applications of the Analytic Hierarchy Process*, Boston: Kluwer Academic Publishers, 1, 46.

Sarkis, J. (2003). "A strategic decision framework for green supply chain management", *Journal of Cleaner Production*, 11(4), 397–409.

Šimelytė, A., Peleckis, K., & Korsakienė, R. (2014). "Analytical network process based on BOCR analysis as an approach for designing a foreign direct investment policy", *Journal of Business Economics and Management*, 15(5), 833–852.

Sun, C., Pan, Y., & Bi, R. (2010). "Study on third-party logistics service provider selection evaluation indices system based on analytic network process with BOCR", *2010 International Conference on Logistics Systems and Intelligent Management (ICLSIM)*.

Talluri, S. & Baker, R. C. (2002). "A multi-phase mathematical programming approach for effective supply chain design", *European Journal of Operational Research*, 141(3), 544–558.

Timor, M. (2011). *Analitik Hiyerarşi Prosesi*. İstanbul: Türkmen Kitabevi. (In Turkish).

Tozanli, O., Kongar, E. & Gupta, S. M. (2019). "A supplier selection model for end-of-life product recovery: an industry 4.0 perspective", in Responsible Manufacturing - Issues Pertaining to Sustainability, Edited by A. Y. Alqahtani, E. Kongar, K. K. Pochampally and S. M. Gupta, CRC Press, 323–344, Chapter 14, ISBN: 978-0815375074.

Uygun, Ö., & Dede, A. (2016). "Performance evaluation of green supply chain management using integrated fuzzy multi-criteria decision making techniques", *Computers & Industrial Engineering*, 102, 502–511.

Uzun, G. & Kabak, M. (2018). "Determining the search and rescue prioritization of coast guard surface vessels by using analytic network process", *Journal of the Faculty of Engineering and Architecture of Gazi University*, 34(2), 819–833.

Vaidya, O. S. & Kumar, S. (2004). "Analytic hierarchy process: an overview of applications", *European Journal of Operational Research (published online)*, 169(1), 1–29.

Vargas, L. G. (1990). "An overview of the analytic hierarchy process and its applications", *European Journal Of Operational Research*, 48(1), 4.

Verma V. K. & Chandra, B., (2017). "Intention to implement green hotel practices: evidence from indian hotel industry", *International Journal of Management Practice*, 11(1), 24.

Wan, S., Xu, G., & Dong, J. (2017). "Supplier selection using ANP and ELECTRE II in interval 2-tuple linguistic environment", *Information Sciences*, 385–386, 19–38.

Weber, C. A., Current R. J. & Benton W. C. (1991). "Vendor selection criteria and methods", *European Journal Of Operational Research*, 50(2), 2–18.

Wijnmalen D. J .D., (2007). "Analysis of benefits, opportunities, costs, and risks (BOCR) with The AHP–ANP: A critical validation", *Mathematical and Computer Modelling*, 46, 892–905.

Wind, Y., & Saaty, T. L. (1980). "Marketing applications of the analytic hierarchy process", *Management Science*, 26(7), 641–658.

Yan, G. (2009). "Research on green suppliers' evaluation based on AHP & genetic algorithm", *2009 International Conference on Signal Processing Systems, IEEE*, 615–619.

15 A Group Decision-Making Approach Using Fuzzy AHP and Fuzzy VIKOR for Green Supplier Evaluation

Gazi Murat Duman[1], Elif Kongar[2], and Surendra M. Gupta[3]

[1]Trefz School of Business, University of Bridgeport, 230 Park Avenue, Mandeville Hall Room 22B, Bridgeport, CT 06604, USA
[2]Departments of Mechanical Engineering and Technology Management, University of Bridgeport, 221 University Avenue, School of Engineering, 141 Technology Building, Bridgeport, CT 06604, USA
[3]Department of Mechanical and Industrial Engineering, Northeastern University, 334 Snell Engineering Center, 360 Huntington Avenue, Boston, MA 02115, USA

1 INTRODUCTION

Today, the success of supply chain management heavily relies on visibility throughout the supply chain. The ability to respond to sudden, unpredicted changes increases visibility in the supply chain. One effective way to increase visibility is to establish trust, synchronization, and collaboration among multiple parties in the supply chain. To build a reliable and sustainable relationship between the vendor and the buyer, an effective supplier selection is necessary in supply chain management. Furthermore, to maintain a cost-efficient production and achieve a certain quality within the competitive environment of the manufacturing industry, the selection of the right supplier is crucial (Gupta, 2016). Regarding the environmentally-conscious manufacturing concept, comprehensive reviews on the issues were provided by Gungor and Gupta, 1999, Ilgin and Gupta, 2010, Gupta, 2013, and Wang and Gupta, 2012.

Cost and quality are still two major criteria in supplier selection. However, increased concerns regarding the environment—coupled with environmental laws and regulations—require these issues to be addressed as well. Therefore, today's supply chain management includes environmental issues as part of their selection criteria in identifying the ideal supplier(s). This process requires the vendors to follow relevant industry standards such as ISO 14000 and satisfy certain pre-determined conditions.

As mentioned previously, the literature focusing on decision-making models for supplier selection offers several quantitative models (de Boer et al., 2001; Igarashi et al., 2013; Wu and Barnes). Among these, however, ones that focus on green supplier selection are quite limited (Govindan et al., 2015).

One of the important criteria in green supplier selection is green purchasing. Min and Galle, 2001 define green purchasing as the act of environmentally-conscious purchasing that reduces the source of waste while promoting recycling, reclamation, and reuse of purchased materials without adversely affecting overall performance (Igarashi et al., 2013; Min and Galle, 2001). Therefore, decision-makers today are required to embed various environmental criteria into their purchasing decisions.

These environmental supplier selection criteria can be quantitative and/or qualitative in nature (Humphreys et al., 2003). Furthermore, the data for criteria used in the decision-making process is likely to be imprecise, necessitating decision-makers to apply fuzzy methods in various stages of the decision-making process. Moreover, business managers tend to use linguistic terms rather than precise values (Shaw et al., 2012). With this motivation, this paper proposes a hybrid multi-criteria decision-making framework that can handle uncertainty. Theoretical background, problem statement, and the proposed methodology are detailed in the following sections. A case study in a US manufacturing company is provided along with the conclusions and discussion regarding future research.

2   METHODOLOGY

Fuzzy Analytic Hierarchy Process (Fuzzy AHP) is a well-studied and efficient method that has been applied to multi-criteria decision-making problems—particularly to supplier selection processes. One relevant study that has been published was by Mangla et al. In their study, the authors employed Fuzzy AHP to determine the priorities of the identified risks in Green Supply Chain practices of four Indian poly product-manufacturing companies. Wang et al., (2012) utilized a two-stage Fuzzy AHP model to assess the risk of implementing green initiatives. Kannan et al., (2013) combined Fuzzy AHP with multi-objective programming for supplier selection and order allocation in a green supply chain. Lee et al., (2009) applied Fuzzy Extended AHP to accommodate vagueness in expert opinions during the green supplier selection process.

Fuzzy Višekriterijumska Optimizacija i kompromisno Rešenje (Fuzzy VIKOR) is another method that is frequently used in the supplier selection problem. The studies that employ VIKOR—those with green focus—are very limited. Out of these, Datta, Samantra, Mahapatra, Banerjee, and Bandyopadhyay (Datta et al., 2012) utilized an interval-valued fuzzy set coupled with the VIKOR method to evaluate the suppliers' environmental performances. In their study, Kuo et al., 2015 applied a combined approach using the Decision-Making Trial and Evaluation Laboratory (DEMATEL) to use the Analytic Network Process (ANP) (known as

DANP) and VIKOR methods to evaluate the environmental performances of suppliers in the electronics industry. Awasthi and Kannan, 2016 applied Nominal Group Technique (NGT) and Fuzzy VIKOR to determine green supplier development program rankings and recommend the best program(s) for implementation. Wu et al., 2019 developed an integrated approach based on the interval type-2 fuzzy best-worst and extended VIKOR methods to green supplier selection. In their paper, Pérez-Velázquez et al., 2020 proposed a green approach utilizing fuzzy inference and VIKOR in supplier selection for photovoltaic module installation.

This study proposes an integrated hierarchical approach that combines Fuzzy AHP and Fuzzy VIKOR methods, forming a holistic solution methodology for the green supplier selection problem in a fuzzy environment. Fuzzy AHP is utilized to determine the importance weights of Customer Requirements (CRs). Following this, Fuzzy VIKOR is applied to rank provided suppliers. Detailed steps of the proposed approach are provided in Figure 15.1.

To determine the importance weights of the CRs, a Fuzzy AHP approach is applied using Chang's extent analysis (Chang, 1996). The detailed steps of this approach are provided in the following:

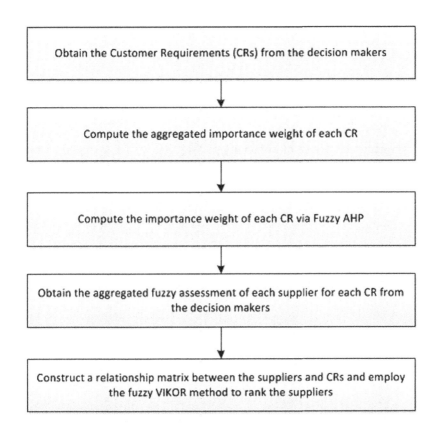

FIGURE 15.1 The Steps of the Proposed Methodology.

i.   Let $X = \{x_1, x_2, x_3, \ldots, x_n\}$' be an object set, whereas $U = \{u_1, u_2, u_3, \ldots, u_m\}$ is a goal set. Each object is taken and extent analysis for each goal is performed, respectively. Therefore, m extent analysis values for each u can be obtained with the following signs: $M_{gi}^1, M_{gi}^2, M_{gi}^3, \ldots, M_{gi}^m$, $i = 1, 2, \ldots, n$, where all the M_{gi}^j, $j = 1, 2, \ldots, m$ are triangular fuzzy numbers. Then, the value of fuzzy synthetic extent with respect to the ith object is defined as:

$$S_i = \sum_{j=1}^{m} M_{gi}^j \otimes \left[\sum_{i=1}^{n} \sum_{j=1}^{m} M_{gi}^j \right]^{-1} \tag{15.1}$$

The first task of the Fuzzy AHP method is to decide the relative importance of each pair of factors in the same hierarchy. By using triangular fuzzy numbers, via pairwise comparison, the fuzzy evaluation matrix $A = (a_{ij})_{nxm}$ is constructed. $\sum_{j=1}^{m} M_{gi}^j$ is obtained by performing the fuzzy addition operation of m extent analysis values for a particular matrix such that:

$$\sum_{j=1}^{m} M_{gi}^j = \left(\sum_{j=1}^{m} l_j, \sum_{j=1}^{m} m_j, \sum_{j=1}^{m} u_j \right), \tag{15.2}$$

and $[\sum_{i=1}^{n} \sum_{j=1}^{m} M_{gi}^j]^{-1}$ is obtained by computing Equation (15.3).

$$\left[\sum_{i=1}^{n} \sum_{j=1}^{m} M_{gi}^j \right]^{-1} = \left(\frac{1}{\sum_{i=1}^{n} \sum_{j=1}^{m} u_i}, \frac{1}{\sum_{i=1}^{n} \sum_{j=1}^{m} m_i}, \frac{1}{\sum_{i=1}^{n} \sum_{j=1}^{m} l_i} \right). \tag{15.3}$$

ii.   Compute the degree of possibility of $S_2 (l_2, m_2, u_2) \geq S_1 (l_1, m_1, u_1)$, where S_2 and S_1 are obtained by Equation (15.1). The degree of possibility between two fuzzy synthetic extents is defined as follows:

$$V (S_2 \geq S_1) = sup_{y \geq x} \left[min \left(\mu_{S_2}(y), \mu_{S_1}(x) \right) \right] \tag{15.4}$$

which can be expressed as:

$$V (S_2 \geq S_1) = hgt (S_1 \cap S_2) = \mu_{S_2}(d) \tag{15.5}$$

$$\mu_{S_2}(d) \begin{cases} 1, & \text{if } m_2 \geq m_1 \\ 0, & \text{if } l_1 \geq u_2 \\ \dfrac{l_1 - u_2}{(m_2 - u_2) - (m_1 - l_1)}, & \text{otherwise} \end{cases}$$

where d is the ordinate of the highest intersection point D between μ_{S1} and μ_{S2} order to compare S_1 and S_2, the values of both $V(S_2 \geq S_1)$ and $V(S_1 \geq S_2)$ are required.

iii. Compute the degree of possibility of a convex fuzzy number to be greater than k convex fuzzy numbers S_i ($i = 1, 2, ..., k$).

$$V(S \geq S_1, S_2, ..., S_k) = V[(S \geq S_1) \text{ and } (S \geq S_2) \text{ and } ... \text{ and } (S \geq S_k)]$$
$$= minV(S \geq S_i) \quad i = 1, 2, ..., k \tag{15.6}$$

iv. Compute the vector W' which is given as follows:

$$W' = (d'(A_1), d'(A_2), ..., d'(A_k))^T \tag{15.7}$$

Assuming that:

$$d'(A_i) = minV(S_i \geq S_j), \quad \text{for } i = 1, 2, ..., k, \ j = 1, 2, ..., k, \ k \neq j \tag{15.8}$$

The normalized vector is indicated by:

$$W = (d(A_1), d(A_2), ..., d(A_k))^T \tag{15.9}$$

where W is a non-fuzzy number calculated for each comparison matrix.

Fuzzy Vikor

i. Assign the linguistic ratings $\tilde{x}_{ij}$ $i = 1, 2, ..., n$ and $j = 1, 2, ..., J$ for alternatives with respect to criteria. The fuzzy linguistic rating ($\tilde{x}_{ij}$) preserves the property that the ranges of normalized triangular fuzzy numbers belong to [0,1]. Let $\tilde{x}_{ij} = (l_{ij}, m_{ij}, u_{ij})$, $\tilde{x}_j^- = (l_j^-, m_j^-, u_j^-)$ and $\tilde{x}_j^+(l_j^+, m_j^+, u_j^+)$

ii. Determine the best rating f_i^+ and the worst rating f_i^- for all criteria. When i is associated with a benefit criterion, then:

$$f_i^+ = max \ \tilde{x}_{ij}, f_i^- = min \ \tilde{x}_{ij},$$

When i is associated with a cost criterion, then:

$$f_i^+ = min \ \tilde{x}_{ij}, f_i^- = max \ \tilde{x}_{ij}, \tag{15.10}$$

iii. Calculate the index S_j, which refers to the separation measure of the jth alternative with the best value. Also, calculate the index R_j, which refers to the separation measure for the jth alternative with the worst value. W_i is the weight of the ith criterion.

$$S_j = \sum_{i=1}^{n} W_i \frac{f_i^+ - x_{ij}}{f_i^+ - f_{ij}}, \quad S_j \in [0, 1]$$

$$R_j = max_i \left[W_i \frac{f_i^+ - x_{ij}}{f_i^+ - f_{ij}} \right], \quad R_j \in [0, 1] \tag{15.11}$$

The solution obtained using R_j has a minimum individual regret function, while the solution obtained using S_j has the maximum group utility.

iv. Compute the Q_j values using the following equation.

$$Q_j = v\frac{(S_j - S^+)}{(S^- - S^+)} + (1 - v)\frac{(R_j - R^+)}{(R^- - R^+)} \tag{15.12}$$

where
$S^+ = max_j S_j, \quad S^- = min_j S_j$
$R^+ = max_j R_j, \quad R^- = min_j R_j$, and v is the strategy weight for the maximization group utility, while $1 - v$ indicates the individual weight regret function. The compromise can be selected with "voting by majority" ($v > 0.5$), with "consensus" ($v = 0.5$), or with "veto" ($v < 0.5$).

v. Rank the alternatives by sorting each S, R, and Q values in increasing order. The result is a set of three ranking lists denoted as $S_{[k]}$, $R_{[k]}$, and $Q_{[k]}$.

vi. Propose the alternative j_1 corresponding to $Q_{[1]}$ (the smallest among Q_j values) as a compromise solution if:

Condition 1: The alternative j_1 has an acceptable advantage, meaning $Q_{[2]} - Q_{[1]} \geq DQ$ where $DQ = 1/(m - 1)$ and m is the number of alternatives.

Condition 2: The alternative j_1 is stable within the decision-making process, meaning it is also the best ranked in $S_{[k]}$ or $R_{[k]}$

If one of the conditions above is not satisfied, then a set of compromised solutions is proposed, which consists of:

- Alternatives j_1 and j_2 where Q_{j2} or $Q_{[2]}$ only if Condition 2 is not satisfied, or
- Alternatives $j_1, j_2, ..., j_k$ if Condition 1 is not satisfied and j_k is determined by the relation $Q_{[k]} - Q_{[1]} < DQ$ for the maximum k, where $Q_{jk} = Q_{[k]}$ (the positions of these alternatives are in closeness).

3 CASE STUDY

The case study is conducted in a leading global manufacturer and distributor of dispensing systems for beauty and personal care on top of home and consumer healthcare needs. The US-based manufacturing company uses plastics as one of its major raw materials. The study focused on evaluating and selecting supplier(s) of plastics used in the plastic injection molding of innovative dispensing pumps. To select the best alternative, 13 potential suppliers were evaluated according to their performance using eight decision criteria. The evaluation was conducted based on the linguistic judgments of three different expert decision-makers who are responsible for the operations, purchasing, and quality control departments in the company. The criteria were defined by the experts are as follows:

- Cost: A measure related to the order and purchasing cost of the raw material for manufacturing. The value also includes different variations of discount options that suppliers could provide regarding the duration of the agreement and the amount of each order.
- Quality: A measure related to the quality of the raw material obtained from the suppliers and its performance in molding and injection processes.
- Logistics Operations: A measure related to the service level of transportation and warehousing, on-time, and full delivery performance operations of the suppliers. The distances to each supplier are also considered in this criterion.
- Service Level: A measure related to the response time to customer requests, flexibility, and past customer service performance of each supplier.
- Financial Position: A measure related to the financial strengths of each supplier for continuous business.
- Organization: A measure related to the company size of the supplier, business existence, and continuity in the industry, in addition to their reputation among the peers.
- Continuous Improvement: A measure related to the continuous improvement in product capability of suppliers, R&D activities, and operational efficiencies.
- Green Image: A measure related to the energy, waste, and natural resources reduction performance of suppliers, in addition to respective CO_2 emission levels, workforce wellness, community services, and responsibility for future generations.

Given the set of above criteria, the decision-makers are asked to rank their preference levels for each criterion. Following this, to prioritize the green image in the

TABLE 15.1
Comparative Linguistic Scale for Ratings of Alternatives and Weights of Criteria

Linguistic Terms	Triangular Fuzzy Number (TFN)
Just equal (EQ)	(1,1,1)
Weak importance of one over another (WI)	(1,1,3)
Fairly preferable (FP)	(1,3,5)
Essential importance of one over another (EI)	(3,5,7)
Strongly preferable (SP)	(5,7,9)
Absolutely preferable (AP)	(7,9,9)

If criteria i has one of the values above assigned in pairwise comparison with criteria j, then criteria j has the reciprocal value when it compared with criteria i.

TABLE 15.2

Pairwise Comparison of Criteria Collected from Decision Makers via Linguistics Terms

	Cost			Quality			Service Level			Logistics Operations		
	DM 1	DM 2	DM 3	DM 1	DM 2	DM 3	DM 1	DM 2	DM 3	DM 1	DM 2	DM 3
Cost	EQ	EQ	EQ	WI	WI	1/WI	FP	FP	FP	FP	EI	EI
Quality	1/WI	1/WI	WI	EQ	EQ	EQ	FP	FP	EI	EI	EI	SP
Service level	1/FP	1/FP	1/FP	1/FP	1/FP	1/EI	EQ	EQ	EQ	1/WI	EQ	1/FP
Logistics operations	1/FP	1/EI	1/EI	1/EI	1/EI	1/SP	WI	EQ	FP	EQ	EQ	EQ
Financial position	1/EI	1/EI	1/SP	1/EI	1/SP	1/SP	1/WI	1/FP	1/FP	1/FP	1/FP	1/WI
Organization	1/AP	1/SP	1/EI	1/SP	1/EI	1/EI	1/FP	1/FP	1/FP	WI	FP	EI
Continuous improvement	1/SP	1/SP	1/EI	1/SP	1/SP	1/SP	1/FP	1/FP	1/FP	1/SP	1/EI	1/EI
Green image	WI	WI	FP	WI	FP	FP	EI	EI	SP	EI	FP	SP

	Financial Position			Organization			Continuous Improvement			Green Image		
	DM 1	DM 2	DM 3	DM 1	DM 2	DM 3	DM 1	DM 2	DM 3	DM 1	DM 2	DM 3
Cost	EI	EI	SP	AP	SP	EI	SP	SP	EI	1/WI	1/WI	1/FP
Quality	EI	SP	SP	SP	EI	EI	SP	SP	SP	1/WI	1/FP	1/FP
Service level	WI	FP	FP	FP	FP	FP	FP	FP	FP	1/EI	1/EI	1/SP
Logistics operations	FP	FP	WI	1/WI	1/FP	1/EI	SP	EI	EI	1/EI	1/FP	1/SP
Financial position	EQ	EQ	EQ	EI	FP	FP	1/EI	1/EI	1/FP	1/SP	1/EI	1/SP
Organization	1/EI	1/FP	1/FP	EQ	EQ	EQ	1/EI	1/EI	1/EI	1/SP	1/SP	1/AP
Continuous improvement	EI	EI	FP	EI	EI	EI	EQ	EQ	EQ	1/SP	1/EI	1/SP
Green image	SP	EI	EI	EI	EI	SP	EI	EI	SP	EQ	EQ	EQ

supplier evaluation process, a Fuzzy AHP approach is employed. The evaluation scale used in Fuzzy AHP is provided in Table 15.1.

The data obtained from the decision-makers for pairwise comparison is provided in Table 15.2.

To obtain the aggregated assessment of each decision-maker, Equation (15.13) is constructed. Here, let $\tilde{x}_j^r = (l_{xj}^r, m_{xj}^r, u_{xj}^r)$ describe a fuzzy number assigned to j^{th} criterion by r^{th} decision-maker ($r = 1,2, ..., k$), then the aggregated assessment given by k decision-makers can be expressed as:

$$\tilde{x}_j = \frac{1}{k}(\tilde{x}_j^1 + \tilde{x}_j^2 + \cdots + \tilde{x}_j^k)$$
(15.13)

TABLE 15.3
The Aggregated results of Pairwise Comparison Collected from Decision Makers via TFN

	Cost	Quality	Service Level	Logistics Operations
Cost	(1,1,1)	(0.78,1,2.34)	(1,2.34,4.34)	(2.34,4.34,6.34)
Quality	(0.56,1,1.67)	(1,1,1)	(1.67,3.67,5.67)	(3.67,5.67,7.67)
Service level	(0.25,0.56,1)	(0.19,0.29,0.78)	(1,1,1)	(0.52,0.78,1)
Logistics operations	(0.17,0.25,0.56)	(0.14,0.19,0.29)	(1,1.67,3)	(1,1,1)
Financial position	(0.14,0.19,0.29)	(0.13,0.17,0.25)	(0.25,0.56,1)	(0.25,0.56,1)
Organization	(0.13,0.16,0.23)	(0.14,0.19,0.29)	(0.2,0.34,1)	(1.67,3,5)
Continuous improvement	(0.13,0.17,0.25)	(0.12,0.15,0.2)	(0.2,0.34,1)	(0.14,0.19,0.29)
Green image	(1,1.67,3.67)	(1,2.34,4.34)	(3.67,5.67,7.67)	(3,5,7)

	Financial Position	Organization	Continuous Improvement	Green Image
Cost	(3.67,5.67,7.67)	(5,7,8.34)	(4.34,6.34,8.34)	(0.29,0.78,1)
Quality	(4.34,6.34,8.34)	(3.67,5.67,7.67)	(5,7,9)	(0.25,0.56,1)
Service level	(1,2.34,4.34)	(1,3,5)	(1,3,5)	(0.14,0.19,0.29)
Logistics operations	(1,2.34,4.34)	(0.23,0.52,0.78)	(3.67,5.67,7.67)	(0.16,0.23,0.52)
Financial position	(1,1,1)	(1.67,3.67,5.67)	(0.17,0.25,0.56)	(0.14,0.19,0.29)
Organization	(0.19,0.29,0.78)	(1,1,1)	(0.15,0.2,0.34)	(0.14,0.19,0.29)
Continuous improvement	(2.34,4.34,6.34)	(3,5,7)	(1,1,1)	(0.14,0.19,0.29)
Green image	(3.67,5.67,7.67)	(3.67,5.67,7.67)	(3.67,5.67,7.67)	(1,1,1)

The aggregated results of pairwise comparison after using TFN are given in Table 15.3. The values of the fuzzy synthetic extent for the criteria, calculated via Equation (15.1) through Equation (15.5), are given in Table 15.4.

The degree of possibilities of the fuzzy values above and corresponding normalized weight vectors can easily be obtained via Equations (15.6) through (15.9). The results are presented in Table 15.5.

TABLE 15.4
The Fuzzy Synthetic Extent Values for the Criteria

S_{Cost}	(0.093, 0.206, 0.458)
$S_{Quality}$	(0.101, 0.224, 0.489)
$S_{Service\ Level}$	(0.026, 0.081, 0.215)
$S_{Logistics\ Operations}$	(0.037, 0.086, 0.212)
$S_{Financial\ Position}$	(0.019, 0.048, 0.117)
$S_{Organization}$	(0.018, 0.039, 0.104)
$S_{Continuous\ Improvement}$	(0.036, 0.083, 0.191)
$S_{Green\ Image}$	(0.104, 0.237, 0.544)

TABLE 15.5
The Degree of Possibilities and Weight Vectors of the Criteria

Criteria	The Degree of Possibility	Weight
Cost	$V (S_{Cost} \geq S_{Quality},\ \ldots, S_{Green\ Image}) = 0.92062$	0.221894
Quality	$V (S_{Quality} \geq S_{Cost}, \ldots, S_{Green\ Image}) = 0.96769$	0.233240
Service level	$V (S_{Service\ Level} \geq S_{Cost}, \ldots, S_{Green\ Image}) = 0.41546$	0.100138
Logistics operations	$V (S_{Logistics\ Ops.} \geq S_{Cost}, \ldots, S_{Green\ Image}) = 0.41649$	0.100387
Financial position	$V (S_{Financial\ Pos.} \geq S_{Cost}, \ldots, S_{Green\ Image}) = 0.06663$	0.016061
Organization	$V (S_{Organization} \geq S_{Cost}, \ldots, S_{Green\ Image}) = 0.00172$	0.000416
Continuous improvement	$V (S_{Cont.\ Imp} \geq S_{Cost}, \ldots, S_{Green\ Image}) = 0.36028$	0.086838
Green image	$V (S_{Green\ Image} \geq S_{Cost}, \ldots, S_{Cont.\ Imp.}) = 1$	0.241026

TABLE 15.6
Linguistic Scale to Evaluate the Ratings of the Suppliers

Linguistic Terms	Triangular Fuzzy Number (TFN)
Very low (VL)	(0,1,2)
Low (L)	(2,3,4)
Medium (M)	(4,5,6)
High (H)	(6,7,8)
Very high (VH)	(8,9,10)

TABLE 15.7

Linguistic Ratings of Each Supplier with Respect to Each Criteria

	Cost			Quality			Logistics Operations			Financial Position		
	DM 1	DM 2	DM 3	DM 1	DM 2	DM 3	DM 1	DM 2	DM 3	DM 1	DM 2	DM 3
Supplier 1	M	H	M	H	M	H	H	M	M	M	L	M
Supplier 2	VH	VH	VH	VH	VH	VH	M	H	H	L	M	H
Supplier 3	H	M	M	VH	H	VH	M	L	M	M	M	H
Supplier 4	H	H	M	VH	H	H	M	H	H	VL	L	L
Supplier 5	VH	VH	VH	M	M	H	H	VH	VH	H	M	M
Supplier 6	VH	H	VH	H	H	M	VH	H	H	M	M	M
Supplier 7	M	VH	VH	H	H	H	VH	VH	H	M	H	H
Supplier 8	VH	H	H	M	H	H	H	VH	VH	H	VH	VH
Supplier 9	VH	VH	VH	H	VH	VH	H	H	H	H	VH	H
Supplier 10	VH	VH	H	M	H	H	VH	H	VH	H	M	M
Supplier 11	H	H	H	H	VH	VH	M	M	H	H	M	H
Supplier 12	H	VH	VH	VH	VH	H	H	VH	H	L	VL	L
Supplier 13	M	H	H	M	H	H	H	H	VH	M	H	VH

	Service Level			Organization			Continuous Improvement			Green Image		
	DM 1	DM 2	DM 3	DM 1	DM 2	DM 3	DM 1	DM 2	DM 3	DM 1	DM 2	DM 3
Supplier 1	M	M	H	H	M	M	H	M	H	M	M	H
Supplier 2	H	H	H	H	VH	VH	VH	VH	H	H	H	H
Supplier 3	H	M	H	M	H	H	H	H	VH	H	H	VH
Supplier 4	VH	H	H	VH	VH	H	M	H	M	H	VH	H
Supplier 5	VH	VH	H	H	H	VH	M	H	H	M	H	H

(Continued)

TABLE 15.7 (Continued)

Supplier 6	L	M	M	H	M	H	M	L	M	M	L	VL
Supplier 7	H	VH	H	H	VH	H	M	L	L	M	L	M
Supplier 8	VH	H	H	H	M	VH	H	M	M	H	M	M
Supplier 9	L	M	M	M	H	H	VL	L	VL	VL	L	M
Supplier 10	L	L	VL	H	M	L	M	H	H	M	H	H
Supplier 11	L	M	M	M	M	L	H	M	H	H	VH	H
Supplier 12	VH	H	H	H	H	M	H	H	H	M	H	M
Supplier 13	VH	H	VH	H	H	VH	H	M	H	H	M	M

The weight vector represents the importance degree of each criterion, which is used as an input in the Fuzzy VIKOR method to evaluate the suppliers. Additional inputs are the fuzzy evaluations of the suppliers with respect to the criteria. These are obtained from the decision-makers in the company using linguistic terms and their corresponding TFN as given in Table 15.6.

The data collected from the decision-makers regarding the assessment of each supplier is provided in Table 15.7.

Equation (15.13) is employed again to aggregate the fuzzy ratings of each supplier from decision-makers. The results are presented in Table 15.8.

TABLE 15.8

Fuzzy Numbers for the Aggregated Ratings of Each Supplier

	Cost	Quality	Logistics Operations	Financial Position
Supplier 1	(4.67,5.67,6.67)	(5.33,6.33,7.33)	(4.67,5.67,6.67)	(3.33,4.33,5.33)
Supplier 2	(8,9,10)	(8,9,10)	(5.33,6.33,7.33)	(4,5,6)
Supplier 3	(4.67,5.67,6.67)	(7.33,8.33,9.33)	(3.33,4.33,5.33)	(4.67,5.67,6.67)
Supplier 4	(5.33,6.33,7.33)	(6.67,7.67,8.67)	(5.33,6.33,7.33)	(1.33,2.33,3.33)
Supplier 5	(8,9,10)	(5.33,6.33,7.33)	(7.33,8.33,9.33)	(4.67,5.67,6.67)
Supplier 6	(7.33,8.33,9.33)	(4.67,5.67,6.67)	(6.67,7.67,8.67)	(4,5,6)
Supplier 7	(6.67,7.67,8.67)	(6,7,8)	(6.67,7.67,8.67)	(5.33,6.33,7.33)
Supplier 8	(6.67,7.67,8.67)	(5.33,6.33,7.33)	(7.33,8.33,9.33)	(7.33,8.33,9.33)
Supplier 9	(8,9,10)	(7.33,8.33,9.33)	(6,7,8)	(6.67,7.67,8.67)
Supplier 10	(7.33,8.33,9.33)	(5.33,6.33,7.33)	(7.33,8.33,9.33)	(4.67,5.67,6.67)
Supplier 11	(6,7,8)	(7.33,8.33,9.33)	(4.67,5.67,6.67)	(5.33,6.33,7.33)
Supplier 12	(7.33,8.33,9.33)	(7.33,8.33,9.33)	(6.67,7.67,8.67)	(1.33,2.33,3.33)
Supplier 13	(5.33,6.33,7.33)	(5.33,6.33,7.33)	(6.67,7.67,8.67)	(5.33,6.33,7.33)
	Service Level	**Organization**	**Continuous Improvement**	**Green Image**
Supplier 1	(4.67,5.67,6.67)	(4.67,5.67,6.67)	(5.33,6.33,7.33)	(4.67,5.67,6.67)
Supplier 2	(6,7,8)	(7.33,8.33,9.33)	(7.33,8.33,9.33)	(6,7,8)
Supplier 3	(5.33,6.33,7.33)	(5.33,6.33,7.33)	(6.67,7.67,8.67)	(6.67,7.67,8.67)
Supplier 4	(6.67,7.67,8.67)	(7.33,8.33,9.33)	(4.67,5.67,6.67)	(6.67,7.67,8.67)
Supplier 5	(7.33,8.33,9.33)	(6.67,7.67,8.67)	(5.33,6.33,7.33)	(5.33,6.33,7.33)
Supplier 6	(3.33,4.33,5.33)	(5.33,6.33,7.33)	(3.33,4.33,5.33)	(2,3,4)
Supplier 7	(6.67,7.67,8.67)	(6.67,7.67,8.67)	(2.67,3.67,4.67)	(3.33,4.33,5.33)
Supplier 8	(6.67,7.67,8.67)	(6,7,8)	(4.67,5.67,6.67)	(4.67,5.67,6.67)
Supplier 9	(3.33,4.33,5.33)	(5.33,6.33,7.33)	(0.67,1.67,2.67)	(2,3,4)
Supplier 10	(1.33,2.33,3.33)	(4,5,6)	(5.33,6.33,7.33)	(5.33,6.33,7.33)
Supplier 11	(3.33,4.33,5.33)	(3.33,4.33,5.33)	(5.33,6.33,7.33)	(6.67,7.67,8.67)
Supplier 12	(6.67,7.67,8.67)	(5.33,6.33,7.33)	(5.33,6.33,7.33)	(5.33,6.33,7.33)
Supplier 13	(7.33,8.33,9.33)	(6.67,7.67,8.67)	(5.33,6.33,7.33)	(4.67,5.67,6.67)

TABLE 15.9

The Weighted Normalized Fuzzy Decision Matrix

	Cost	Quality	Logistics Operations	Financial Position
Supplier 1	(0.222,0.222,0.222)	(0.155,0.164,0.171)	(0.064,0.068,0.072)	(0.007,0.008,0.009)
Supplier 2	(0.129,0.14,0.148)	(0.233,0.233,0.233)	(0.073,0.076,0.079)	(0.009,0.01,0.01)
Supplier 3	(0.222,0.222,0.222)	(0.214,0.216,0.218)	(0.046,0.052,0.057)	(0.01,0.011,0.011)
Supplier 4	(0.194,0.199,0.202)	(0.194,0.199,0.202)	(0.073,0.076,0.079)	(0.003,0.004,0.006)
Supplier 5	(0.129,0.14,0.148)	(0.155,0.164,0.171)	(0.1,0.1,0.1)	(0.01,0.011,0.011)
Supplier 6	(0.141,0.151,0.158)	(0.136,0.147,0.155)	(0.091,0.092,0.093)	(0.009,0.01,0.01)
Supplier 7	(0.155,0.164,0.171)	(0.175,0.181,0.187)	(0.091,0.092,0.093)	(0.012,0.012,0.013)
Supplier 8	(0.155,0.164,0.171)	(0.155,0.164,0.171)	(0.1,0.1,0.1)	(0.016,0.016,0.016)
Supplier 9	(0.129,0.14,0.148)	(0.214,0.216,0.218)	(0.082,0.084,0.086)	(0.015,0.015,0.015)
Supplier 10	(0.141,0.151,0.158)	(0.155,0.164,0.171)	(0.1,0.1,0.1)	(0.01,0.011,0.011)
Supplier 11	(0.173,0.18,0.185)	(0.214,0.216,0.218)	(0.064,0.068,0.072)	(0.012,0.012,0.013)
Supplier 12	(0.141,0.151,0.158)	(0.214,0.216,0.218)	(0.091,0.092,0.093)	(0.003,0.004,0.006)
Supplier 13	(0.194,0.199,0.202)	(0.155,0.164,0.171)	(0.091,0.092,0.093)	(0.012,0.012,0.013)

	Service Level	Organization	Continuous Improvement	Green Image
Supplier 1	(0.064,0.068,0.072)	(0.064,0.068,0.072)	(0.063,0.066,0.068)	(0.169,0.178,0.185)
Supplier 2	(0.082,0.084,0.086)	(0.1,0.1,0.1)	(0.087,0.087,0.087)	(0.217,0.22,0.222)
Supplier 3	(0.073,0.076,0.079)	(0.073,0.076,0.079)	(0.079,0.08,0.081)	(0.241,0.241,0.241)
Supplier 4	(0.091,0.092,0.093)	(0.1,0.1,0.1)	(0.055,0.059,0.062)	(0.241,0.241,0.241)
Supplier 5	(0.1,0.1,0.1)	(0.091,0.092,0.093)	(0.063,0.066,0.068)	(0.193,0.199,0.204)
Supplier 6	(0.046,0.052,0.057)	(0.073,0.076,0.079)	(0.039,0.045,0.05)	(0.072,0.094,0.111)
Supplier 7	(0.091,0.092,0.093)	(0.091,0.092,0.093)	(0.032,0.038,0.043)	(0.121,0.136,0.148)
Supplier 8	(0.091,0.092,0.093)	(0.082,0.084,0.086)	(0.055,0.059,0.062)	(0.169,0.178,0.185)
Supplier 9	(0.046,0.052,0.057)	(0.073,0.076,0.079)	(0.008,0.017,0.025)	(0.072,0.094,0.111)
Supplier 10	(0.018,0.028,0.036)	(0.055,0.06,0.064)	(0.063,0.066,0.068)	(0.193,0.199,0.204)
Supplier 11	(0.046,0.052,0.057)	(0.046,0.052,0.057)	(0.063,0.066,0.068)	(0.241,0.241,0.241)
Supplier 12	(0.091,0.092,0.093)	(0.073,0.076,0.079)	(0.063,0.066,0.068)	(0.193,0.199,0.204)
Supplier 13	(0.1,0.1,0.1)	(0.091,0.092,0.093)	(0.063,0.066,0.068)	(0.169,0.178,0.185)

TABLE 15.10

Ranking of Suppliers

Suppliers	S_j	Ranking	R_j	Ranking	Q_j	Ranking
Supplier 1	0.3092	8	0.1166	8	0.2085	8
Supplier 2	0.2407	4	0.1387	12	0.2227	9
Supplier 3	0.1402	1	0.0723	2	0.0319	1
Supplier 4	0.2228	3	0.0904	4	0.1106	3

(Continued)

TABLE 15.10 (Continued)

Suppliers	S_j	Ranking	R_j	Ranking	Q_j	Ranking
Supplier 5	0.3571	10	0.1387	12	0.2788	11
Supplier 6	0.5266	13	0.1458	13	0.3757	13
Supplier 7	0.3756	11	0.0875	3	0.1779	4
Supplier 8	0.3003	7	0.1166	8	0.2042	6
Supplier 9	0.4421	12	0.1387	12	0.3198	12
Supplier 10	0.3416	9	0.1166	8	0.2240	10
Supplier 11	0.1826	2	0.0684	1	0.0441	2
Supplier 12	0.2962	6	0.1184	9	0.2061	7
Supplier 13	0.2730	5	0.1166	8	0.1910	5

The normalized fuzzy decision matrix is obtained using Equation (15.10). In our model, the first criterion is defined as the cost criteria and the others are defined as the benefit criteria. The weighted normalized decision matrix can be obtained via Equation (15.11), multiplying the normalized decision matrix with the weights of the criteria which are computed through the Fuzzy AHP approach as presented in Table 15.9.

S_j, R_j, and Q_j values are computed by selecting "consensus" ($v = 0.5$) and the results are presented in Table 15.10.

Given these results, we observed that Condition 1 is not satisfied ($Q_{[2]}-Q_{[1]} \geq DQ$, $DQ = 0.083$), whereas Condition 2 is. Both Supplier 3 and 11 are compromised solutions. It is suggested that compromise candidate Supplier 3 is the best alternative, owing to its closeness to the best candidate.

4 CONCLUSION

Fuzzy AHP and Fuzzy VIKOR methods are both well-studied for the supplier selection problem. However, these methods are rarely applied jointly in supplier evaluation and selection models with a green focus. Both models suffer from various limitations. A hybrid model utilizing both, however, highlights the advantages of each while lessening their respective shortcomings. For instance, Fuzzy AHP performs well in pairwise comparison. On the other hand, many alternatives and decision-makers in the Fuzzy AHP model requires large numbers of pairwise comparisons. Therefore, when the agility of this process is concerned, the Fuzzy VIKOR approach performs superior.

Taking advantage of both methods, this study proposed a real-world case study for supplier evaluation and selection. The assessment measures and related ratings—with respect to the main criteria—are directly obtained from the experts using linguistic terms. In the future, the data set can be expanded to include quantitative measures for additional technical sub-criteria.

REFERENCES

Awasthi, A., and Kannan, G.: 'Green supplier development program selection using NGT and VIKOR under fuzzy environment', *Computers & Industrial Engineering*, 2016, 91, pp. 100–108.

Chang, D.-Y.: 'Applications of the extent analysis method on fuzzy AHP', *European Journal of Operational Research*, 1996, 95, (3), pp. 649–655.

Datta, S., Samantra, C., Mahapatra, S. S., Banerjee, S., and Bandyopadhyay, A.: 'Green supplier evaluation and selection using VIKOR method embedded in fuzzy expert system with interval-valued fuzzy numbers', *International Journal of Procurement Management*, 2012, 5, (5), pp. 647–678.

de Boer, L., Labro, E., and Morlacchi, P.: 'A review of methods supporting supplier selection', *European Journal of Purchasing & Supply Management*, 2001, 7, (2), pp. 75–89.

Govindan, K., Rajendran, S., Sarkis, J., and Murugesan, P.: 'Multi criteria decision making approaches for green supplier evaluation and selection: A literature review', *Journal of Cleaner Production*, 2015, 98, pp. 66–83.

Gupta, S. M.: 'Lean manufacturing, green manufacturing and sustainability', *Journal of Japan Industrial Management Association*, 2016, 67, (2E), pp. 102–105.

Gupta, S. M.: *Reverse Supply Chains: Issues and Analysis* (CRC Press, 2013).

Gungor, A., and Gupta, S. M.: 'Issues in environmentally conscious manufacturing and product recovery: a survey', *Computers & Industrial Engineering*, 1999, 36, (4), pp. 811–853.

Humphreys, P. K., Wong, Y. K., and Chan, F. T. S.: 'Integrating environmental criteria into the supplier selection process', *Journal of Materials Processing Technology*, 2003, 138, (1–3), pp. 349–356.

Igarashi, M., de Boer, L., and Fet, A. M.: 'What is required for greener supplier selection? A literature review and conceptual model development', *Journal of Purchasing and Supply Management*, 2013, 19, (4), pp. 247–263.

Ilgin, M. A., and Gupta, S. M.: 'Environmentally conscious manufacturing and product recovery (ECMPRO): A review of the state of the art', *Journal of Environmental Management*, 2010, 91, (3), pp. 563–591.

Kannan, D., Khodaverdi, R., Olfat, L., Jafarian, A., and Diabat, A.: 'Integrated fuzzy multi criteria decision making method and multi-objective programming approach for supplier selection and order allocation in a green supply chain', *Journal of Cleaner Production*, 2013, 47, pp. 355–367.

Kuo, T. C., Hsu, C.-W., and Li, J.-Y.: 'Developing a green supplier selection model by using the DANP with VIKOR', *Sustainability*, 2015, 7, (2), pp. 1661–1689.

Lee, A. H. I., Kang, H.-Y., Hsu, C.-F., and Hung, H.-C.: 'A green supplier selection model for high-tech industry', *Expert Systems with Applications*, 2009, 36, (4), pp. 7917–7927.

Mangla, S. K., Kumar, P., and Barua, M. K.: 'Risk analysis in green supply chain using fuzzy AHP approach: A case study', *Resources, Conservation and Recycling*, 2015, 104, pp. 375-390.

Min, H., & Galle, W. P.: 'Green purchasing practices of US firms', *International Journal of Operations & Production Management*, 2001, 21(9), pp. 1222–1238.

Pérez-Velázquez, A., Oro-Carralero, L. L., and Moya-Rodríguez, J. L.: 'Supplier selection for photovoltaic module installation utilizing fuzzy inference and the VIKOR method: a green approach', *Sustainability*, 2020, 12, (6), pp. 2242

Shaw, K., Shankar, R., Yadav, S. S., and Thakur, L. S.: 'Supplier selection using fuzzy AHP and fuzzy multi-objective linear programming for developing low carbon supply chain', *Expert Systems with Applications*, 2012, 39, (9), pp. 8182–8192.

Wang, H.-F., and Gupta, S. M.: *Green Supply Chain Management: Product Life Cycle Approach*, McGraw-Hill Education, NewYork (2012).

Wang, X., Chan, H. K., Yee, R. W. Y., and Diaz-Rainey, I.: 'A two-stage fuzzy-AHP model for risk assessment of implementing green initiatives in the fashion supply chain', *International Journal of Production Economics*, 2012, 135, (2), pp. 595–606.

Wu, C., and Barnes, D.: 'An integrated model for green partner selection and supply chain construction', *Journal of Cleaner Production*, 2016, 112, pp. 2114–2132.

Wu, Q., Zhou, L., Chen, Y., and Chen, H.: 'An integrated approach to green supplier selection based on the interval type-2 fuzzy best-worst and extended VIKOR methods', *Information Sciences*, 2019, 502, pp. 394–417.

16 An Improved Extension of Weighted Hierarchical Fuzzy Axiomatic Design

Sustainable Route Selection Problem in Intermodal Transportation Networks

Kemal Subulan and Adil Baykasoğlu
Faculty of Engineering, Department of Industrial
Engineering, Dokuz Eylül University, Buca, Izmir, 35397,
Turkey

1 INTRODUCTION

In recent years, sustainable route selection problems in intermodal transportation systems have received increasing attention from both researchers and practitioners in the global logistics sector. On the other hand, most of the previous research in literature only focused on single-mode route selection problems that generally involve road freight transport options. Actually, sustainable route selection problems in intermodal transportation systems are much more complicated than unimodal, since the computational efficiency required to select the best path will considerably deteriorate by increasing the number of traffic and transportation modes, hubs, terminals, etc. (Qu and Chen 2008). Moreover, multimodal logistics networks are commonly characterized by dynamically-changing conditions and multiple modes of simultaneous transportation operations. Furthermore, most global logistics companies have several conflicting and non-commensurable goals/criteria, such as the minimization of total costs, transit times, risks/damages and environmental effects, and the maximization of reliability, safety, and social benefits/contributions, etc. in selecting the best transport route options.

For the aforementioned reasons, a set of non-dominated or non-inferior solutions may arise instead of a single optimal solution. In such a case, decision analysts should select the most sustainable way of transportation that is also the best compromise. Therefore, the route selection in multimodal/intermodal transportation systems become a complex multiple criteria decision-making (MCDM) problem for road users and intermodal operators. However, there is a lack of studies in literature focused on quantitative modeling of social indicators, along with the economic and environmental criteria in route selection problems. Consequently, sustainable route selection in intermodal transportation systems is still a hot research topic for logistics and transportation professionals.

Based on these motivations, this research presents an improved extension of the weighted hierarchical fuzzy axiomatic design (WFAD) approach to select the best transport route in an intermodal logistics network by considering several sustainability factors. In fact, the main aim of this research is to provide a decision aid for the intermodal routing operators to evaluate different transport route options and sustainability factors in a more complete and systematic manner, while considering the economic, environmental, social, and risky aspects of the problem. The reason for applying the AD methodology and its weighted fuzzy extension as an MCDM approach is that it can consider the decision-maker's minimum and maximum requirements and, therefore, provide a further comparison of alternatives in their level of competency to satisfy these **requirements** (Shahbandarzadeh et al. 2011). It should also be highlighted that AD methodology enables intermodal operators to assess various transport route options with respect to several qualitative and quantitative criteria, simultaneously. Additionally, some of the sustainability criteria may not be valid for a few transport route options in an intermodal transportation system and, therefore, they may not involve any information related to these functional requirements (FRs) or criteria. In that situation, the information contents of such transport routes can be expressed as unavailable for that FR and will take a zero value because there will be no information needed for FR for this type of transport route options. This is an important property of AD methodology that makes it a more appropriate MCDM method for intermodal route selection problems; some transport modes may not include any information on the sustainability criteria.

In this research, conventional crisp AD methodology is extended by incorporating both negative- and positive-valued information contents that correspond to over and under achievements on the FRs, respectively. In addition, the classical WFAD method is modified by considering the criteria hierarchy, and new formulations are proposed to calculate the weighted information contents under different circumstances. Finally, an evaluation procedure is also presented to deal with the negative- and positive-valued information contents while ranking the alternative transport routes. To show the practicality and applicability of the proposed extensions, a real-life application of an international logistics company is presented to select the road, marine, and railway transport routes from Turkey to Europe. To do this, a comprehensive questionnaire survey is first performed with a group of company experts to specify the FR/criteria, design parameters, and system ranges of the alternative transport routes.

Then, the proposed improved extensions of crisp and fuzzy AD methods are applied, and the results are reported. Afterward, the importance weights of the sustainability criteria are also linguistically evaluated by company experts—the financial advisor, drayage, network, terminal, and intermodal operators of the logistics company. After applying the normalization and defuzzification processes on the hierarchical fuzzy weights, the WFAD method is applied with the proposed extensions, and results are compared to each other. According to the computational results, economic, timing, and reliability factors are specified as the most essential sustainability criteria while selecting the best transport route option on the intermodal logistics network. It is also figured out that apart from the direct road or combined marine and road freight transport options, intermodal transportation may provide the most sustainable way of freight transportation to the logistics company. Additionally, Trieste seaport, Munich, and Cologne railway stations should be the most frequently used terminals/hubs on these best intermodal routes.

The rest of this chapter is organized as follows: a brief literature survey on route selection problems in single and multimodal transportation systems is presented in Section 16.2. Some preliminaries and fundamentals of the crisp, fuzzy, and weighted AD methodology are given in Section 16.3. Furthermore, some shortcomings of the conventional AD methodology are discussed in that section and then, the improved extensions are proposed. In Section 16.4, a real-life application in an international logistics company is presented to select the most sustainable way of road, marine, and railway transport operations from Turkey to Europe.

2 A BRIEF LITERATURE REVIEW ON SUSTAINABLE ROUTE SELECTION PROBLEM

An overview of the recent articles where sustainable route selection problems and applications in single and multimodal transportation systems were studied can be summarized as follows:

Qu and Chen (2008) proposed a hybrid MDCDM method that combines the fuzzy analytical hierarchy process (AHP) and artificial neural network (ANN) theory for the route selection problem of multimodal transportation networks. To handle the relationship between the criteria and the performance of the alternatives, they applied an improved ANN with error backpropagation. Sawadogo and Anciaux (2009) employed an approach based on the ELECTRE method for the multimodal route selection problem within a green supply chain network. They considered the economic, environmental, and social criteria (i.e. noise pollution and social costs related to accidents) to help decision-makers choose the best route in an intermodal transportation system. An integrated GIS and MCDM approach was proposed by Farkas (2009) for the sustainable route/site selection problem of a metro-rail network. A hierarchical decision tree model was also presented to incorporate economic, institutional, and social perspectives as well as environmental objectives. A decision support system (DSS) based on a comprehensive AHP model was developed by Özceylan (2010) to select the best transportation mode in intermodal logistics operations. A methodology was developed by Pahlavani and Delavar (2014) using a locally linear neuro-fuzzy model that is trained with an incremental tree-based learning algorithm for a multi-criteria route selection problem. They also considered the driver's preferences and classified any route based on these preferences. An integrated quantitative risk assessment, AHP, and data envelopment analysis (DEA) methodology was developed by Kengpol et al. (2014) to evaluate the transportation risk and select an optimal multimodal transportation route. Hamurcu and Eren (2016) implemented the Analytic Network Process (ANP) and

Technique for Order Preference by Similarity to Ideal Solution (TOPSIS) methods for the route selection problem of a planned monorail transport system in Turkey. Wang and Yeo (2018) implemented an integrated Fuzzy Delphi and Fuzzy ELECTRE-I methods to select the most efficient intermodal route for cargo transportation from Korea to Central Asia. They have concluded that total cost, reliability, transportation capability, total time, and security are the most important factors for logistics companies in selecting intermodal transport routes. A hierarchical framework based on the grey Decision Making Trial and Evaluation Laboratory (DEMATEL) and ANP was developed by Kumar and Anbanandam (2020) to analyze the interrelationships and prioritize factors that influence a sustainable intermodal freight transport system.

Finally, some of the previously reviewed articles can be summarized and classified as in Table 16.1, with respect to different transportation modes, sustainability criteria (cost, time, environment, risk, security, etc.), and the relevant solution methods used in the route selection problems.

3 OVERVIEW OF AXIOMATIC DESIGN (AD) METHODOLOGY

Suh (2001a) first introduced the AD theory and its principles to yield a systematic search process for the design space and achieve the best solution among several design alternatives (Kulak and Kahraman 2005a; Çakır 2018). The main aim of the AD methodology is to provide designers or decision-makers with more creative/innovative designs by reducing the random search and iterative trial-and-error processes while determining the best design among several design alternatives (Shahbandarzadeh et al. 2011). In detail, two design axioms (i.e. independence and information) constitute the main concepts of AD methodology (Suh 2001b).

TABLE 16.1
Summary of the Literature Review on Transportation Route Selection Problem.

	Transportation						Cost			Time						Risk									Security		Environment			Method
	Road	Marine	Railway	Pipeline	Airline	Intermodal	Transit cost	Infrastructure	Insurance	Transit time	Distance	# of trips	Load/unload	Customs wait	Border crossing	Equipment breakdown	Infrastructure level	Traffic	Reliability	Cabotage	Flexibility	Transport productivity	Accessibility	Accidents	Theft	Country risks	CO2 emission	Energy & fuel consumption	NOx emission	
Ko (2009)	x						x			x	x					x	x	x					x		x	x				Fuzzy AHP
Murat & Kulak (2005)	x							x			x								x	x					x		x			Information Axiomatic Design
Moon et al. (2015)			x				x			x		x	x						x	x										Fuzzy TOPSIS
Banai (2006)		x					x			x									x	x	x		x							AHP
Özceylan (2010)	x	x	x	x	x	x	x			x										x	x		x			x				AHP
Ahmed & Asmael (2009)	x						x			x						x	x		x											GIS based AHP & DEM model
Hamurcu & Eren (2016)			x							x	x		x								x		x							TOPSIS & AHP
Owczarzak & Zak (2015)	x	x					x												x	x	x					x				ELECTRE III - IV
Nosal & Solecka (2014)	x	x								x	x								x	x						x				AHP Ranking method
Alkubaisi (2014)			x				x							x					x				x		x	x				GIS & GPS based MADM method
Odeyale et al. (2014)	x	x	x				x							x					x	x	x		x			x				Fuzzy AHP
Farkas (2009)			x							x				x									x		x	x				GIS based Expert System
Yedla & Shrestha (2003)	x									x	x															x				AHP
Kazan & Çiftci (2013)	x	x	x	x	x					x		x	x	x					x							x				PROMETHEE & AHP
Jacyna & Wasiak (2015)	x	x								x	x			x											x	x			x	Ranking & MAJA Method
This study	x	x	x				x	x	x	x	x	x	x	x	x		x	x	x	x	x				x	x	x	x	x	Fuzzy Hierarchical Axiomatic Design

According to the first axiom, independence of the functional requirements (FR) has to be satisfied in the design process. To maintain independence of the FRs that characterize the design objectives, decomposition of a complex design problem into several sub-problems and provision of independent mappings between these problems and design solutions can be useful (Kulak and Kahraman 2005b). On the other hand, the second information axiom specifies that a design solution that already satisfied the independence axiom and has the smallest information content among the others is the best alternative. In other words, the information axiom presents the choice of the best design option that has the minimum information required. Thus, this axiom provides the basis for the decision-making process in case of multiple attributes/FRs and alternatives (Suh 2001c). For that reason, AD methodology was utilized by several design applications in different fields and its usefulness was tested via several comparisons in MCDM literature (Babic 1999; Kulak and Kahraman 2005b; Özel and Özyörük 2007; Kahraman and Cebi 2009; Celik et al. 2009; Cicek and Celik 2010; Kannan et al. 2015; Büyüközkan and Göçer 2017; Chen et al. 2018).

3.1 Crisp AD Method and the Proposed Extensions

Under a crisp decision-making environment, we have certain information about the system and design ranges for the FRs. In detail, we have just a probability density function, $f(FR_j)$ for any FR_j. As mentioned previously for the information axiom, information is defined in terms of information content (I_j), which is a basic form of the probability of fulfilling a given FR_j. This probability (p_j) of achieving the FR_j in its lower and upper bounded design range $[dr^l, dr^u]$ can be calculated as in Equation (16.1), when FR_j is a continuous random variable (Kulak and Kahraman 2005b; Çakır 2018):

$$p_j = \int_{dr^l}^{dr^u} f\left(FR_j\right) dFR_j \tag{16.1}$$

In fact, the probability of success for the FR_j is specified by what the designer desires to accomplish in terms of tolerance or design range and what the system can deliver (i.e. system range). According to the conventional AD methodology, a feasible or acceptable design solution arises within a common range with an overlap/intersection between the designer-specified range and system range, as displayed in Figure 16.1. In this research, since interval values are used for the tolerances and system ranges, we utilized uniform distribution for the FRs. Therefore, Equation (16.1) can also be replaced by Equation (16.2) when FR_j has a uniform probability density function.

$$p_j = Common\ range/System\ range \tag{16.2}$$

Based on the probability of success for the FR_j in Equations (16.1) and (16.2), its information content, I_j can be expressed as in the following Eq. (16.3).

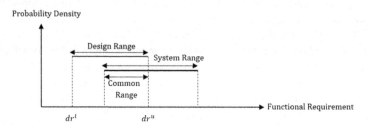

FIGURE 16.1   Design, System, and Common Ranges in Case of a Uniform Probability Density Function of an FR.

$$I_j = log_2\left(\frac{1}{p_j}\right) or\ log_2\left(\frac{System\ range}{Common\ range}\right) \tag{16.3}$$

Unfortunately, the above formulation of the conventional AD methodology may suffer from some shortcomings and, therefore, may cause questionable results in terms of its validity in practical applications. For instance, when an alternative has a much better performance than the desired range of regarding FR, it is put to failure and excluded from the comparison process since it shares no common area with that FR. However, such an alternative should find more chances of being selected. Another drawback is that conventional AD methodology gives the same selection chance to the alternatives which do not share a common area with the design range of an FR and have better performances to each other (Shahbandarzadeh et al. 2011).

To overcome this problematic issue, we modified Equation (16.3) in the following Equation (16.4) to compute the information content of alternatives for benefit type criteria/FRs. Similarly, information contents for cost type criteria/FRs can also be obtained by using Equation (16.5). Indeed, the first three conditions in Equations (16.4)–(16.5) correspond to the calculation of information contents in the conventional crisp AD methodology. The last two conditions given in Equations (16.4)–(16.5) are considered by the proposed modification and displayed in Figure 16.2.

$$I_j = \begin{cases} Infinite & if\ sr^u \le dr^l \\ log_2\left(\frac{sr^u - sr^l}{sr^u - dr^l}\right) & if\ dr^l < sr^u\ \&\ dr^l > sr^l \\ 0 & if\ dr^l = sr^l\ \&\ dr^u = sr^u \\ log_2\left(\frac{1}{mid\,[sr^l, sr^u] - mid\,[dr^l, dr^u]}\right) & if\ sr^l < dr^u\ \&\ 1 < dr^l < sr^l\ or\ dr^u \le sr^l \\ log_2\,(mid\,[sr^l,\ sr^u] - mid\,[dr^l,\ dr^u]) & if\ sr^l < dr^u\ \&\ dr^l < sr^l \le 1\ or\ dr^u \le sr^l \end{cases} \tag{16.4}$$

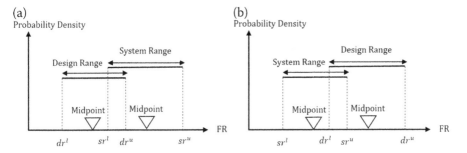

FIGURE 16.2 Depiction of the Last Conditions in Equations (16.4)–(16.5) for (a) Benefit and (b) Cost Type Criteria.

$$I_j = \begin{cases} Infinite & if\ dr^u \le sr^l \\ log_2\left(\frac{sr^u - sr^l}{dr^u - sr^l}\right) & if\ sr^l < dr^u\ \&\ sr^l > dr^l \\ 0 & if\ dr^l = sr^l\ \&\ dr^u = sr^u \\ log_2\left(\frac{1}{mid\,[dr^l, dr^u] - mid\,[sr^l, sr^u]}\right) & if\ dr^l < sr^u\ \&\ 1 < sr^l < dr^l\ or\ sr^u \le dr^l \\ log_2(mid\,[dr^l, dr^u] - mid\,[sr^l, sr^u]) & if\ dr^l < sr^u\ \&\ sr^l < dr^l \le 1\ or\ sr^u \le dr^l \end{cases}$$

$$(16.5)$$

In these last conditions, midpoints of the system and design ranges are utilized to calculate information contents. According to this reformulation, the information content tends to take bigger negative values as the performance of the alternative improves. By performing such a modification on the crisp AD method, higher priority can be given to the alternatives that have better performances (the first problematic issue), and these alternatives can be more logically compared to each other (the second problematic issue).

Furthermore, since there are several FRs (or multiple criteria) available in a design problem, the overall information content, I_o, is computed in conventional AD methodology as in Equation (16.6) by summing all the information contents related to the entire FRs ($j \in M$). If one or more probabilities of success are zero, I_o will approach infinity and the designed system may never work. In the contrasting case, I_o will be equal to zero when all the probabilities are one for the FRs. In that situation, the highest probability of success leads to minimal information content—therefore, the best design solution can be obtained by the conventional AD methodology.

$$I_o = \sum_{j=1}^{M} I_j \qquad (16.6)$$

On the other hand, since the proposed improved extension of the AD methodology may cause negative valued information contents for some FRs, summation of all the positive- and negative-valued information contents may cause some questionable and misleading results. For instance, after summing all the information contents, an inferior alternative with high total positive information content may look superior due to its negative-valued information contents. For this reason, we proposed the following evaluation procedure in Equations (16.7) – (16.8) to rank the alternatives after calculating both total positive- and negative-valued information contents.

$$IP_o = \sum_{j=1}^{M\setminus\{N\}} I_j \quad \text{and} \quad IN_o = \sum_{j=1}^{M\setminus\{P\}} I_j \tag{16.7}$$

where $M\setminus\{N\}$ and $M\setminus\{P\}$ correspond to sets of FRs that have positive- and negative-valued information contents, respectively. According to Equation (16.8), the best alternative has the least positive and the most negative-valued information contents.

$$\begin{cases} \textit{Rank the alternatives in an ascending order of } IP_o \\ \textit{Rank the alternatives in an descending order of } IN_o \\ \textit{Select the best alternative with minimum } IP_o \textit{ and } IN_o \\ \textit{For a pair of alternatives}\,(A,\,B),\,\textit{If } IPA_o = IPB_o, \\ \textit{select minimum}\,(INA_o,\,INB_o) \end{cases} \tag{16.8}$$

3.2 Fuzzy Axiomatic Design (FAD) Method

Under uncertain environments, there may be incomplete information about the design and system ranges. In other words, they may have some sort of ambiguity or vagueness. In this case, fuzzy data can be obtained for the FRs in terms of linguistic evaluations, fuzzy sets, or fuzzy numbers. The fuzzy multi-attribute axiomatic design approach was first proposed by Kulak and Kahraman (2005a) to select the best transportation company problem with linguistic assessments. In a similar manner, evaluations of the transportation routing alternatives with respect to the reliability/risk criteria for the present sustainable intermodal route selection problem are defined linguistically in this research. The linguistic terms are first converted into their corresponding fuzzy numbers by using numerical approximation systems. Due to the ease of use and simplicity, we utilized the triangular fuzzy numbers (TFNs), as in most of the available studies in MCDM literature (Çakır 2018). In contrast to the crisp AD methodology where a probability density function is used, a membership function of the triangular fuzzy numbers is constructed for the given FR. Therefore, the common region is the intersection area of the predefined triangular membership functions for the design and system ranges as indicated by Figure 16.3 (Kulak and Kahraman 2005a, 2005b). Hence, information content can be calculated as in Equation (16.9) by the conventional FAD method for a linguistically evaluated qualitative type FR. It should be highlighted here that this calculation can also be used for the quantitative type FRs whose design and system range values are uncertain and represented by fuzzy numbers.

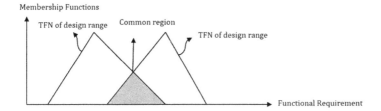

FIGURE 16.3 Design, System, and Common Ranges under a Fuzzy Decision-Making Environment.

$$I_j = log_2 \left(\frac{TFN \; of \; System \; Range}{Common \; Region} \right) \quad \forall \; j \in M \qquad (16.9)$$

However, the conventional FAD method may also suffer from some shortcomings. For instance, an axiomatic and fundamental fact that the higher the performance of an alternative is in a benefit type criteria, the more utility it provides—and vice versa for the cost type criteria— may be violated by the conventional FAD method. Moreover, an alternative that performs a higher performance than the design requirement—a fuzzy number far from the fuzzy number of the design range—does not share a common area. Therefore, its information content will have an unfavorable infinite value that may cause the elimination of such an efficient alternative from the comparison process. Another drawback of the conventional FAD method is that the same grade or selection chance is given to the alternatives that have equal distances from the design range of an FR. For instance, when the design range of an FR is specified as "Good" by the decision analyst, both the "Fair" and "Very Good" assessments will have the same grade or selection chance, while the higher chance should be given to an alternative which is linguistically evaluated as "Very Good". In other words, performances of the alternatives should be logically comparable in each FR so that the better their performance, the better evaluation or higher selection chance they receive (Shahbandarzadeh et al. 2011). Based on these facts, Shahbandarzadeh et al. (2011) developed an improved extension of the FAD method where information contents are computed **in such a way that**, as the performance of the alternatives improves, the information contents tend to have bigger negative values. On the other hand, this gradual improvement should not have a significant influence since the decision analyst is already satisfied with the performances close to the design ranges of the FR. For all these reasons, we applied this improved extension of the FAD method by (Shahbandarzadeh et al. 2011). According to their modifications, information contents of the alternatives can be calculated with Equations (16.10)–(16.13) for the benefit and cost type criteria when the triangular fuzzy numbers are used in the evaluation process. As in the crisp AD methodology section, the last conditions in Equations (16.10) and (16.12) are also demonstrated in Figure 16.4. If there is a common area with respect to these last conditions in Equations (16.10) and (16.12), it can be calculated by Equations (16.11) and (16.13). If there is no common area, this value can be taken as zero. In detail, these last conditions can provide larger negative-valued information contents as the performance of the alternatives gets better than the design range of an FR.

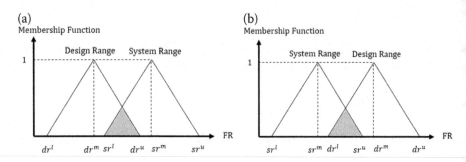

FIGURE 16.4 Depiction of TFNs in the Last Conditions of Equations (16.10)–(16.12) for (a) Benefit and (b) Cost Type Criteria.

$$I_j = \begin{cases} Infinite & if\ sr^u \le dr^l \\ log_2\left(\dfrac{(sr^u - sr^l)\cdot(sr^u - sr^m) + (sr^u - sr^l)\cdot(dr^m - dr^l)}{(sr^u - dr^l)^2}\right) & if\ dr^l < sr^u\ \&\ dr^l > sr^l \\ 0 & if\ dr^l = sr^l\ \&\ dr^u = sr^u \\ log_2\left(1 - \dfrac{sr^m - dr^m - dr^u + sr^l + common\ area}{2\cdot(sr^u - dr^l)}\right) & if\ sr^l < dr^u\ \&\ dr^l < sr^l\ or\ dr^u \le sr^u \end{cases}$$

(16.10)

$$common\ area = \frac{(dr^u - sr^l)^2}{sr^m - sr^l - dr^m + dr^u}$$

(16.11)

$$I_j = \begin{cases} Infinite & if\ dr^u \le sr^l \\ log_2\left(\dfrac{(sr^u - sr^l)\cdot(dr^u - dr^m) + (sr^u - sr^l)\cdot(sr^m - sr^l)}{(dr^u - sr^l)^2}\right) & if\ sr^l < dr^u\ \&\ sr^l > dr^l \\ 0 & if\ dr^l = sr^l\ \&\ dr^u = sr^u \\ log_2\left(1 - \dfrac{dr^m - sr^m + dr^l - sr^u + common\ area}{2\cdot(dr^u - sr^l)}\right) & if\ dr^l < sr^u\ \&\ sr^l < dr^l\ or\ sr^u \le dr^l \end{cases}$$

(16.12)

$$common\ area = \frac{(sr^u - dr^l)^2}{dr^m - dr^l - sr^m + sr^u}$$

(16.13)

3.3 WEIGHTED FAD (WFAD) METHOD AND THE PROPOSED EXTENSIONS

In most real-life applications, a decision analyst tends to assign different weights for each criterion or FR. To deal with these, Kulak and Kahraman (2005b) proposed the following Equation (16.14) to compute information contents in the weighted crisp or fuzzy AD methods. For each design alternative $i \in I$ and $FR_j\ j \in M$, information contents are calculated by using these criteria weights (w_j) under three circumstances.

$$WI_{ij} = \begin{cases} \left[\left[log_2\left(\dfrac{1}{p_{ij}}\right) \right] \right]^{1/w_j} & 0 \leq I_{ij} < 1 \\[2ex] \left[\left[log_2\left(\dfrac{1}{p_{ij}}\right) \right] \right]^{w_j} & I_{ij} > 1 \\[2ex] w_j & I_{ij} = 1 \end{cases} \qquad (16.14)$$

When the criteria hierarchy is taken into consideration, the above formulation can be modified as in Equation (16.15) by considering the number of levels in the criteria hierarchy. In detail, the final importance weight of an FR can be determined by multiplying all the weights in the sequential levels in the criteria hierarchy.

$$WI_{ij} = \begin{cases} \left[\left[log_2\left(\dfrac{1}{p_{ij}}\right) \right] \right]^{1/\Pi_{k \in K} w_{kj}} & 0 \leq I_{ij} < 1 \\[2ex] \left[\left[log_2\left(\dfrac{1}{p_{ij}}\right) \right] \right]^{\Pi_{k \in K} w_{kj}} & I_{ij} > 1 \\[2ex] \Pi_{k \in K} w_{kj} & I_{ij} = 1 \end{cases} \qquad (16.15)$$

where k represents the number of levels in criteria hierarchy for the FR_j. However, the negative valued information contents are not considered by this conventional WFAD method in Equation (16.14). For this reason, Shahbandarzadeh et al. (2011) proposed the formulation in Equation (16.16) to deal with the weighted extension of the FAD method. If the information contents of the alternatives are calculated using the last conditions in Equations (16.4), (16.5), (16.10), and (16.12), Equation (16.16) should be employed to concern the negative-valued information contents in both crisp and fuzzy AD methods.

$$WI_{ij} = -\left[\left| log_2\left(\dfrac{1}{p_{ij}}\right) \right|^{1-w_i} \right] \qquad -1 < I_{ij} < 0 \qquad (16.16)$$

In this research, we also extended the above formulation in Equation (16.17) to take other circumstances and the criteria hierarchy into account.

$$WI_{ij} = \begin{cases} -\left[\left| log_2\left(\dfrac{1}{p_{ij}}\right) \right|^{1-\left(\Pi_{k \in K} w_{kj}\right)} \right] & -1 < I_{ij} \leq 0 \\[3ex] -\left[\left| log_2\left(\dfrac{1}{p_{ij}}\right) \right|^{1+\left(\Pi_{k \in K} w_{kj}\right)} \right] & I_{ij} < -1 \\[3ex] -\left(1 + \Pi_{k \in K} w_{kj} \right) & I_{ij} = -1 \end{cases} \qquad (16.17)$$

Finally, we used the linguistic variables to assess the importance weights of the criteria/FRs. After that, corresponding TFNs of the linguistic variables are converted into crisp values by using a defuzzification procedure of Hsieh et al. (2004), Opricovic and Tzeng (2003). This procedure aims to locate the Best Non-fuzzy Performance (BNP) value. Usage of the Center of Area (COA) method offers the simplest and the most practical way to compute BNP without the need for preferences of any decision analyst (Wu et al. 2009). The BNP value of the TFN for a given FR_j can be found by using Equation (16.18) and making use of its lower, middle, and upper values (LR_j, MR_j, UR_j). After performing this defuzzification process, the crisp weights are normalized according to Equation (16.19) to satisfy that the summation of all criteria/FRs weights is equal to '1' within each level of criteria hierarchy.

$$BNP_j = LR_j + \left[\left(UR_j - LR_j \right) + \left(MR_j - LR_j \right) \right]/3 \quad \forall \, j \in M \qquad (16.18)$$

$$w_{kj} = BNP_j / \sum_{j \in M_k} BNP_j \quad \forall \, k \in K \qquad (16.19)$$

where M_k represents the subsets of FRs belonging to the kth level of criteria hierarchy. For comparison of the computational results, we implemented both the proposed improved extensions of unweighted and weighted FAD methods to select the best routing alternatives in a sustainable intermodal transportation network. Briefly, the application procedure of the proposed WFAD method is summarized in Figure 16.5.

4   A REAL-LIFE APPLICATION IN AN INTERNATIONAL LOGISTICS COMPANY

To show the applicability and practicality of the proposed improved extension of WFAD methodology, a real-life case study is presented for a sustainable route selection problem of a large-sized international logistics company in Turkey.

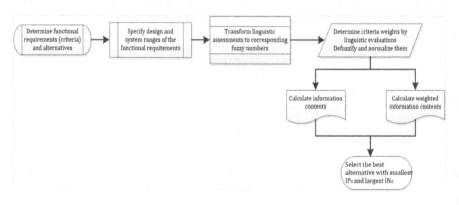

FIGURE 16.5   Application Procedure of the WFAD Methodology.

4.1 Description of Route Selection Problem in Intermodal Transportation

The logistics company provides various transportation models and services to its customers, including direct road freight transport, combined marine and road transport, and intermodal transportation. It performs international transportation services between Turkey and all over Europe in both export/import directions. In the direct road freight transport, loads are transported from Turkey to the consolidation centers of the company via highways and delivered again from these centers to the final destinations in Europe. The combined marine and road transport incorporate two distinct modes of transport and serves as an alternative to direct road transport for particularly heavy products because weight limitations are imposed on European highways. The loads are first transferred from the seaports in Turkey to the seaports in Europe by Ro-Ro vessels and then delivered from these ports to the relevant destinations via highways. Intermodal service is a transportation option wherein three different transport modes are utilized (i.e., maritime, railway, and road transports). Trailers or containers arriving by Ro-Ro vessels to European seaports are loaded on block or public trains, and then transferred to the destination rail stations and finally delivered to the customers by road freight transport (Baykasoğlu and Subulan 2016; Baykasoğlu and Subulan 2019; Baykasoğlu et al. 2019a, 2019b). All these routing alternatives and the intermodal logistics network configuration of the logistics company are displayed in Figure 16.6.

The company experts seek to reduce on-road mileage by using intermodal services and reduce carbon dioxide, hydrocarbon, particle, nitrogen emissions, and fuel/electricity consumption. Therefore, in addition to the economic dimension of freight transportation, selection of the more environmental-friendly way of transportation is intended. Furthermore, customer satisfaction, reliability, and risks of the transportation services are also considered by the company experts to focus on the

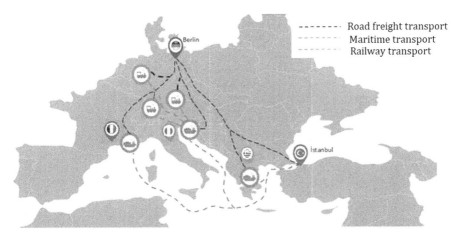

FIGURE 16.6   Alternative Transportation Routes on the Intermodal Logistics Network.

social aspects of the examined problem. In detail, timely delivery of loads/freights is related to the transportation service level of the company, and therefore, customer demands should be satisfied as soon as possible. Moreover, infrastructure levels and equipment potentials of the seaports, rail stations, and hubs/consolidation centers are also considered while selecting the more reliable transportation routes. Feasibility and flexibility of the routing schedules are other critical issues, especially in selecting more reliable maritime and railway transport routes. Additionally, company experts have also focused their attention on the cabotage limitations, political risks, security precautions, and damages/risks during the transportation operations. For that reason, they aim to select the best routing alternative under the sustainability factors. Briefly, three pillars of sustainability for the route selection problem in an intermodal transport network are depicted in Figure 16.7.

A detailed hierarchy of the route selection problem is also obtained, as in Figure 16.8, from the company experts by performing a comprehensive questionnaire survey. The company experts consist of drayage, terminal and network operators, fleet planners, and financial advisors who work in the business development department of the company. First, they have determined the FRs (main and sub-criteria) and seven alternative transportation routes between Istanbul and Berlin as in Table 16.2 below. As it is clearly seen in Figure 16.6 and Table 16.2, the trailers or containers can be directly shipped from Istanbul to Berlin via road freight transport.

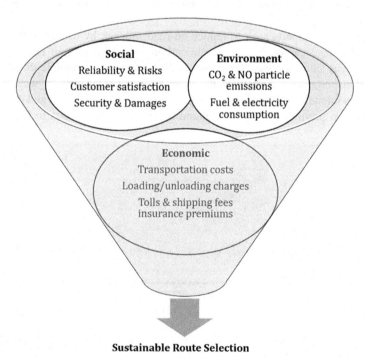

FIGURE 16.7 Three pillars of sustainability for route selection problem in an intermodal transportation network.

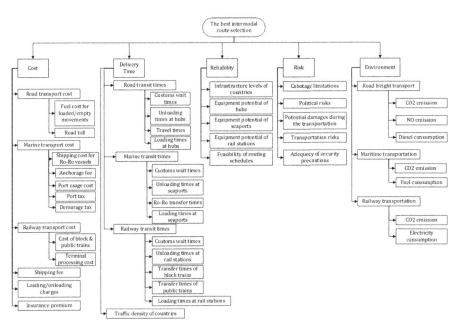

FIGURE 16.8 Hierarchy of sustainable route selection criteria in an intermodal transportation network.

TABLE 16.2
Alternative Routes on the Intermodal Transportation Network

Route No	Terminals/Ports/Hubs/Stations	Modes of Transport
1	Istanbul – Berlin	Direct road freight transport
2	Istanbul – Lavrio Port – Berlin	Combined marine and road transport
3	Istanbul – Trieste Port – Berlin	Combined marine and road transport
4	Istanbul – Toulon Port – Berlin	Combined marine and road transport
5	Istanbul – Trieste Port – Munich Station - Berlin	Intermodal transportation
6	Istanbul – Trieste Port – Cologne Station – Berlin	Intermodal transportation
7	Istanbul – Trieste Port – Ludwigshafen Station – Berlin	Intermodal transportation

Alternatively, they are first shipped from Haydarpasa port of Istanbul to Trieste, Lavrio, and Toulon ports via company-owned Ro-Ro vessels, and then transferred to the final destination in Berlin via road freight transport. Otherwise, arriving freights to Trieste port can be loaded on block or public trains that offer full trainload and less-than trainload services at Munich, Cologne, and Ludwigshafen railway stations. After that, they are consolidated, and then transferred by trucks to final destinations in Berlin. It should be emphasized here that only the Trieste port has railway connections on the intermodal logistics network. Moreover, instead of block trains, there are only public train services from Trieste port to Munich railway station. Similarly, there are only block train services from Trieste port to Cologne and Ludwigshafen rail stations.

4.2 DESIGN AND SYSTEM RANGE DATA FOR ALTERNATIVE TRANSPORT ROUTES

Application of the FAD and WFAD methodologies begin with the specification of functional requirements (criteria) and design parameters based on the company experts' preferences. After that, system design ranges of the alternative routes are specified and the relationships between them are used to select the best routing alternative by considering multiple transportation modes. The company experts have reached a group consensus about design ranges of the FRs and the system range data were generated by evaluating all the routing alternatives in Table 16.3. Except for the traffic density sub-criteria, all the cost, delivery time, and environmental criteria of the alternative routes took crisp numerical values. In fact, traffic densities of the countries on the related transport routes were evaluated linguistically. Moreover, linguistic assessments of the qualitative type criteria—such as risk and reliability—were also carried out, as in Table 16.3. Then, these linguistic assessments were transformed into the corresponding TFNs by using the numerical approximation systems for tangible/intangible criteria as given in Figure 16.9. For instance, traffic density, cabotage limitations, political risks, and damages during the transportation operations were considered as tangible FRs, whereas the infrastructure level of the order countries, security precautions, and equipment potential of the ports, hubs, and railway stations were taken as intangible FRs. It should be noted here that similar to the terminal processing costs at the railway stations and port usage costs, the processing costs at the hubs or consolidation centers should also be taken into consideration in the context of road transport cost criteria.

4.3 COMPARISON OF THE RESULTS FOR INFORMATION CONTENTS

In this section, the information contents of each alternative transport route were first calculated in each FR without criteria weights by using Equations (16.3) – (16.13). Indeed, criteria weights were temporarily excluded in the applications of crisp and FAD methodologies. In other words, unweighted information contents were first computed. After that, these weights would be considered while implementing the WFAD methodology. For the sake of brevity, information content calculations were shown below for the randomly selected alternative routes and FRs. For instance, since the fuel cost of loaded/empty vehicle movements is a cost type FR that is targeted to minimize, information contents of the randomly selected routes —1, 3, and 7—can be calculated using Equation (16.5) as in the following Equations (16.20) – (16.22).

TABLE 16.3

Design and System Range Data for FRs and Routing Alternatives

Functional Requirements (FR)	Unit	Design Ranges	Routing Alternatives						
			Route No. 1	Route No. 2	Route No. 3	Route No. 4	Route No. 5	Route No. 6	Route No. 7
		Design Ranges				System Ranges			
Cost									
Road toll	(€)	280–450	440–590	425–565	200–350	250–400	175–285	175–285	175–285
Anchorage fee	(€)	20–40	N.A	30–40	16–32	37–45	16–32	16–32	16–32
Port usage cost	(€)	85–140	N.A	95–155	60–90	130–175	60–90	60–90	60–90
Port tax	(€)	140–300	N.A	100–170	290–370	210–270	290–370	290–370	290–370
Demurrage tax	(€)	45–60	N.A	55–78	47–62	50–75	35–50	58–76	51–68
Terminal processing cost	(€)	240–300	N.A	N.A	N.A	N.A	260–315	260–315	260–315
Shipping fee	(€)	890–1500	1370–2200	1050–1500	790–995	900–1400	850–1100	765–920	975–1250
Loading/unloading charges	(€)	240–470	170–250	450–650	300–500	425–570	440–510	450–580	465–600
Fuel costs of loaded/empty moves	(€)	700–1400	2631–2780	2300–2864	1298–1600	1276–1500	700–850	680–750	752–922
Shipping cost for Ro-Ro vessels	(€)	460–850	N.A	465–520	750–800	840–900	750–800	750–800	750–800
Cost of block and public trains	(€)	500–800	N.A	N.A	N.A	N.A	600–750	730–900	700–850
Insurance premium	(€)	56–65	61–68	58–67	45–59	62–73	58–72	39–57	56–69
Delivery Time	(hr.)	6–14	12–20	12–20	3–10	12–20	N.A	N.A	N.A

(Continued)

TABLE 16.3 (Continued)

Functional Requirements (FR)		Routing Alternatives						
		Route No. 1	Route No. 2	Route No. 3	Route No. 4	Route No. 5	Route No. 6	Route No. 7
Customs wait times in road transport	(hr.) 7–11	N.A	4–9	6–8	5–9	6–8	6–8	6–8
Customs wait times in marine transport	(hr.) 4–9	N.A	N.A	N.A	N.A	5–8	6–11	4–8
Customs wait times in railway transport	(hr.)	N.A	N.A	N.A	N.A	N.A	N.A	N.A
Unloading times at hubs	(hr.) 10–16	14–18	13–18	14–19	10–15	8–12	9–13	15–18
Unloading times at seaports	(hr.) 35–45	N.A	32–37	35–41	39–46	41–49	31–40	37–46
Unloading times at rail stations	(hr.) 21–29	N.A	N.A	N.A	N.A	23–30	20–27	18–24
Travel times in road transport	(hr.) 10–25	24–28	23–30	10–15	15–19	6–12	7–14	7–13
Ro-Ro transfer times	(hr.) 40–70	N.A	30–45	60–65	85–89	60–63	59–64	57–61
Transfer times of block trains	(hr.) 20–30	N.A	N.A	N.A	N.A	N.A	22–26	19–23
Transfer times of public trains	(hr.) 20–25	N.A	N.A	N.A	N.A	22–26	N.A	N.A
Loading times at hubs	(hr.) 20–29	20–25	18–26	23–29	25–32	19–31	20–25	22–28
Loading times at seaports	(hr.) 40–47	N.A	38–41	39–43	42–48	39–43	39–43	39–43

(Continued)

TABLE 16.3 (Continued)

Functional Requirements (FR)	Unit	Design Range	Routing Alternatives (System Range)						
			Route No. 1	Route No. 2	Route No. 3	Route No. 4	Route No. 5	Route No. 6	Route No. 7
Loading times at rail stations	(hr.)	22–32	N.A	N.A	N.A	N.A	20–28	23–33	23–29
Traffic density of countries	–	Low	High	High	Medium	Low	Very Low	Very Low	Very Low
Reliability									
Infrastructure levels of countries	–	Very Good	Fair	Good	Very Good	Good	Excellent	Excellent	Excellent
Equipment potential of hubs	–	Very Good	Good	Good	Very Good	Very Good	Very Good	Very Good	Very Good
Equipment potential of seaports	–	Very Good	N.A	Good	Good	Very Good	Very Good	Very Good	Very Good
Equipment potential of rail stations	–	Very Good	N.A	N.A	N.A	N.A	Very Good	Excellent	Good
Feasibility/flexibility of route schedules	–	High	Very Low	Low	High	Medium	High	High	High
Risk									
Cabotage limitations	–	Low	N.A	High	Low	Medium	Low	Low	Low
Political risks	–	Low	Medium	Medium	Low	Low	Low	Very Low	Low
Potential damages during transport	–	Low	Very High	High	Medium	Medium	Low	Low	Low
	–	Very Low	High	Medium	Low	Low	Low	Very Low	Low

(Continued)

TABLE 16.3 (Continued)

Functional Requirements (FR)		Routing Alternatives						
		Route No. 1	Route No. 2	Route No. 3	Route No. 4	Route No. 5	Route No. 6	Route No. 7
Transportation/fleet risks								
Security precautions at the countries	–	Low	Low	High	Medium	High	High	Medium
Environment								
CO_2 emission by road transport (kg)	6–18	17–25	17–23	7–10	12–16	4–8	5–9	4–9
CO_2 emission by marine transport (kg)	30–50	N.A	20–29	39–52	35–51	39–52	39–52	39–52
CO_2 emission by rail transport (kg)	0.9–1.2	N.A	N A	N.A	N.A	1.1–1.3	1.1–1.4	0.7–1.0
NO emission by road transport (g/kWh)	20–35	42–54	38–45	20–32	30–37	15–22	14–21	14–22
Diesel consumption in road transport (l/km)	190–400	710–745	740–810	350–480	360–545	180–190	165–185	150–175
Fuel consumption in marine transport (t)	58–78	N.A	25–45	70–84	71–91	70–84	70–84	70–84
Electricity consumption in rail transport (kW)	800000–900000	N.A	N.A	N.A	N.A	794000–915000	826000–910400	786000–864000

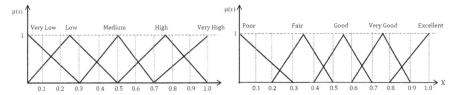

FIGURE 16.9 The numerical approximation systems for tangible/intangible FRs (Source: Çakır, 2018).

$$dr^u = 1400 \leq sr^l = 2631, \quad I_1 = Infinite \tag{16.20}$$

$$log_2\left(\frac{sr^u - sr^l}{dr^u - sr^l}\right) = log_2\left(\frac{1600 - 1298}{1400 - 1298}\right) = 1.57 \tag{16.21}$$

$$log_2\left(\frac{1}{mid\,[dr^l, dr^u] - mid\,[sr^l, sr^u]}\right) = \left(\frac{1}{(700 + 1400)/2 - (752 + 922)/2}\right)$$
$$= -7.734 \tag{16.22}$$

In a similar manner, all information contents were calculated and reported in Table 16.4 for the total costs, delivery times, and the environmental impact criteria. Contrary to the crisp design and system range data, FAD methodology and its Equations (16.10) – (16.13) were used to deal with the linguistic assessments of qualitative type criteria under uncertainty.

For instance, because the infrastructure level of the countries on the transportation routes is a benefit type criterion, information contents of the randomly chosen routes-1, 4, and 6 were calculated using Equations (16.10) and (16.11) as in Equations (16.23) and (16.26). Since the design range of this FR was defined as "Very Good" and the performance of the routing alternative-1 was expressed as "Fair", which has not shared a common area with the design requirement, the information content will take an infinite positive value. Similarly, the performance of the routing alternative-4 was specified as "Good" which is worse than the design requirement, the information content will also take a positive value as is clearly seen in Equation (16.24). On the other hand, since the performance of the routing alternative-6 was specified as "Excellent" that leads to a better design option, there exists a common area in Equation (16.25), and the related information content will take a negative value (see in Equation 16.26).

$$sr^u(Fair)0.5 = \leq dr^l = 0.6(Very\ Good), \quad I_1 = Infinite \tag{16.23}$$

TABLE 16.4

Results of the Information Contents for Alternative Routes in an Intermodal Transportation Network.

FRs	Criteria	Sub-criteria	Information Contents (I_i)						
			Route No. 1	Route No. 2	Route No. 3	Route No. 4	Route No. 5	Route No. 6	Route No. 7
COST	Road freight transport cost	Fuel cost for loaded/empty movements	INF	INF	1.57	0.853	-8.103	-8.388	-7.734
		Road toll	3.906	2.485	-6.491	-5.321	-7.076	-7.076	-7.076
	Marine transport cost	Shipping cost for RO-RO vessels	N.A	-7.344	-1.0	2.584	-1.0	-1.0	-1.0
		Anchorage fee	N.A	0	-2.584	1.415	-2.584	-2.584	-2.584
		Port usage cost	N.A	0.415	-5.228	2.169	-5.228	-5.228	-5.228
		Port tax	N.A	-6.409	3.0	-0.584	3.0	3.0	3.0
		Demurrage tax	N.A	2.201	0.206	1.322	-3.322	3.169	0.917
	Railway transport cost	Cost of block and public trains	N.A	N.A	N.A	N.A	0	1.28	0.584
		Terminal processing costs	N.A	N.A	N.A	N.A	0.459	0.459	0.459
		Shipping rate/fee	2.674	0	-8.241	-0.263	-7.781	-8.461	-0.932
		Loading/unloading charges	-7.179	3.321	0.234	1.688	1.222	2.7	4.754
		Insurance premium	0.807	0.362	-3.087	1.874	1.0	-3.643	0.531
DELIVERY TIME	Road transit times	Waiting times at customs	2	2	-1.807	2	N.A	N.A	N.A
		Loading times at hubs	0.807	0.263	0	-1.321	0.415	0	-0.222
		Unloading times at hubs	1.0	0.736	1.322	-0.263	-1.584	-1.0	-0.584
		Highway travel times	2.0	1.807	0	0	-3.087	-2.807	-2.906
	Marine transit times	Waiting times at customs	N.A	-1.321	-1.0	-1.0	-1.0	-1.0	-1.0

(*Continued*)

TABLE 16.4 (Continued)

FRs	Criteria	Sub-criteria	Information Contents (I_i)						
			Route No. 1	Route No. 2	Route No. 3	Route No. 4	Route No. 5	Route No. 6	Route No. 7
		Loading times at seaports	N.A	−2.0	−1.321	0.263	−1.321	−1.321	−1.321
		Unloading times at seaports	N.A	−2.459	0	0.222	0	0	0
		Transfer times via RO-RO vessels	N.A	−3.643	1.0	2.0	1.0	1.0	1.0
	Rail transit times	Waiting times at customs	N.A	N.A	N.A	N.A	0	0.736	0
		Loading times at rail stations	N.A	N.A	N.A	N.A	−1.584	0.152	0
		Unloading times at rail stations	N.A	N.A	N.A	N.A	0.222	−0.584	−2.0
		Transfer times via block trains	N.A	N.A	N.A	N.A	N.A	0	−2.0
		Transfer times via public trains	N.A	N.A	N.A	N.A	0.415	N.A	N.A
	Traffic density of countries		INF	INF	2.169	0	−0.174	−0.174	−0.174
FRs	Criteria	Sub-criteria	Information Contents (I_i)						
			Route No. 1	Route No. 2	Route No. 3	Route No. 4	Route No. 5	Route No. 6	Route No. 7
RELIABILITY	Infrastructure levels of the countries on the transport route		INF	3.169	0	3.169	−0.364	−0.364	−0.364
	Equipment potentials of hubs/consolidation centers		3.169	3.169	0	0	0	0	0
	Equipment potentials of seaports		N.A	3.169	3.169	0	0	0	0
	Equipment potentials of rail stations		N.A	N.A	N.A	N.A	0	−0.364	3.169

(Continued)

TABLE 16.4 (Continued)

FRs	Criteria	Sub-criteria	Information Contents (I_i)						
			Route No. 1	Route No. 2	Route No. 3	Route No. 4	Route No. 5	Route No. 6	Route No. 7
		Feasibility and flexibility of route schedules	INF	INF	0	2.7	0	0	0
RISK		Cabotage limitations	N.A	INF	0	2.169	0	0	0
		Political risks at the countries on the transport route	2.169	2.169	0	0	0	-0.174	0
		Potential damages to freights during transportation operations	INF	INF	2.169	2.169	0	0	0
		Transportation and fleet risks	INF	INF	1.611	1.611	1.611	0	1.611
		Security precautions at the countries on the transport route	INF	INF	0	2.7	0	0	2.7
ENVIRONMENT	Environmental impact by road freight transport	CO₂ emission by road freight transport	3.0	2.584	-1.807	0	-2.584	-2.321	-2.459
		NO emission by road freight transport	INF	INF	0	0.485	-3.169	-3.321	-3.247
		Diesel consumption	INF	INF	1.378	2.209	-6.781	-6.906	-7.049
	Environmental impact by marine transport	CO₂ emission by marine transport	N.A	-3.954	0.241	0.093	0.241	0.241	0.241
		Fuel consumption by RO-RO vessels	N.A	-5.044	0.807	1.514	0.807	0.807	0.807
	Environmental impact by railway transport services	CO₂ emission by railway transport	N.A	N.A	N.A	N.A	1.0	1.584	-2.321
		Electricity consumption by block/public trains	N.A	N.A	N.A	N.A	0.19	0.189	-14.61
Total positive information contents (*PI$_i$*)			INF	INF	18.876	35.209	11.582	15.317	19.773
Total negative information contents (*NI$_i$*)			-7.179	-32.17	-32.56	-8.752	-56.75	-56.72	-64.81
Ranking of the alternative routes			7	6	3	5	1	2	4

$$I_4 = log_2\left(\frac{(sr^u - sr^l).\ (sr^u - sr^m) + (sr^u - sr^l).\ (dr^m - dr^l)}{(sr^u - dr^l)^2}\right)$$

$$= log_2\left(\frac{(0.7 - 0.4).\ (0.7 - 0.55) + (0.7 - 0.4).\ (0.75 - 0.6)}{(0.7 - 0.6)^2}\right) = 3.169$$

$$(16.24)$$

$$common\ area = \frac{(dr^u - sr^l)^2}{sr^m - sr^l - dr^m + dr^u} = \frac{(0.9 - 0.8)^2}{1 - 0.8 - 0.75 + 0.9} = 0.02857$$

$$(16.25)$$

$$I_6 = log_2\left(1 - \frac{sr^m - dr^m - dr^u + sr^l + common\ area}{2.\ (sr^u - dr^l)}\right)$$

$$= log_2\left(1 - \frac{1 - 0.75 - 0.9 + 0.8 + 0.02857}{2.\ (1 - 0.6)}\right) = -0.364 \quad (16.26)$$

It should be emphasized here that if any FR is not valid for a routing alternative, its information content was expressed as N.A (not available) in Tables 16.3 and 16.4 and its value was taken as zero because there will be no information need on that FR for this routing alternative. For instance, since the railway transportation services were not available on routes 1, 2, 3, and 4, their information contents for the railway transit times and costs of block/public trains will be equal to zero. After calculating all the information contents for each FR of each alternative route, total positive- and negative-valued information contents were computed as given in Table 16.4. According to the proposed evaluation procedure in Equations (16.7) and (16.8), final rankings of the alternative transport routes were also specified at the bottom of Table 16.4. It is obviously seen from these rankings that the intermodal routing alternatives-5, 6—and the combined marine and road transport alternative-3—are the most sustainable way of freight transportation since they have the smallest positive and the largest negative -valued information contents. The results of these information contents have shown that the logistics company must use the Trieste seaport and Munich and Cologne rail stations while satisfying the import/export freight demands of its customers between Istanbul and Berlin. Because these route options can fulfill the design requirements defined by the company experts with minimal information contents under the sustainability issues.

So far, we have applied the improved extension of crisp and FAD methodologies and the results were obtained without considering criteria weights. In the WFAD method, the criteria weights should be first determined before calculating the weighted information contents of the alternative transport routes. As mentioned before, some linguistic variables given in Figure 16.10 were used to assess the importance weights of the criteria/FRs. In detail, a group of company experts have made linguistic evaluations of the formerly given main and sub-criteria in Figure 16.8. It should be noted here that these evaluations were structured in a hierarchical manner as seen in Table 16.5.

$\mu(x)$

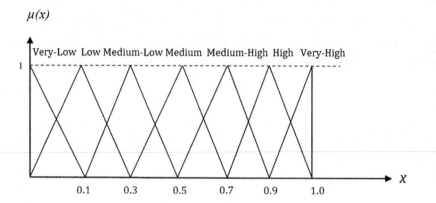

FIGURE 16.10 The corresponding triangular fuzzy numbers for linguistic evaluations of the FRs importance.

After obtaining these linguistic evaluations from the company experts, their relevant TFNs in Figure 16.10 were transformed into the crisp equivalent values by using a defuzzification method proposed by Hsieh et al. (2004), Opricovic and Tzeng (2003). For each main and sub-criteria, the BNP value was computed via the center of area method as in Equation (16.18). For instance, the importance of the potential damages/risks during the transportation operations was linguistically evaluated as "Medium High" by the company experts. The BNP value for that FR was also exemplified in Equation (16.27).

$$BNP = 0.5 + [(0.9 - 0.5) + (0.7 - 0.5)]/3 = 0.7 \qquad (16.27)$$

After performing this defuzzification process, the crisp weights within the same level of criteria hierarchy were normalized using Equation (16.19). This normalization process ensures that the summation of all the weights is equal to '1' within the same level of criteria hierarchy. After applying the defuzzification and normalization processes, the final crisp weights of criteria/FRs can be obtained in Table 16.6. It should be emphasized here that the multiplication of sequential weights within different hierarchy levels gives the final weight of an FR and the summation of all these finals weights will also be equal to 1.0.

While determining the weighted information contents of the alternative transport routes, the hierarchical crisp criteria weights given in Table 16.6 and the formerly calculated information contents in Table 16.4 were utilized. It should be noted that the criteria hierarchy was considered in the calculation of weighted information contents. To do this, the proposed formulations in Equations (16.15) and (16.17) were used. The calculated positive weighted information contents were exemplified in Equations (16.28)–(16.30). For instance, the weighted information content of the fuel consumption cost sub-criteria can be calculated in Equation (16.28) for combined marine and road freight transport option (Alternative transport route-3). In Equation (16.29), weighted information content of the insurance premium for the alternative intermodal transport route-5 was illustrated. The weighted information

TABLE 16.5

Linguistic Evaluations for the Importance Weights of the FRs/Criteria.

Functional Requirements (FRs)			Linguistic variables	
Cost	Road freight transport cost	Fuel cost for loaded/empty movements	Medium	Very high
		Road toll	Medium-high	
	Marine transport cost	Shipping cost for Ro-Ro vessels	High	
		Anchorage fee	High	
		Port usage cost	Medium-high	
		Port tax	High	
		Demurrage tax	Very high	
	Railway transport cost	Cost of block and public trains	Medium	
		Terminal processing cost	Very high	
	Shipping fee		Very high	
	Loading/unloading charges		Very high	
	Insurance premium		Medium-high	
Delivery Time	Road transit times	Customs wait times in road transport	Very high	Very high
		Unloading times at hubs	Very high	
		Travel times in road transport	Very high	
		Loading times at hubs	Very high	
	Marine transit times	Customs wait times in marine transport	Very high	
		Unloading times at seaports	Very high	
		Ro-Ro transfer times	Very high	
		Loading times at seaports	High	
	Rail transit times	Customs wait times in railway transport	High	
		Loading times at rail stations	High	
		Transfer times of block trains	Medium	
		Transfer times of public trains	High	
		Unloading times at rail stations	High	
	Traffic density of countries		Medium	
Reliability	Infrastructure levels of countries on the route		Medium-high	Very high
	Equipment potential of hubs		Medium-high	
	Equipment potential of seaports		High	
	Equipment potential of rail stations		High	
	Feasibility of routing schedules		High	
Risk	Cabotage limitations		High	High
	Political risks		Very high	
	Potential damages during the transportation process		Medium-high	
	Transportation/fleet risks		High	
	Adequacy of security precautions		Very high	

(Continued)

TABLE 16.5 (Continued)

Functional Requirements (FRs)			Linguistic variables	
Environment	Environmental impact of road freight transport	CO_2 emission by road transport	High	Medium-high
		NO emission by road transport	Medium-high	
		Diesel consumption in road transport	Medium	
	Environmental impact of marine transport	CO_2 emission by marine transport	Medium-low	
		Fuel consumption in marine transport	Very high	
	Environmental impact of railway transport	CO_2 emission by rail transport	Medium-high	
		Electricity consumption in rail transport	Medium	

content of the fuel consumption of Ro-Ro vessels in marine transport was also computed in Equation (16.30) for the intermodal transport route-5 again.

$$I_{13} = 1.57 \Rightarrow WI_{13} = \left[log_2 \left(\frac{1}{P_{13}} \right) \right]^{w_1 * w_2 * w_3} = (1.57)^{0.659 * 0.191 * 0.216} = 1.012 \qquad (16.28)$$

$$I_{(12)5} = 1.0 \Rightarrow WI_{(12)5} = w_1 * w_2 = 0.138 * 0.216 = 0.029 \qquad (16.29)$$

$$I_{(41)5} = 0.807 \Rightarrow WI_{(41)5} = \left[log_2 \left(\frac{1}{P_{(41)5}} \right) \right]^{\frac{1}{w_1 * w_2 * w_3}} = (0.807)^{\frac{1}{0.5 * 0.339 * 0.157}} = 0.0003 \qquad (16.30)$$

In addition to these positive weighted information content calculations, some of the negative-valued weighted information contents were also exemplified in Equations (16.31)–(16.33). For instance, weighted information contents of the loading and unloading times at rail stations were calculated in Equations (16.31) and (16.32) for the intermodal alternative transport routes 5 and 6, respectively. Finally, one can compute the weighted information content of waiting times at customs in marine transport route 4 as in Equation (16.33). All other weighted information contents were calculated in a similar manner for each FR of each transport route option and reported in Table 16.7.

$$WI_{(22)5} = - \left[\left| log_2 \left(\frac{1}{P_{(22)5}} \right) \right|^{1 + (w_1 * w_2 * w_3)} \right] = -[|-1.584|^{1 + (0.191 * 0.263 * 0.216)}] = -1.591$$

$$(16.31)$$

TABLE 16.6

Hierarchical Crisp Weights of the Criteria/FRs after the Defuzzification and Normalization Processes.

	Functional Requirements (FRs)		Weights		
Cost	Road freight transport cost	Fuel cost for loaded/empty movements	0.659	0.191	0.216
		Road toll	0.341		
	Marine transport cost	Shipping cost for Ro-Ro vessels	0.236	0.191	
		Anchorage fee	0.171		
		Port usage cost	0.211		
		Port tax	0.211		
		Demurrage tax	0.171		
	Railway transport cost	Cost of block and public trains	0.527	0.191	
		Terminal processing cost	0.473		
	Shipping fee			0.191	
	Loading/unloading charges			0.099	
	Insurance premium			0.138	
Delivery Time	Road transit times	Customs wait times in road transport	0.284	0.293	0.216
		Unloading times at hubs	0.284		
		Travel times in road transport	0.284		
		Loading times at hubs	0.147		
	Marine transit times	Customs wait times in marine transport	0.264	0.293	
		Unloading times at seaports	0.264		
		Ro-Ro transfer times	0.364		
		Loading times at seaports	0.236		
	Rail transit times	Customs wait times in railway transport	0.213	0.263	
		Loading times at rail stations	0.191		
		Transfer times of block trains	0.191		
		Transfer times of public trains	0.191		
		Unloading times at rail stations	0.213		
	Traffic density of countries			0.152	
Reliability	Infrastructure levels of countries on the route			0.175	0.216
	Equipment potential of hubs			0.175	
	Equipment potential of seaports			0.217	

(*Continued*)

TABLE 16.6 (Continued)

	Functional Requirements (FRs)		Weights		
	Equipment potential of rail stations		0.217		
	Feasibility of routing schedules		0.217		
Risk	Cabotage limitations		0.198	0.194	
	Political risks		0.221		
	Potential damages during the transportation process		0.160		
	Transportation/fleet risks		0.198		
	Adequacy of security precautions		0.221		
Environment	Environmental impact of road freight transport	CO_2 emission by road transport	0.406	0.419	0.157
		NO emission by road transport	0.141		
		Diesel consumption in road transport	0.453		
	Environmental impact of marine transport	CO_2 emission by marine transport	0.500	0.339	
		Fuel consumption in marine transport	0.500		
	Environmental impact of railway transport	CO_2 emission by rail transport	0.500	0.242	
		Electricity consumption in rail transport	0.500		

$$WI_{(23)6} = -\left[\left| log_2\left(\frac{1}{P_{(23)6}}\right) \right|^{|1-(w_1 * w_2 * w_3)|} \right] = -[|-0.584|^{1-(0.213*0.263*0.216)}]$$

$$= -0.587 \tag{16.32}$$

$$I_{(17)4} = -1 \Rightarrow WI_{(17)4} = -(1 + w_1 * w_2 * w_3) = -(1 + 0.264 * 0.293 * 0.216)$$

$$= -1.016 \tag{16.33}$$

In Figure 16.11, the most sustainable ways of freight transportation were demonstrated on the intermodal logistics network. In other words, the best three transport routes obtained from the crisp, FAD, and WFAD methodologies were presented in that figure. According to Tables 16.4 and 16.7, weighted and unweighted information contents lead to very close rankings for the alternative transport routes. The specified weights within the criteria hierarchy took similar values since most of the FRs were evaluated as "High", "Very High" or "Medium High" by the company experts. Both results in terms of the weighted/unweighted information contents have shown that intermodal and combined marine and road transport options are the most sustainable ways of freight transportation between Turkey and Europe. In

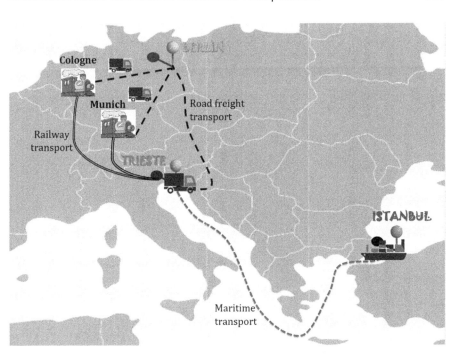

FIGURE 16.11 The most sustainable transport routes on the intermodal logistics network.

detail, the combined marine and road transport via Trieste seaport was ranked as the third routing option by the crisp and FAD methodologies, whereas it has been later ranked as fourth by the WFAD method. Additionally, it was also depicted in Figure 16.11 that Trieste seaport, Munich, and Cologne railway stations should be the most frequently used terminals/hubs on these sustainable transport routes.

5 CONCLUDING REMARKS

In this research, an improved extension of the weighted hierarchical FAD method is proposed and its application to a real-life sustainable route selection problem in an intermodal transportation network is also presented. In detail, the conventional crisp AD method is extended by generating negative-valued information contents (overachievement) as well as positive-valued information contents (under-achievement). Moreover, the conventional WFAD method is also modified by incorporating the criteria hierarchy and the negative-valued information contents. Finally, an evaluation procedure is proposed by considering both total positive- and negative-valued information contents for the FRs. In the application study, the crisp and fuzzy data for the design and system ranges are first obtained from the company experts by performing a comprehensive questionnaire survey. Then, the computational results in terms of both weighted and unweighted information contents are generated and compared with each other. The company experts have announced that the best transport routes selected by the proposed improved extension of the

TABLE 16.7

Results of Weighted Information Contents for Alternative Routes in an Intermodal Transportation Network.

FRs	Criteria	Sub-criteria	Weighted Information Contents (W_l)						
			Route No. 1	Route No. 2	Route No. 3	Route No. 4	Route No. 5	Route No. 6	Route No. 7
COST	Road freight transport cost	Fuel cost for loaded/empty movements	INF	INF	1.012	0.0028	−8.577	−8.887	−8.176
		Road tolls	1.019	1.013	−6.664	−5.447	−7.273	−7.273	−7.273
	Marine transport cost	Shipping cost for RO-RO vessels	N.A	−7.487	−1.009	1.009	−1.009	−1.009	−1.009
		Anchorage fee	N.A	0	−2.601	1.002	−2.601	−2.601	−2.601
		Port usage cost	N.A	0	−5.303	1.006	−5.303	−5.303	−5.303
		Port tax	N.A	−6.513	1.009	−0.586	1.009	1.009	1.009
		Demurrage tax	N.A	1.005	0	1.002	−3.35	1.008	0
	Railway transport cost	Cost of block and public trains	N.A	N.A	N.A	N.A	0	1.005	0
		Terminal processing costs	N.A	N.A	N.A	N.A	0	0	0
		Shipping rate/fee	1.041	0	−8.99	−0.278	−8.468	−9.24	−0.934
		Loading/unloading charges	−7.488	1.026	0	1.011	1.004	1.021	1.034
		Insurance premium	0.0007	0	−3.192	1.019	0.029	−3.786	0
DELIVERY TIME	Road transit times	Waiting times at customs	1.0125	1.0125	−1.826	1.0125	N.A	N.A	N.A
		Loading times at hubs	0	0	0	−1.324	0	0	−0.225
		Unloading times at hubs	0.018	0	1.005	−0.269	−1.597	−1.018	−0.589
		Highway travel times	1.0125	1.01	0	0	−3.15	−2.859	−2.962
	Marine transit times	Waiting times at customs	N.A	−1.327	−1.017	−1.017	−1.017	−1.017	−1.017

(Continued)

TABLE 16.7 (Continued)

FRs	Criteria	Sub-criteria	Weighted Information Contents (WI_f)						
			Route No. 1	Route No. 2	Route No. 3	Route No. 4	Route No. 5	Route No. 6	Route No. 7
		Loading times at seaports	N.A	-2.021	-1.326	0	-1.326	-1.326	-1.326
		Unloading times at seaports	N.A	-2.496	0	0	0	0	0
		Transfer times via RO-RO vessels	N.A	-3.753	0.023	1.016	0.023	0.023	0.023
	Rail transit times	Waiting times at customs	N.A	N.A	N.A	N.A	0	0	0
		Loading times at rail stations	N.A	N.A	N.A	N.A	-1.591	0	0
		Unloading times at rail stations	N.A	N.A	N.A	N.A	0	-0.587	-2.016
		Transfer times via block trains	N.A	N.A	N.A	N.A	N.A	0	-2.015
		Transfer times via public trains	N.A	N.A	N.A	N.A	0	N.A	N.A
	Traffic density of countries		INF	INF	1.026	0	-0.184	-0.184	-0.184

FRs	Criteria	Sub-criteria	Weighted Information Contents (WI_f)						
			Route No. 1	Route No. 2	Route No. 3	Route No. 4	Route No. 5	Route No. 6	Route No. 7
RELIABILITY	Infrastructure levels of the countries on the transport route		INF	1.045	0	1.045	-0.378	-0.378	-0.378
	Equipment potentials of hubs/consolidation centers		1.045	1.045	0	0	0	0	0
	Equipment potentials of seaports		N.A	1.055	1.055	0	0	0	0
	Equipment potentials of rail stations		N.A	N.A	N.A	N.A	N.A	-0.381	1.055
	Feasibility and flexibility of route schedules		INF	INF	0	1.048	0	0	0

(Continued)

TABLE 16.7 (Continued)

FRs	Criteria	Sub-criteria	Weighted Information Contents (WI_i)						
			Route No. 1	Route No. 2	Route No. 3	Route No. 4	Route No. 5	Route No. 6	Route No. 7
RISK		Cabotage limitations	N.A	INF	0	1.03	0	0	0
		Political risks at the countries on the transport route	1.033	1.033	0	0	0	-0.187	0
		Potential damages to freights during transportation operations	INF	INF	1.024	1.024	0	0	0
		Transportation and fleet risks	INF	INF	1.018	1.018	1.018	0	1.018
		Security precautions at the countries on the transport route	INF	INF	0	1.044	0	0	1.044
ENVIRONMENT	Environmental impact by road freight transport	CO_2 emission by road freight transport	1.029	1.025	-1.84	0	-2.65	-2.374	-2.519
		NO emission by road freight transport	INF	INF	0	0	-3.2	-3.358	-3.282
		Diesel consumption	INF	INF	1.009	1.024	-7.179	-7.315	-7.471
	Environmental impact by marine transport	CO_2 emission by marine transport	N.A	-4.1	0	0	0	0	0
		Fuel consumption by RO-RO vessels	N.A	-5.265	0.0003	1.011	0.0003	0.0003	0.0003
	Environmental impact by railway transport services	CO_2 emission by railway transport	N.A	N.A	1.009	N.A	0.019	1.009	-2.358
		Electricity consumption by block/public trains	N.A	N.A	N.A	N.A	0	0	-15.37
Total positive weighted information contents (WPI_i)			INF	INF	8.181	16.32	3.102	5.075	5.183
Total negative weighted information contents (WNI_i)			-7.488	-32.96	-33.77	-8.92	-58.85	-59.08	-67.01
Ranking of the alternative routes			7	6	4	5	1	2	3

hierarchical WFAD method can be accepted as the most sustainable and applicable options by the logistics company. On the other hand, while realizing a selected transport route, an unexpected event may occur and, therefore, some parts of the route may not be realized. In such a case, rerouting is needed, which necessitates the resolution of the problem. A dynamic and online decision support system that incorporates the proposed extensions can be developed in the future. Furthermore, the development of an interval type-2 hierarchical WFAD method is also scheduled as a future work.

ACKNOWLEDGMENT

We would like to offer many thanks to the business development engineers in Ekol Logistics, Inc. who provided insight and expertise that greatly assisted this research.

REFERENCES

Ahmed, N. G., & Asmael, N. M. (2009). A GIS-Assisted optimal urban route selection based on multi criteria approach. *The Iraqi Journal for Mechanical and Material Engineering*, 2, 557–567.

Alkubaisi, M. I .T. (2014). Predefined evaluating criteria to select the best tramway route. *Journal of Traffic and Logistics Engineering*, 2(3), 211–217.

Babic, B. (1999). Axiomatic design of flexible manufacturing systems. *International Journal of Production Research*, 37(5), 1159–1173.

Banai, R. (2006). Public transportation decision-making: A case analysis of the Memphis light rail corridor and route selection with analytic hierarchy process. *Journal of Public Transportation*, 9 (2), 1–24.

Baykasoğlu, A., & Subulan, K. (2016). A multi-objective sustainable load planning model for intermodal transportation networks with a real-life application. *Transportation Research Part E*, 95, 207–247.

Baykasoğlu, A., & Subulan, K. (2019). A fuzzy-stochastic optimization model for the intermodal fleet management problem of an international transportation company. *Transportation Planning and Technology*, 42(8), 777–824.

Baykasoglu, A., Subulan, K., Tasan, A.S., & Dudakli, N. (2019a). A review of fleet planning problems in single and multimodal transportation systems. *Transportmetrica A: Transport Science*, 15(2), 631–697.

Baykasoglu, A., Subulan, K., Taşan, A.S., Dudakli, N., Turan, M., Çelik, E., & Ülker, Ö. (2019b). Development of a web-based decision support system for strategic and tactical sustainable fleet management problems in intermodal transportation networks. *Lean and Green Supply Chain Management*, Springer, Cham, pp. 189–230.

Büyüközkan, G., & Göçer, F. (2017). Application of a new combined intuitionistic fuzzy MCDM approach based on axiomatic design methodology for the supplier selection problem. *Applied Soft Computing*, 52, 1222–1238.

Çakir, S. (2018). An integrated approach to machine selection problem using fuzzy SMART-fuzzy weighted axiomatic design. *Journal of Intelligent Manufacturing*, 29, 1433–1445.

Celik, M., Cebi, S., Kahraman, C., & Er, I.D. (2009). Application of axiomatic design and TOPSIS methodologies under fuzzy environment for proposing competitive strategies on Turkish container ports in maritime transportation network. *Expert Systems with Applications*, 36(3), 4541–4557.

Chen, W., Goh, M., & Zou, Y. (2018). Logistics provider selection for omni-channel environment with fuzzy axiomatic design and extended regret theory. *Applied Soft Computing*, 71, 353–363.

Cicek, K., & Celik, M. (2010). Multiple attribute decision-making solution to material selection problem based on modified fuzzy axiomatic design-model selection interface algorithm. *Materials & Design*, 31(4), 2129–2133.

Farkas, A. (2009). Route/site selection of urban transportation facilities: An integrated GIS/MCDM approach. *7th International Conference on Management, Enterprise and Benchmarking*, (June 2010), 169–184.

Hamurcu, M., & Eren, T. (2016). A multi-criteria decision-making for monorail route selection in Ankara. *International Journal of Industrial Electronics and Electrical Engineering*, 4(5), 121–125.

Hsieh, T.Y., Lu, S.T., & Tzeng, G.H. (2004). Fuzzy MCDM approach for planning and design tenders selection in public office buildings. *International Journal of Project Management*, 22(7), 573–584.

Jacyna, M., & Wasiak, M. (2015). Multicriteria decision support in designing transport systems. In *Tools of Transport Telematics*(pp. 11–23). Springer.

Kahraman, C., & Cebi, S. (2009). A new multi-attribute decision making method: Hierarchical fuzzy axiomatic design. *Expert Systems with Applications*, 36(3), 4848–4861.

Kannan, D., Govindan, K., & Rajendran, S. (2015). Fuzzy axiomatic design approach based green supplier selection: a case study from Singapore. *Journal of Cleaner Production*, 96(1), 194–208.

Kazan, H., & Çiftci, C. (2013). Transport path selection: Multi-criteria comparison. *International Journal of Operations and Logistics Management*, 2(4), 33–48.

Kengpol, A., Tuammee, S., & Tuominen, M. (2014). The development of a framework for route selection in multimodal transportation. *The International Journal of Logistics Management*, 25(3), 581–610.

Ko, H. J. (2009). A DSS approach with Fuzzy AHP to facilitate international multi-modal transportation network. *KMI International Journal of Maritime Affairs and* Fisheries, 1, 51–70.

Kulak, O., & Kahraman, C. (2005a). Fuzzy multi-attribute selection among transportation companies using axiomatic design and analytic hierarchy process. *Information Sciences*, 170, 191–210.

Kulak, O., & Kahraman, C. (2005b). Multi-attribute comparison of advanced manufacturing systems using fuzzy vs. crisp axiomatic design approach. *International Journal of Production Economics*, 95, 415–424.

Kumar, A., & Anbanandam, R. (2020). Analyzing interrelationships and prioritising the factors influencing sustainable intermodal freight transport system: A grey-DANP approach. *Journal of Cleaner Production*, 252, Article 119769.

Moon, D. S., Kim, D. J., & Lee, E. K. (2015). A study on competitiveness of sea transport by comparing international transport routes between Korea and EU. *The Asian Journal of Shipping and* Logistics, 31(1), 1–20.

Murat, Y. Ş., & Kulak, O. (2005). Route selection in transportation networks using information axiom. *Pamukkale University Engineering College Journal of Engineering Sciences*, 11(3), 425–435.

Nosal, K., & Solecka, K. (2014). Application of AHP method for multi-criteria evaluation of variants of the integration of urban public transport. *Transportation Research Procedia*, 3, 269–278.

Odeyale, S. O., Alamu, O. J., & Odeyale, E. O. (2014). Performance evaluation and selection of best mode of transportation in Lagos state metropolis. *International Journal for Traffic and Transport Engineering*, 4(1), 76–89.

Opricovic, S., & Tzeng, G.H. (2003). Defuzzification within a multicriteria decision model. *International Journal of Uncertainty, Fuzziness and Knowledge-Based Systems*, 11(5), 635–652.

Owczarzak, L., & Zak, J. (2015). Design of passenger public transportation solutions based on autonomous vehicles and their multiple criteria comparison with traditional forms of passenger transportation. *Transportation Research Procedia*, 10, 472–482.

Özceylan, E. (2010). A decision support system to compare the transportation modes in logistics. *International Journal of Lean Thinking*, 1(1), 58–83.

Özel, B., & Özyörük, B. (2007). Supplier selection with fuzzy axiomatic design. *Journal of the Faculty of Engineering and Architecture of Gazi University*, 22(3), 415–423.

Pahlavani, P., & Delavar, M. R. (2014). Multi-criteria route planning based on a driver's preferences in multi-criteria route selection. *Transportation Research Part C*, 40, 14–35.

Qu, L., & Chen, Y. (2008). A hybrid MCDM method for route selection of multimodal transportation network. *Lecture Notes in Computer Science*, 5263, 374–383.

Sawadogo, M., & Anciaux, D. (2009). Intermodal transportation within the green supply chain: An approach based on the ELECTRE method. *International Conference on Computers & Industrial Engineering*, (July 2009), pp. 839–844.

Shahbandarzadeh, H., Jafarnejad, A., & Raeesi, R. (2011). Using a modified extension of mixed fuzzy and crisp axiomatic design method in vendor selection for supply chain. *Journal of Industrial Management*, 3(7), 37–54.

Suh, N. P. (2001a). Design of thinking design machine. *Annals of the CIRP*, 39(1), 145–149.

Suh, N. P. (2001b). Designing-in of quality through axiomatic design. *IEEE Transactions on Reliability*, 44(2), 256–264.

Suh, N.P. (2001c). *Axiomatic Design—Advances and Applications*. Oxford University Press.

Vinodh, S., Kamala, V., & Jayakrishna, K. (2015). Application of fuzzy axiomatic design methodology for selection of design alternatives. *Journal of Engineering, Design and Technology*, 13(1), 2–22.

Wang, Y., & Yeo, G.T. (2018). Intermodal route selection for cargo transportation from Korea to Central Asia by adopting fuzzy Delphi and fuzzy ELECTRE I methods. *Maritime Policy & Management*, 45(1), 3–18.

Wu, H.Y., Tzeng, G.H., & Chen, Y.H. (2009). A fuzzy MCDM approach for evaluating banking performance based on Balanced Scorecard. *Expert Systems with Applications*, 36(6), 10135–10147.

Yedla, S., & Shrestha, R.M. (2003). Multi-criteria approach for the selection of alternative options for environmentally sustainable transport system in Delhi. *Transportation Research Part A*, 37(8), 717–729.

17 Minimizing Delays on Single-Line Train Scheduling

Muzaffer Alım[1], Çağrı Koç[2], and Saadettin Erhan Kesen[3]

[1]Technology Faculty, Batman University, Batı Raman Kampüsü, Yenişehir Mah., Batman, 72000, Turkey
[2]Department of Business Administration, Social Sciences University of Ankara, Hükümet Meydanı, No:2, Ulus, Ankara, Turkey
[3]Industrial Engineering Department, Konya Technical University, Alaeddin Keykubat Kampüsü, Selçuklu, Konya, Turkey

1 INTRODUCTION

The excessive demand for transportation and the higher costs of energy sources has greatly increased the importance of public transportation. In addition, the increasing potential environmental problems caused by personal transport reveal the need to direct people to public transport. In line with all these, railway transportation that is effective not only for passengers but also for freight transportation stands, out with its economy, safety, high capacity, and cleanliness in terms of CO_2 emissions (Wang et al. 2014).

Railway transportation has always been given great importance in Turkey since its foundation; it has made large financial investments in the infrastructure system. With the investments in recent years on high-speed trains, the railways have started to appeal to a larger audience and gain more importance. As of 2018, the Turkish Republic State Railways (TCDD) — which is the monopolist public railways corporation — has control over railway systems containing a total of 12,740 km of lines, 11,527 km of which are conventional and 1,213 km of which are high-speed line. In this, 101 million passengers were transported in 2018 and 5,560 million passenger-km were produced accordingly. The occupancy rate in passenger trains remained at 65% (TCDD 2018a).

In 2018, despite all the efforts, the share of railways in passenger transportation holds only 2.5% and the ratio of freight is at 3.7% in total (TCDD 2018b). It is essential to make railways more attractive to increase this ratio that is below the European average. Moreover, the railway is a greener and cleaner type of transportation mode than the road in environmental impacts. The average CO_2 emission (g/tonne-km) is 45 for diesel-electric powered railways, 123 for road freight transportation. The numbers show that railway transport produces almost three times less CO_2 emission than road transport. A similar effect can be observed on the emission of nitrous oxide (NO_x) — road transport produces 75% more NO_x as compared to railways (Wee et al. 2005). Railway transportation is much more advantageous in case of the required occupancy rate, provided not only in terms of emissions but also energy consumption efficiency (Skrucany et al. 2017).

All aforementioned indicators show the importance of railway transportation in terms of both economic and environmental impacts (Wang et al. 2014, Lin et al. 2017). However, it has been seen that the share of the railways in passenger transportation is very low and the occupancy rate of trains is not at the desired level. As its reasons are examined, we see that one of the most important is the high travel times on the railways. The length of these is not only related to the speed of trains, but also to exceptionally long delays. One of the biggest causes arise from a single line. For trains that use the same line at the same time period, to continue on their way, they must yield to the train coming from the opposite direction. Thus, one of the trains waits at a station for another to come so that the line opens. When we consider the fact that approximately 95% of the total railways in Turkey is a single line, its huge impact needs more attention. It is possible to invest in double lines — a definitive solution to the problem — but it brings forbiddingly high financial costs. For this reason, scheduling on the single line needs to be done in the most appropriate way to use on hand resources more efficiently.

The train scheduling problem has been widely studied in literature and aims to find the arrival/departure times of trains at/from every station under some operational constraints (Xu et al. 2019b). We refer readers to the surveys of Wang et al. (2011), Fang et al. (2015), and Scheepmaker et al. (2017) for up-to-date coverage on the train schedule.

The train scheduling problem on a single line is to decide how long the trains wait at which station to minimize waiting times of the trains for meeting or overtaking. Szpigel (1973) introduced the problem and formulated it with mathematical modeling. Higgins et al. (1996) developed an integer mathematical model to minimize delays and operating costs of trains. Higgins et al. (1997) reviewed the application of different heuristic methods including tabu search, genetic algorithm, and local search into the single line train scheduling problem. Dessouky and Leachman (1995) argued that each delay in the system affects the plans of all other system elements due to this huge interaction (i.e. ripple effect, the analytical solutions are inadequate). Thus, they proposed a simulation methodology to analyze the different railway networks including double and single-line systems. Gafarov et al. (2015) revised the train schedule between two stations, where the train speed is considered equal as a special case of the single machine equal processing time scheduling. They proposed polynomial-time solution algorithms. Liu and Kozan (2009) modeled the problem as blocking the parallel-machine job shop scheduling problem. Additionally, various solution methods such as simulated-annealing (Yue et al. 2017), genetic algorithm and artificial neural networks (Dündar and Şahin 2013), heuristics based on decomposition of the problem (Corman et al. 2017), neighborhood search (Samà et al. 2017), and ant colony optimization (Sahana et al. 2014) have been used in literature.

Based on real-life applications, some studies have discussed the train scheduling problem under different settings and constraints. Zhou and Zhong (2007) investigated cases when a capacity limit at stations exists. Gafarov et al. (2015) assumed that the speed of the trains on the single line is the same. Unlike this study, the situation where trains have different speeds, — even speeds are considered as decision variables — have also been examined (Luan et al. 2018, Yang et al. 2018). Heydar et al. (2013) investigated the optimization of the cyclic timetable of the railway that contains two types of trains — express and local. With an increase in the use of high-speed trains, the attention on train scheduling optimization shifted from conventional lines to high-speed trains (Yu et al. 2018, Chen et al. 2017, Yang et al. 2016). Some studies not only considered operational optimization, but also the customer side. Corman et al. (2017) addressed the problem by assessing passenger perspectives on delay management and Parbo et al. (2016) provided a review of passenger views on trains schedules.

Recent studies have addressed the problem from different perspectives as a result of the increased importance and effectiveness of railway transport and advanced technology. In this context, researchers focused on the integration of the problem with others of the same environment and the inclusion of different constraints. Xu et al. (2019a) considered the integration of train scheduling and locomotive assignment problem with the constraint of limited locomotives. Similarly, some other studies — such as the integration of the problem with speed profile optimization of trains (Yang et al. 2018) and fleet maneuvers scheduling (Wang et al. 2018) — examined different decision variables simultaneously. Samà et al. (2017) combined the routing of trains with scheduling decisions and proposed some meta-heuristic methods for the solution.

Increased CO_2 emission encourages researchers to consider the train schedule by also considering the minimization of its environmental impacts. The most important motivation here is that transportation is responsible for 22% of total CO_2 emissions worldwide (Miandoab et al. 2020). Lin et al. (2017) compared the environmental impacts of highways and railways. Some studies (Miandoab et al. 2020, Lin et al. 2017) aimed to reduce the negative environmental impacts of railways minimize the carbon emission in the objective function, while others minimize the amount of fuel/energy consumption (Higgins et al. 1996, Montrone et al. 2016). In addition, Li et al. (2013) formulated the multi-objective model to minimize energy consumption, carbon emission, and total passenger travel time simultaneously. Yang et al. (2018) highlighted the integration of schedule and speed optimization to achieve more energy-efficient transportation for a metro-line. Based on their case, there is a possibility of reducing energy consumption by 4.52%.

All the above-mentioned studies show that railway transportation is superior to other modes of transportation in many aspects. For this reason, it is important to improve railway operations, thereby increasing railway utilization and meeting passengers' expectations. In this study, a mathematical model is proposed to minimize delays resulting from a single line. The model is tested numerically on a specified line and potential improvements are demonstrated by comparisons made between the proposed and the existing train schedules.

2 PROBLEM DEFINITION

The desired occupancy rate must be achieved to obtain the expected benefit from railway transport. This is accomplished by improving the service quality and minimizing delays.

In railways, the waiting times of trains at a particular station generally consist of three categories. The standard waiting is for loading and unloading of passengers. This is a requirement for passenger trains at the stations on the route. Similarly, freight trains also wait for the products to be loaded/unloaded. Emergency waiting is the second type and it emerges from a breakdown or incident. The train waits at the station until the emergency case is tackled and the line opens. The third type is waiting for a meeting with another train coming from the opposite direction, or overtaking a train in the same direction on a single line.

When trains from opposite directions use the same train line at the same time, one of them needs to wait at the station until the other train arrives. With the arrival of another train at the planned station for a meeting, the line becomes clear for the waiting train to continue its way. This is called the meeting of trains and it can take place only at stations with multiple lines. An illustration of the trains meeting is presented in Figure 17.1.

In some situations, trains moving in the same direction on the same line may also have to meet because the train ahead may be slow, which is more valid for freight trains. In this case, it will be more convenient if the faster train coming from behind passes the slow train. For this to happen, the slow train ahead waits at the designated station. The other train arrives and leaves that station earlier. This event is known as the overtaking of trains. This situation is represented in Figure 17.2.

For both situations, one of the trains needs to wait for another to arrive at the designated station. For the single line, this waiting type is almost inevitable but can be minimized with better planning.

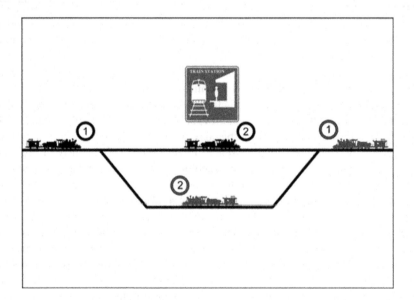

FIGURE 17.1 Meeting of Trains.

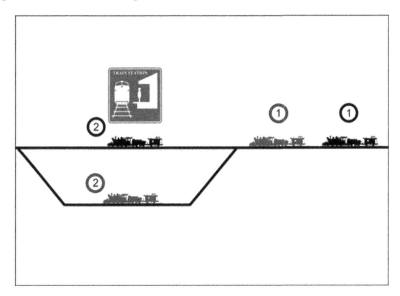

FIGURE 17.2   Overtaking of Trains.

3   MATHEMATICAL FORMULATION

We now present our mathematical formulation to minimize the waiting times due to the meeting and overtaking of trains with its relevant assumptions and notation.

3.1   ASSUMPTIONS

- In the model, it was considered how long the trains travel between the two consecutive stations instead of their speeds. These travel duration are assumed to be constant but can vary for different trains.
- The operation time for shunting is neglected. This is because the train arriving at the station later enters on a different line from the waiting train. Thus, the line can be opened for the waiting train.
- Standard waiting time (loading/unloading) of trains varies depending on the station and the trains. These waits, however, are assumed to be constant for different journeys.
- All the railways, excluding the stations, are assumed to be single line. Thus, meeting/overtaking can only take place at stations.
- We assume that there are enough lines at the stations for multiple trains' meeting/overtaking operations.
- It is predetermined which train pairs, moving in opposite directions, will need to meet but the location of this meeting is not known in advance.
- The index numbers of the trains are given according to their departure from the origin station. For overtaking, only the train with a big index number can pass a train with a smaller index number.

TABLE 17.1

Notation

Parameters

N	Number of stations
n	Number of inbound trains
m	Number of outbound trains
i, c	index of inbound trains, $i,c = \{1,2,...,n\}$
f, t	index of outbound trains, $f,t = \{n+1, n+2, ... ,n+m\}$
j	index of station, $j = \{1,2, ... ,N\}$
$w_i(j)$	standard waiting time of train i at station j
$p_i(k, j)$	travelling time of train i from station k to j
d_i	departure time of train i from the first station
$a_i(j)$	arriving time of train i to station j
$d_i(j)$	departure time of train i from station j
$I_i(j)$	waiting time of train i at station j due to single line
$org(i)$	origin station of train I
$des(i)$	destination station of train I
$\theta(i,f)$	the set of possible stations for meeting of trains i and f
$\varphi(i,c)$	the set of possible stations for overtaking of trains i and c
$\beta(i,f)$	the set of pairs of trains i and f to meet
M	a large number

3.2 Notation

The notation used in the mathematical formulation includes parameters as the input of the mathematical model and decision variables as the output of the model.

Parameters

Table 17.1 presents the parameters used in the formulation.

Decision Variables

The decision variables are all binary and determine at which station meeting/overtaking occurs.

$$x_{\text{if}}(j) = \begin{cases} 1, & \text{\textit{if train i arrives to station j before f}} \\ & \text{\textit{and meeting occurs,}} \\ 0, & \text{\textit{otherwise}} \end{cases} \qquad (17.1)$$

$$x_{\text{ic}}(j) = \begin{cases} 1, & \text{\textit{if train i arrives to station j before train c}} \\ & \text{\textit{and overtaking occurs,}} \\ 0, & \text{\textit{otherwise}} \end{cases} \qquad (17.2)$$

$$y_{ic}(j) = \begin{cases} 1, & \begin{array}{l} \textit{if there is overtaking between train c and i} \\ \textit{at station j,} \end{array} \\ 0, & \textit{for the stations before overtaking occurs} \end{cases} \quad (17.3)$$

3.3 MATHEMATICAL FORMULATION

The integer programming formulation of the problem is as follows.

$$Min\ z = \sum_{j=0}^{N} \left(\sum_{i=1}^{n} l_i(j) + \sum_{f=n+1}^{n+m} l_f(j) \right) \quad (17.4)$$

$$a_i(j) = d_i(j-1) + p_i(j-1,j) \quad \forall\ (i),\ org(i)+1 \le j \le dest(i) \quad (17.5)$$

$$a_f(j) = d_f(j+1) + p_f(j+1,j) \quad \forall\ (f),\ dest(f)+1 \le j \le org(f)-1 \quad (17.6)$$

$$d_i(j) = a_i(j) + w_i(j) + l_i(j) \quad \forall\ (i),\ org(i)+1 \le j \le dest(i)-1 \quad (17.7)$$

$$d_f(j) = a_f(j) + w_f(j) + l_f(j) \quad \forall\ (f),\ dest(f)+1 \le j \le org(f)-1 \quad (17.8)$$

$$\sum_{j \in \theta(i,f)} [X_{if}(j) + X_{fi}(j)] = 1 \quad \forall\ i,f \in \beta \quad (17.9)$$

$$\sum_{j \in \Phi(i,c)} X_{ic}(j) \le 1 \quad \forall\ c,i \quad (17.10)$$

$$\sum_{j \in \Phi(t,f)} X_{ft}(j) \le 1 \quad \forall\ t,f \quad (17.11)$$

$$Y_{ic}(j) = \sum_{k=0}^{j} X_{ic}(k) \quad \forall\ c,i,j \quad (17.12)$$

$$Y_{ft}(j) = \sum_{k=j}^{N} X_{ft}(k) \quad \forall\ t,f,j \quad (17.13)$$

$$d_i(org(i)) = d_i \quad \forall\ i \quad (17.14)$$

$$d_f(org(f)) = d_f \quad \forall\ f \quad (17.15)$$

$$a_f(j) \le d_i(j) + \left(1 - X_{if}(j)\right) \times M \quad \forall\ i,f \in \theta(i,f) \quad (17.16)$$

$$a_i(j) \leq a_f(j) + (1 - X_{if}(j)) \times M \quad \forall \ i, f \in \theta(i, f) \tag{17.17}$$

$$a_i(j) \leq d_f(j) + (1 - X_{fi}(j)) \times M \quad \forall \ i, f \in \theta(i, f) \tag{17.18}$$

$$a_f(j) \leq a_i(j) + (1 - X_{fi}(j)) \times M \quad \forall \ i, f \in \theta(i, f) \tag{17.19}$$

$$d_c(j) \leq d_i(j) + (1 - X_{ic}(j)) \times M \quad \forall \ i, c \in \Phi(i, c) \tag{17.20}$$

$$a_i(j) \leq a_c(j) + (1 - X_{ic}(j)) \times M \quad \forall \ i, c \in \Phi(i, c) \tag{17.21}$$

$$d_t(j) \leq d_f(j) + (1 - X_{ft}(j)) \times M \quad \forall \ t, f \in \Phi(t, f) \tag{17.22}$$

$$a_f(j) \leq a_t(j) + (1 - X_{ft}(j)) \times M \quad \forall \ t, f \in \Phi(t, f) \tag{17.23}$$

$$d_i(j) \leq d_c(j) + Y_{ic}(j) \times M \quad \forall \ i, c \in \Phi(i, c) \tag{17.24}$$

$$d_c(j) \leq d_i(j) + (1 - Y_{ic}(j)) \times M \quad \forall \ i, c \in \Phi(i, c) \tag{17.25}$$

$$d_f(j) \leq d_t(j) + Y_{ft}(j) \times M \quad \forall \ t, f \in \Phi(t, f) \tag{17.26}$$

$$d_t(j) \leq d_f(j) + (1 - Y_{ft}(j)) \times M \quad \forall \ t, f \in \Phi(t, f) \tag{17.27}$$

In this formulation, the objective function (17.4) minimizes the sum of the total delays due to meeting and overtaking of trains. Constraints (17.5) and (17.6) state the arrival times of the trains to each station by using the departure time from the previous stations and travel time. Constraints (17.7) and (17.8) determine the departure times based on the arrival times, standard waiting time, and delays due to a single line at the station. Constraint (17.9) ensures the meeting of the train pairs that must meet. Constraints (17.10) and (17.11) show that overtaking is not necessary and may occur only once. Constraints (17.12) and (17.13) set the binary variable "y" value to "1" for the station where the overtaking takes place and the following stations. Constraints (17.14) and (17.15) determine the departure times from initial stations. Constraint (17.16) forces that train i cannot depart until the train f arrives at the station. Constraint (17.17) shows that train i comes before train f if there is a meeting, at which train i need to wait. Constraints (17.18) and (17.19) are valid if the train f comes earlier to the station and waits for train i. Constraints (17.20) and (17.21) ensure that train i cannot depart before c and train i comes to the station earlier if there is an overtaking of c to i. Constraints (17.22) and (17.23) are valid for the overtaking of outbound trains. Constraints (17.24) and (17.25) provide that train c departs after i, unless there is an overtaking between these trains. In case of an overtaking, train c starts to depart before train i. Similar conditions hold for outbound trains by constraints (17.26) and (17.27).

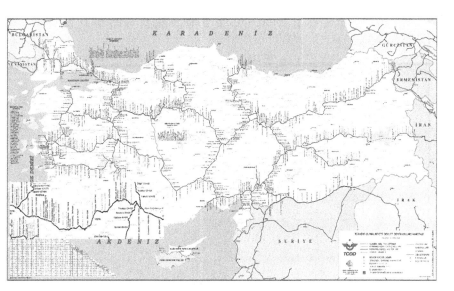

FIGURE 17.3 Railway Map between Manisa and Afyon.

TABLE 17.2
Departure/Destination stations and times of trains

Train	FROM Station	Time	TO Station	Time
1	Manisa	03:15	Afyon	14:35
2	Manisa	04:50	Uşak	13:45
3	Manisa	08:03	Uşak	12:34
4	Manisa	10:25	Afyon	20:34
5	Manisa	12:15	Alaşehir	14:15
6	Manisa	12:32	Afyon	21:57
7	Uşak	20:02	Afyon	23:24
8	Manisa	14:47	Afyon	01:30
9	Manisa	18:05	Alaşehir	20:04
10	Alaşehir	05:30	Manisa	07:25
11	Afyon	05:15	Manisa	15:53
12	Afyon	07:18	Uşak	10:33
13	Alaşehir	15:10	Manisa	17:05
14	Uşak	11:05	Manisa	19:08
15	Uşak	16:00	Manisa	20:35
16	Afyon	13:50	Manisa	23:37
17	Afyon	15:40	Manisa	02:19
18	Afyon	17:55	Manisa	04:39

TABLE 17.3

Traveling Time between Consecutive Stations

Stations	Train																	
	1	2	3	4	5	6	7	8	9	10	11	12	13	14	15	16	17	18
0–1	14	17	12	15	12	15	–	15	12	12	17	–	12	17	13	16	17	18
1–2	16	18	12	17	12	17	–	17	12	11	15	–	11	17	13	15	15	16
2–3	16	15	11	14	10	14	–	14	10	10	14	–	10	15	10	14	14	15
3–4	12	14	9	13	9	14	–	14	9	8	11	–	8	12	8	11	11	12
4–5	22	22	19	21	19	21	–	12	19	19	23	–	19	23	19	23	23	23
5–6	27	28	28	26	28	26	–	26	28	27	26	–	27	26	27	25	26	25
6–7	22	18	21	22	21	22	–	22	21	20	19	–	20	20	15	19	19	19
7–8	17	19	8	18	–	18	–	18	–	–	15	–	–	14	8	15	15	14
8–9	20	51	15	19	–	19	–	19	–	–	18	–	–	17	12	18	18	18
9–10	51	12	36	51	–	50	–	51	–	–	43	–	–	43	32	43	43	43
10–11	14	24	13	13	–	12	–	13	–	–	15	–	–	14	13	15	15	15
11–12	18	28	17	19	–	19	–	19	–	–	30	–	–	30	17	30	30	30
12–13	25	29	18	25	–	26	–	24	–	–	33	–	–	31	17	33	33	32
13–14	24	21	19	26	–	26	–	25	–	–	28	–	–	31	18	28	28	28
14–15	18	–	13	17	–	17	–	17	–	–	22	–	–	22	13	21	22	21
15–16	56	–	–	54	–	56	59	57	–	–	57	57	–	–	–	63	57	57
16–17	18	–	–	19	–	19	20	19	–	–	21	23	–	–	–	21	21	22
17–18	38	–	–	28	–	28	26	27	–	–	23	23	–	–	–	23	23	14
18–19	20	–	–	16	–	18	16	18	–	–	17	17	–	–	–	17	17	17
19–20	31	–	–	30	–	20	32	31	–	–	32	32	–	–	–	32	32	32
20–21	28	–	–	27	–	27	27	28	–	–	27	27	–	–	–	27	27	27
21–22	4	–	–	4	–	4	4	4	–	–	4	4	–	–	–	4	4	4

TABLE 17.4
Comparison of the results with current case

Train	Current Case		Model Results		Change
	Travel	Delay	Travel	Delay	
1	628	98	569	39	59
2	535	81	472	18	63
3	271	0	272	1	-1
4	609	94	536	21	73
5	119	0	126	7	-7
6	555	47	548	40	7
7	202	0	202	0	0
8	642	91	619	68	23
9	119	0	119	0	0
10	115	0	115	0	0
11	638	73	504	29	44
12	195	0	308	45	-45
13	115	0	122	7	-7
14	483	111	449	77	34
15	275	16	264	5	11
16	587	16	616	45	-29
17	639	46	593	0	46
18	644	69	632	54	12
Total	**7371**	**742**	**7066**	**459**	**283**

4 CASE STUDY

The mathematical model introduced in Section 3 is applied to a real case and its effectiveness is measured. The chosen case is a single line railway between the cities of Manisa and Afyon, consisting of a 356 km line, 22 stations, and 18 in-bound/outbound trains. The selected line is shown in Figure 17.3. In the line, although there are more stations, those where trains do not stop and the stations without multiple lines are disregarded. Table 17.2 presents the first status of the trains with their origin and destination stations including the times.

As can be seen from Table 17.2, the trains from 1 to 9 are traveling from Manisa (0) to Afyon (21), or some other stations in between these two cities, but they are all traveling in the same direction. The opposite direction trains are indexed from 10 to 18 and their travel journey is from Afyon to Manisa. Trains from opposite directions must meet mutually to continue their way. The aim of this application is to minimize time loss due to meetings. For the determined scenario, the travel times between the stations (Table 17.3) and the standard waiting times of the trains at the stations are considered based on the real data.

Based on real data, the traveling time, departure times from the origin stations, and the standard waiting times of trains are taken as input to run the mathematical model. Furthermore, it is determined which trains from opposite directions at the current schedule and the same meetings are planned for the model. It is only aimed to determine the optimal stations for these meetings to occur. There is no restriction for the overtaking of trains and the decision has been made by the model. The mathematical model has been solved using CPLEX and optimal meeting/overtaking stations are determined for the selected train line. The results of the model and the current case are compared in Table 17.4.

In Table 17.4, the total travel duration of each train and the delay due to a single line are presented for the current case and model. The delays due to meeting/ overtaking has been reduced from 742 to 459 minutes and achieved improvement is approximately 38%.

When two different schedules are compared, there is a big difference between model results and the current situation with regards to the overtaking of trains. In the current plan, overtaking takes place only between trains 2 and 3, while it occurs between trains 8-9 and 11-12 in the model results. For the meeting cases, all plan meeting is unsurprisingly the same for the current and model schedules. However, the stations in the schedule differ.

5   CONCLUSIONS

This study considered the train scheduling problem on a single line and intended to use railways in a more efficient way with the aid of a developed mathematical model. Previous studies in literature have been examined and the importance of railway transport has been underlined.

The proposed integer mathematical model is applied to a real train line that is about 364 km and lies between the cities of Manisa and Afyon. The single line delays are reduced by about 38% compared to the current schedule.

Although this study is intended for a specific line, a serious improvement has been achieved. The model has a greater improvement potential as a result of the application to the railways across the country. However, addressing such a large problem size will increase the complexity to a greater extent and the solution time of the model. For this reason, the development of bespoke heuristic methods will be a promising future study for the problem.

Moreover, instant position lining of trains can be possible with the advent of developing technology. In this way, all train positions can be known continuously, and, in case of any unexpected event, the model can be run, and the schedule can be updated instantly. So, there will be an advantage of intervening even the slightest delay by fast planning. In addition, current or potential delays can be reported to passengers abruptly. All these can help railway transport increase customer satisfaction and cause passengers to prefer trains more frequently.

REFERENCES

Chen, F., Shen, X., Wang, Z., and Yang, Y. (2017). An evaluation of the low-carbon effects of urban rail based on mode shifts. *Sustainability*, *9*(3), 401.

Corman, F., D'Ariano, A., Marra, A. D., Pacciarelli, D., and Samà, M. (2017). Integrating train scheduling and delay management in real-time railway traffic control. *Transportation Research Part E: Logistics and Transportation Review*, *105*, 213–239.

Dessouky, M. M., and Leachman, R. C. (1995). A simulation modeling methodology for analyzing large complex rail networks. *Simulation*, *65*(2), 131–142.

Dündar, S., and Şahin, İ. (2013). Train re-scheduling with genetic algorithms and artificial neural networks for single-track railways. *Transportation Research Part C: Emerging Technologies*, *27*, 1–15.

Fang, W., Yang, S., and Yao, X. (2015). A survey on problem models and solution approaches to rescheduling in railway networks. *IEEE Transactions on Intelligent Transportation Systems*, *16*(6), 2997–3016.

Gafarov, E. R., Dolgui, A., and Lazarev, A. A. (2015). Two-station single-track railway scheduling problem with trains of equal speed. *Computers & Industrial Engineering*, *85*, 260–267.

Heydar, M., Petering, M. E., and Bergmann, D. R. (2013). Mixed integer programming for minimizing the period of a cyclic railway timetable for a single track with two train types. *Computers & Industrial Engineering*, *66*(1), 171–185.

Higgins, A., Kozan, E., and Ferreira, L. (1996). Optimal scheduling of trains on a single line track. *Transportation research part B: Methodological*, *30*(2), 147–161.

Higgins, A., Kozan, E., and Ferreira, L. (1997). Heuristic techniques for single line train scheduling. *Journal of heuristics*, *3*(1), 43–62.

Li, X., Wang, D., Li, K., and Gao, Z. (2013). A green train scheduling model and fuzzy multi-objective optimization algorithm. *Applied Mathematical Modelling*, *37*(4), 2063–2073.

Lin, B., Liu, C., Wang, H., and Lin, R. (2017). Modeling the railway network design problem: A novel approach to considering carbon emissions reduction. *Transportation Research Part D: Transport and Environment*, *56*, 95–109.

Liu, S. Q., and Kozan, E. (2009). Scheduling trains as a blocking parallel-machine job shop scheduling problem. *Computers & Operations Research*, *36*(10), 2840–2852.

Luan, X., Wang, Y., De Schutter, B., Meng, L., Lodewijks, G., and Corman, F. (2018). Integration of real-time traffic management and train control for rail networks-Part 1: Optimization problems and solution approaches. *Transportation Research Part B: Methodological*, *115*, 41–71.

Miandoab, M. H., Ghezavati, V., and Mohammaditabar, D. (2020). Developing a simultaneous scheduling of passenger and freight trains for an inter-city railway considering optimization of carbon emissions and waiting times. *Journal of Cleaner Production*, *248*, 119303.

Montrone, T., Pellegrini, P., Nobili, P., and Longo, G. (2016), June. Energy consumption minimization in railway planning. In *2016 IEEE 16th International Conference on Environment and Electrical Engineering (EEEIC)* (1–5). IEEE.

Parbo, J., Nielsen, O. A., and Prato, C. G. (2016). Passenger perspectives in railway timetabling: a literature review. *Transport Reviews*, *36*(4), 500–526.

Sahana, S. K., Jain, A., and Mahanti, P. K. (2014). Ant colony optimization for train scheduling: an analysis. *International Journal of Intelligent Systems and Applications*, *6*(2), 29.

Samà, M., Corman, F., and Pacciarelli, D. (2017). A variable neighbourhood search for fast train scheduling and routing during disturbed railway traffic situations. *Computers & Operations Research*, *78*, 480–499.

Scheepmaker, G. M., Goverde, R. M., and Kroon, L. G. (2017). Review of energy-efficient train control and timetabling. *European Journal of Operational Research*, *257*(2), 355–376.

Skrucany, T., Kendra, M., Skorupa, M., Grencik, J., and Figlus, T. (2017). Comparison of chosen environmental aspects in individual road transport and railway passenger transport. *Procedia Engineering*, *192*, 806–811.

Szpigel, B. (1973). Optimal train scheduling on a single line railway. *Journal of Operations Research*, 72, 344–351.

TCDD. (2018a). 2018 Faaliyet Raporu. TCDD. URL:http://www.tcddtasimacilik.gov.tr/uploads/images/3/Strateji/Faaliyet_Raporlari/faaliyet_raporu_2018.pdf.

TCDD. (2018b). 2018 İstatistik Yıllığı. TCDD. URL: http://www.tcddtasimacilik.gov.tr/uploads/images/Strateji/TCDD-T-2018-istatistik-yilligi.pdf.

Wang, Y., Ning, B., Cao, F., De Schutter, B., and Van den Boom, T. J. (2011), July. A survey on optimal trajectory planning for train operations. In *Proceedings of 2011 IEEE International Conference on Service Operations, Logistics and Informatics* (589–594). IEEE.

Wang, Y., D'Ariano, A., Yin, J., Meng, L., Tang, T., and Ning, B. (2018). Passenger demand oriented train scheduling and rolling stock circulation planning for an urban rail transit line. *Transportation Research Part B: Methodological*, *118*, 193–227.

Wang, Y. F., Li, K. P., Xu, X. M., and Zhang, Y. R. (2014). Transport energy consumption and saving in China. *Renewable and Sustainable Energy Reviews*, *29*, 641–655.

Wee, B. V., Janse, P., and Brink, R. V. D. (2005). Comparing energy use and environmental performance of land transport modes. *Transport Reviews*, *25*(1), 3–24.

Xu, X., Li, K., and Lu, X. (2019a). Simultaneous locomotive assignment and train scheduling on a single-track railway line: A simulation-based optimization approach. *Computers & Industrial Engineering*, *127*, 1336–1351.

Xu, X., Li, K., Yang, L., and Gao, Z. (2019b). An efficient train scheduling algorithm on a single-track railway system. *Journal of Scheduling*, *22*(1), 85–105.

Yang, L., Qi, J., Li, S., and Gao, Y. (2016). Collaborative optimization for train scheduling and train stop planning on high-speed railways. *Omega*, *64*, 57–76.

Yang, S., Wu, J., Yang, X., Sun, H., and Gao, Z. (2018). Energy-efficient timetable and speed profile optimization with multi-phase speed limits: Theoretical analysis and application. *Applied Mathematical Modelling*, *56*, 32–50.

Yu, X., Lang, M., Gao, Y., Wang, K., Su, C. H., Tsai, S. B., Huo, M., Yu, X., and Li, S. (2018). An empirical study on the design of China high-speed rail express train operation plan—From a sustainable transport perspective. *Sustainability*, *10*(7), 2478.

Yue, Y., Han, J., Wang, S., and Liu, X. (2017). Integrated train timetabling and rolling stock scheduling model based on time-dependent demand for urban rail transit. *Computer-Aided Civil and Infrastructure Engineering*, *32*(10), 856–873.

Zhou, X., and Zhong, M. (2007). Single-track train timetabling with guaranteed optimality: Branch-and-bound algorithms with enhanced lower bounds. *Transportation Research Part B: Methodological*, *41*(3), 320–341.

18 Sustainable Scheduling and Application in Aviation Industry

İbrahim Zeki Akyurt
Turkish Airlines Flight Academy, Aydın, 09120, Turkey

1 INTRODUCTION

Scheduling is the operational problem that deeply affects daily productivity in the industry. These problems include production scheduling problems, nurse scheduling problems, project scheduling problems, timetable scheduling problems, and so on. Although scheduling problems are comprehensive and complex, growing concerns about sustainable scheduling is critical to businesses. Operational level planning can be done using scheduling. It is also the most effective way to reduce energy costs and sustainability issues. Most sustainable planning research deals with environmental issues such as electricity consumption or carbon emissions, even though there are many other environmental and social indicators. However, increased manufacturers are realizing that solutions for the environment are not limited, such as the implementation of socially responsible, economically sound, management programs (Akbar and Irohara, 2018).

Many studies have been published about sustainable scheduling and application in literature using various methods. Sustainable scheduling analysis (Baruah and Burns, 2006), a sustainable machine scheduling problem (Wang, 2017), and a review for scheduling of sustainable manufacturing (Akbar and Irohara, 2018). Yue and You (2013) proposed a mixed-integer linear fractional program for sustainable scheduling of batch processes under economic and environmental criteria. Baker and Baruah (2009) presented fixed-priority scheduling of sporadic task sets that is sustainable under a variety of scheduling parameter relaxations. Fang et al. (2011) provided a multi-objective mixed-integer programming for a new shop scheduling in sustainable manufacturing. Musavi and Bozorgi-Amiri (2017) investigated a sustainable hub location-scheduling problem using multi-objective mixed-integer linear programming optimizing. De et al. (2016) addressed a particle algorithm for sustainable integrated dynamic ship routing and scheduling optimization. De et al. (2017) proposed two novel algorithms that include Non-Dominated Sorting Genetic Algorithm II (NSGA-II) and Multi-Objective Particle Swarm Optimization (MOPSO) for sustainable ship routing and scheduling with draft restrictions. Xu et al. (2016) presented a multi-objective joint model of energy consumption and production efficiency. Hammad et al. (2010) considered traffic scheduling problems for energy sustainable vehicular. Zhang et al. (2017) examined sustainable scheduling of cloth production processes using combined multi-objective genetic and tabu algorithm. Gong et al. (2015) applied a generic mixed-integer linear programming for sustainable manufacturing which includes job scheduling on a single machine.

The aim of this study is to propose a new sustainable mathematical model for the pilot scheduling problem for a flight school. Flight scheduling is one of the most vital problems in flight schools since it is affected by various conditions such as regulations, rules, standards, resources, etc. Besides, a pilot candidate takes the courses for 18 months, which needs a sustainable approach. Therefore, it is hard to find suitable schedules. This study concerns the sustainable scheduling approach of the candidate and flight instructor scheduling.

2 SUSTAINABLE PILOT SCHEDULING MODEL

A detailed scheduling problem that includes rules for flight training is examined. The problem includes flight instructor and the working rules for pilots, candidate pilots, and the rules regarding the training of these pilots, weather conditions and flight permits for the ground base, and, finally, the rules for airplanes and aircraft. All these factors are illustrated in Figure 18.1. The rules about flight instructor include certificates, permission plans, flight hours, flight hours' restrictions on training aircraft. Candidate pilots are responsible for the tasks of pilots, their completion status, and daily duty limits. The ground square creates a limitation by considering obstacles such as flight times and airport traffic. Finally, the aircraft have restrictions such as maintenance conditions, the number of aircraft that can fly simultaneously, and the available aircraft capacity at hand.

There are three objectives in the problem. The first objective is to ensure that flight instructors are available for continuous flight to provide them with training in different missions. The second objective is to achieve the highest number of

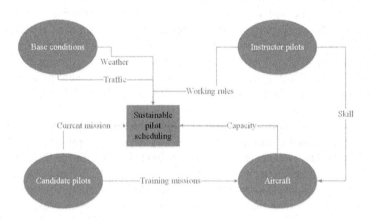

FIGURE 18.1   The Sustainable Pilot Scheduling Problem.

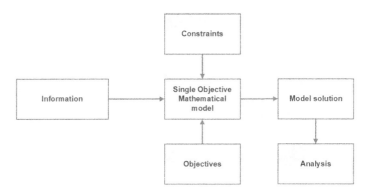

FIGURE 18.2 The Methodology of the Study.

assignments that will follow the rules. The last objective is to evaluate these two goals together. The flow of the methodology that will occur according to the objective of the problem can be seen in Figure 18.2.

The assumptions of the sustainable pilot scheduling problem are given as follows:

i. All model parameters are known in advance.
ii. Maintenance of the aircraft is known in advance.
iii. Repeated practices are negligible in the model.
iv. Each single objective model has the same weight as the goal programming model.
v. The working schedule for the instructors is known in the beginning.
vi. The number of available flights for the base conditions are known in advance.

Accordingly, the notation of the mathematical model is given in Table 18.1.

In the modeling, two objectives z_1, and z_2 are modeled as a single objective model. The first objective function (1) aims to maximize assigning different phases to the flight instructor. This objective allows a sustainable plan for the flight instructor, as it maximizes assignments to different tasks. The second single objective function (2) deals with assigning the maximum number of missions available. Finally, the third objective function (3) focuses these two single objective functions into one objective.

$$\max z_1 = \sum_i^I \sum_t^{T-1} \sum_k^K \sum_v^K |k * y_{ikt} - v * y_{ivt+1}| \tag{18.1}$$

$$\max z_2 = \sum_i^I \sum_j^J \sum_k^K \sum_t^T x_{ijkt} \tag{18.2}$$

$$\max z = z_1 + z_2 \tag{18.3}$$

TABLE 18.1
The Notation of the Mathematical Model

Sets

I	Flight instructors $(1, ..., I)$
J	Pilot candidates $(1, ..., J)$
K	Flight mission phases $(1, ..., K)$
U	Type of aircraft $(1, ..., U)$
T	Time periods $(1, ..., T)$

Parameters

S_{ui}	Skill availability for pilot i for aircraft type u
Q_{ut}	Number of available u type aircraft at period t
O_{uk}	The necessity of u type aircraft in the phase k
B_{jk}	The number of succeeded missions for candidate j in phase k
D_k	The number of total missions in phase k
P_{it}	Available working schedule for the pilot i at period t
A_{kt}	Resource usage ratio for phase k at period t
z_1	The objective value of the maximum number of instructor assignments of different phases model
z_2	The objective value of the maximum assignment model

Decision Variables

x_{ijkt}	$\begin{cases} 1 & \text{if the candidate pilot } j \text{ assigned to instructer } i \\ & \text{for a mission in phase } k \text{ at period } t \\ 0 & \text{otherwise} \end{cases}$
y_{ikt}	$\begin{cases} 1 & \text{if the Instructer } I \\ & \text{for a mission in phase } k \text{ at period } t \\ 0 & \text{otherwise} \end{cases}$
f_t, g_t	Binary variables for avoiding assignment of 1st and 2nd phases in a period
s_1^+, s_1^-	Positive and negative deviations from the first objective z_1
s_2^+, s_2^-	Positive and negative deviations from the second objective z_2
s_3^+, s_3^-	Positive and negative deviations from the third objective z_3

The constraints of all the mathematical models are given in Equations (18.4)–(18.17) as follows:

$$\sum_{i}^{I} x_{ijkt} \leq 1 \; \forall j \in J, \; \forall k \in K, \; \forall t \in T \tag{18.4}$$

$$\sum_{j}^{J} \sum_{k}^{K} x_{ijkt} \leq P_{it} \; \forall i \in I, \forall t \in T \tag{18.5}$$

$$\sum_{i}^{I}\sum_{j}^{J}\sum_{k}^{K} x_{ijkt} * O_{uk} \leq Q_{ut} \quad \forall u \in U, \ \forall t \in T \tag{18.6}$$

$$\sum_{i}^{I}\sum_{j}^{J}\sum_{k}^{K} x_{ijkt} * A_{kt} \leq 1 \quad \forall t \in T \tag{18.7}$$

$$\sum_{i}^{I}\sum_{t}^{T} x_{ijkt} + B_{jk} \leq D_k \quad \forall j \in J, \ \forall k \in K \tag{18.8}$$

$$\sum_{i}^{I}\sum_{t}^{T} x_{ijkt} + B_{jk} - D_k \leq -1 * \sum_{i}^{I}\sum_{t}^{T} x_{ijkt} \quad \forall j \in J, \ \forall k \in K \tag{18.9}$$

$$x_{ijkt} = 0 \quad \forall i \in I, \forall j \in J, \ \forall k \in K, \ \forall t \in T \ \ if \ \ D_k - B_{jk} \leq 0 \tag{18.10}$$

$$x_{ij1l} \leq x_{ij1t} \quad \forall i \in I, \forall j \in J, \ \forall l \in T, \ \forall t \in T \ \ if \ t < l \tag{18.11}$$

$$\sum_{i}^{I}\sum_{t}^{T} x_{ij1t} \leq M * f_t \quad \forall t \in T \tag{18.12}$$

$$\sum_{i}^{I}\sum_{t}^{T} x_{ij2t} \leq M * g_t \quad \forall t \in T \tag{18.13}$$

$$f_t + g_t \leq 1 \quad \forall t \in T \tag{18.14}$$

$$y_{ikt} \leq \sum_{j}^{J} x_{ijkt} \quad \forall i \in I, \ \forall k \in K, \ \forall t \in T \tag{18.15}$$

$$x_{ijkt} \in \{0, 1\} \quad \forall i \in I, \forall j \in J, \ \forall k \in K, \ \forall t \in T \tag{18.16}$$

$$y_{ikt} \in \{0, 1\} \quad \forall i \in I, \ \forall k \in K, \ \forall t \in T \tag{18.17}$$

$$f_t \in \{0, 1\}, g_t \in \{0, 1\} \quad \forall t \in T \tag{18.18}$$

Constraint (4) ensures that a student can only be assigned to one instructor in a period for a phase. Constraint (5) forces assignment not less than the available number of pilot instructors in a period. Constraint (6) ensures the availability of aircraft in the schedule and constraint (7) forces consideration of base availability. Constraints (8) and (9) provide the completion of phases. Constraint (9) ensures to start the next phase before the current phase is succeeded. Constraint (10) avoids assigning a mission if there are no available missions in a phase. Constraint (11) is a case-specific constraint that provides that the same instructor should complete the first phase. Constraints (12)–(14) provide that there are only first or second phase candidate pilots in a day. Constraint (15) calculates whether an instructor pilot is assigned to a mission or not. Constraints (16)–(18) refer to the valid ranges of variables.

3   CASE STUDY

In the case study, a flight school of about 100 students and 30 instructors are considered. In this school, flight education consists of five different phases and three different types of aircraft. A candidate pilot must pass 16, 24, 39, 31, and 15 missions in each phase, respectively. There are three pilot instructors having only one aircraft skill, seven flight instructors having two aircraft skills, and the rest of the instructors having all skills. According to the Directorate General of Civil Aviation rules (http://web.shgm.gov.tr/en), a flight instructor has a limitation to flight missions. Therefore, there is a working schedule obeyed—flight times, holidays, and off days.

Each mission has different flight times and aircraft type needs as summarized in Table 18.2. Figure 18.3 shows the number of missions is passed at the start of the planning period. Each pilot candidate has its own phase missions; therefore, some of the candidates have similar completed missions. Finally, flights are affected by the traffic, weather, and sunlight. Fifteen days are for missions, provided by the forecasts.

4   EXPERIMENTAL RESULTS

The mathematical models proposed for the 15-day planning problem examined are solved. Accordingly, Figure 18.4 shows the variation of workloads on the flight instructor between models. Although in the first model, the maximum scheduling model allows pilots to have a balanced shift assignment, there are significant differences between the pilots when the sustainable development of the second model—and pilot skills—is considered. However, it is seen that the second model makes fewer assignments. It can be said that a more balanced appointment was made in the model that considers both objectives. Taking sustainable and maximum efficiency objectives together has provided a more convenient scheduling model.

TABLE 18.2
Aircraft and Flight Phases

Phases/Aircraft Type	1	2	3	4	5
Cessna	1	1	1	0	0
DA-40	0	0	0	1	0
DA-42	0	0	0	0	1

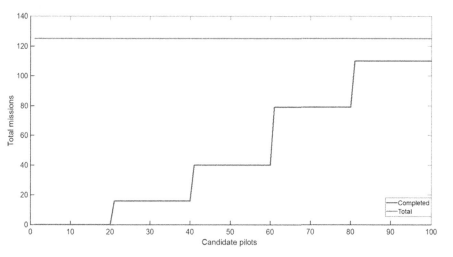

FIGURE 18.3 The Current Progress of the Candidate Pilots.

Figure 18.5 shows the number of charts made according to the periods. The same results are seen for the first model and the third model. In the second model, fewer assignments are observed. Accordingly, taking a combination of sustainable and maximum efficiency objectives in terms of the relevant criteria has revealed a more appropriate scheduling model.

The number of assignments made for students is shown in Figure 18.6. The first model and the third model create similar assignments. In the second model, fewer pilot assignments are obtained. Considering the sustainable and efficient model in common, it is seen that it makes different assignments. Taking a combination of sustainable and maximum-efficient objectives, the relevant criteria has revealed a more appropriate scheduling model.

5 CONCLUSION AND DISCUSSION

Sustainability emerges as an increasingly critical issue after the 1990s. This situation has made environmental and social issues important, not only with the economic factors (cost or profit) in decision problems arising in supply chains. Focusing on economic decisions with minimum cost, minimum environmental damage, and maximum social benefit forces sustainable solutions with multiperspective views.

Sustainability is one of the most important issues not only for production companies but also for service companies. With sustainability, it becomes possible to consider economic, environmental, and social conditions. In this study, an exemplary case study is applied to a flight academy for sustainable flight scheduling. In the proposed model, a unique mathematical scheduling model has been introduced by including the restrictions specific to the flight school. The mathematical model is solved for three different purpose situations and the results are examined. Accordingly, quite different results are obtained only from an economic or social

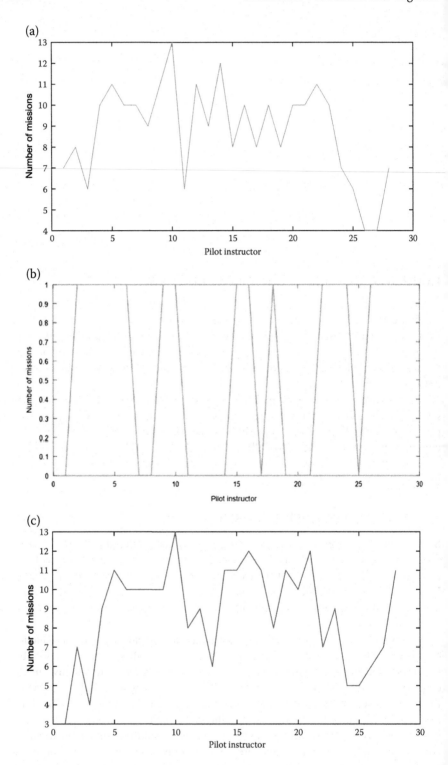

FIGURE 18.4   Results of the Models for Flight Instructors (a) First Model, (b) Second Model, and (c) Third Model.

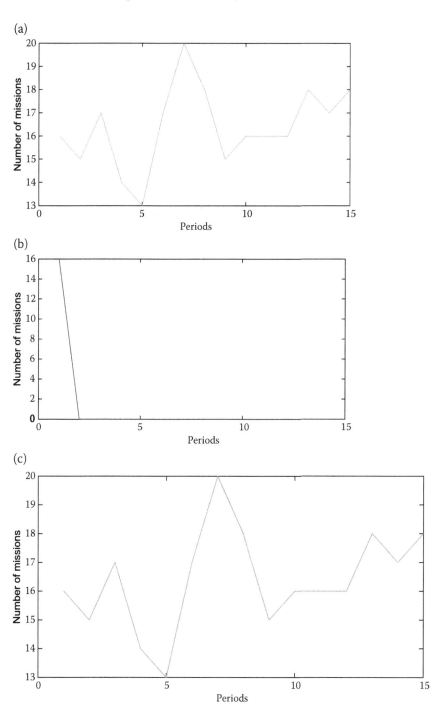

FIGURE 18.5 Results of the Models for Planning Periods (a) First Model, (b) Second Model, and (c) Third model.

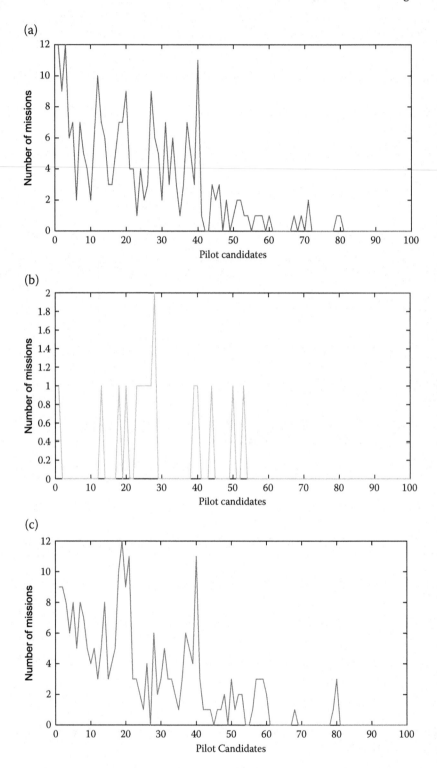

FIGURE 18.6 Results of the Models for Candidate Pilots (a) First Model, (b) Second Model, and (c) Third Model.

perspective. However, taking both criteria together has allowed a more consistent schedule model. In future studies, it is suggested to introduce functions for this purpose—intuitive approaches to the case of problem coverage and different sustainability objectives.

REFERENCES

Akbar, M., and Irohara, T. (2018). "Scheduling for sustainable manufacturing: A review." *Journal of Cleaner Production*, 205, 866–883.

Baker, T. P., and Baruah, S. K. (2009), July. "Sustainable multiprocessor scheduling of sporadic task systems." In *2009 21st Euromicro Conference on Real-Time Systems* (pp. 141–150). IEEE.

Baruah, S., and Burns, A. (2006), December. "Sustainable scheduling analysis." In *2006 27th IEEE International Real-Time Systems Symposium (RTSS'06)* (pp. 159–168). IEEE.

De, A., Mamanduru, V. K. R., Gunasekaran, A., Subramanian, N., and Tiwari, M. K. (2016). "Composite particle algorithm for sustainable integrated dynamic ship routing and scheduling optimization." *Computers & Industrial Engineering*, 96, 201–215.

De, A., Choudhary, A., and Tiwari, M. K. (2017). "Multiobjective approach for sustainable ship routing and scheduling with draft restrictions." *IEEE Transactions on Engineering Management*, 66(1), 35–51.

Directorate General of Civil Aviation, http://web.shgm.gov.tr/en, Access Date: 20.05.2020.

Fang, K., Uhan, N., Zhao, F., and Sutherland, J. W. (2011). "A new shop scheduling approach in support of sustainable manufacturing." In *Glocalized Solutions for Sustainability in Manufacturing* (pp. 305–310). Springer, Berlin, Heidelberg.

Gong, X., De Pessemier, T., Joseph, W., and Martens, L. (2015). "An energy-cost-aware scheduling methodology for sustainable manufacturing." *Procedia CIRP*, 29, 185–190.

Hammad, A. A., Badawy, G. H., Todd, T. D., Sayegh, A. A., and Zhao, D. (2010), December. "Traffic scheduling for energy sustainable vehicular infrastructure." In *2010 IEEE Global Telecommunications Conference GLOBECOM 2010* (pp. 1–6). IEEE.

Musavi, M., and Bozorgi-Amiri, A. (2017). "A multi-objective sustainable hub location-scheduling problem for perishable food supply chain." *Computers & Industrial Engineering*, 113, 766–778.

Wang, J. (2017). "Sustainable machine scheduling problem to minimize carbon emission." *Control and Decision* 32(6):1063–1068.

Xu, W., Shao, L., Yao, B., Zhou, Z., and Pham, D. T. (2016). "Perception data-driven optimization of manufacturing equipment service scheduling in sustainable manufacturing." *Journal of Manufacturing Systems*, 41, 86–101.

Yue, D., and You, F. (2013). "Sustainable scheduling of batch processes under economic and environmental criteria with MINLP models and algorithms." *Computers & Chemical Engineering*, 54, 44–59.

Zhang, R. (2017). "Sustainable scheduling of cloth production processes by multi-objective genetic algorithm with Tabu-enhanced local search." *Sustainability*, 9(10), 1754.

Author Index

Subject Index

(The numbers refer to Chapter Numbers)

Printed in the United States
by Baker & Taylor Publisher Services